The Epic History of Biology

The Epic History of Biology

Anthony Serafini

PLENUM PRESS • NEW YORK AND LONDON

Library of Congress Cataloging-in-Publication Data

Serafini, Anthony.
The epic history of biology / Anthony Serafini.
p. cm.
Includes bibliographical references and index.
ISBN 0-306-44511-5
1. Biology--History. 2. Medicine--History. 3. Biologists.
4. Medical scientists. I. Title.
QH305.S48 1993
574'.09--dc20 93-27895
CIP

ISBN 0-306-44511-5

Plenum Press is a division of Plenum Publishing Corporation
233 Spring Street, New York, N.Y. 10013

Printed in the United States of America

Preface

Of all the sciences, biology doubtless touches our lives in the most direct way. With new developments in recombinant DNA, cloning, new reproductive technologies, and environmental concerns, biology has taken on a significance perhaps greater than at any point in history.

Over the past one hundred fifty years or so, science has become increasingly subdivided. Correspondingly, the myth has appeared that it has become increasingly arcane and impenetrable to all but a handful of specialists. True, there have been a number of popular and accessible books on, for example, physics. The complexities of physics are, however, easily matched by the myriad subtleties and complexities of biology. Further, the history of biology is so immense that it is well-nigh impossible to cover all within the space limits set by the publisher. For these reasons, difficult decisions had to be made. I have spent more time on molecular biology, for example, than a classical biologist perhaps might like. I have all but passed over the Orient, since the history of eastern biology is a complete story in itself.

Biology has, therefore, a difficult and complex history. Biology should, however, be something any educated person can grasp. To this end, I have tried to keep notes to a minimum. I have tried to lay out in the most readable fashion possible the principal developments in the biological sciences. I proceed from the viewpoint of the historian according to what is today called narrative history, a method I believe best suited for clarity and continuity in a book like this.

The progress of any science cannot be fully measured outside the cultural and historical context, the zeitgeist, in which it flowers. I have paid some attention not only to ideas, therefore, but also to the lives and the times of the persons who created them. Though I observe the classic general time divisions of history—keeping in mind that the boundaries between the various historical epochs are neither crystal-sharp nor etched in stone—I have tried to note special forward surges in biology: the advent of the inductive method with the work of Bacon and Descartes in the 17th century, with the accompanying postulate that the universe can be understood in terms of general laws; the impact of the entrance of the microscope; and the importance of the computer as it relates to scientific inquiry.

There is another question I have had to confront: the vague boundaries between the sciences. The history of science shows that no one science develops in total isolation from another; the progress of genetics, for example, has over the past century depended heavily on mathematics, biochemistry, and even computer science. Thus I note the work of people who, though not strictly biologists, have nevertheless contributed tremendously to biology. Among these are Charles Lyell, Linus Pauling, and Dorothy Wrinch.

ACKNOWLEDGMENTS

Crossover among the sciences creates an an impediment for the historian. How can he or she hope to grasp fully all the complex interrelations among the various sciences, let alone among the branches of the strictly biological sciences? (Just as the sciences themselves are continually evolving, so too are the relationships among them.) To this end I have sought the advice of specialists in a variety of branches of science. Here I owe my thanks to the following people for assistance, helpful conversations, and advice at various stages over the years in my research and writing on the biological sciences. First, despite the passage of years, I am indebted to the faculty of Cornell University for their fine tutelage, although the field has changed and evolved considerably since I received my baccalaureate in zoology in 1965. Next, I thank Professor Robert Olby, Department of Philosophy, The University of Leeds; Professor Paul Joss, Department of Physics, MIT; Professor John Archibald Wheeler, University of Texas at Austin; Professor P. W. Anderson, Princeton University; Professor Verner Shoemaker, Emeritus, Washington University; Professor Jane Armstrong, Department of Biology, Centenary College of New Jersey; Professor Anne Misziak, Department of Chemistry, Centenary College of New Jersey; Professor George Wilder, Department of Biology, Cleveland State University; Professor Garland Allen, Department of Biology, Washington University; Professor Edmund Erde, University of Medicine and Dentistry of New Jersey; Professor Alice Levine Baxter, Loomis–Chaffee School, Windsor, Connecticut; Barbara G. Beddal; H. Tristram Engelhardt, Jr., Center for Ethics, Baylor College of Medicine; Professor Israel Scheffler, Department of Philosophy, Harvard University; Professor Janet Fleetwood, Medical College of Pennsylvania; Dr. Dorothy Prisco, Office of the Academic Vice-President, Wesley College of Delaware; Professor John Edsall, Biological Laboratories, Harvard University; Professor Everett Mendelsohn, Department of the History of Science, Harvard University; Professor Dennis Senchuk, Department of Philosophy, Indiana University. I am indebted also to Professor Philip J. Pauly, Department of History, Rutgers University; Dr. Mary Jane Capozzolli-Ingui; Professor L. Pearce Williams, John Stambaugh Professor of the History of Science at Cornell University; James Reed, Dean and Professor of History, Rutgers University; my editor, Naomi Gross, for tactful and meticulous editing; my wife Tina and my daughter Alina Serafini; and many others. Also, since I am not a historian by training, I wish to express my profound thanks to Professor David Weir and Raymond Frey of the Department of History at Centenary College of New Jersey for their advice on writing and researching history.

Contents

CHAPTER 1. THE BEGINNINGS 1

CHAPTER 2. MESOPOTAMIA 9

CHAPTER 3. GREEK MEDICINE 15

CHAPTER 4. ROME 45

CHAPTER 5. THE MIDDLE AGES 55

CHAPTER 6. THE RENAISSANCE 59

CHAPTER 7. PESTILENCE IN THE RENAISSANCE 69

CHAPTER 8. THE AGE OF VESALIUS 75

CHAPTER 9. THE HARVEY ERA 89

CHAPTER 10. THE AGE OF NEWTON *103*

CHAPTER 11. THE MICROSCOPE AND LEEUWENHOEK *111*

CHAPTER 12. THE MEETING OF BIOLOGY AND CHEMISTRY *117*

CHAPTER 13. RAY AND THE EMERGENCE OF CELL THEORY *125*

CHAPTER 14. THE AGE OF LINNAEUS AND THE ENLIGHTENMENT *139*

CHAPTER 15. LAMARCK AND HIS SYSTEM *171*

CHAPTER 16. THE RISE OF PALEONTOLOGY *185*

CHAPTER 17. BIOLOGY IN THE VICTORIAN ERA *195*

CHAPTER 18. DARWIN AND HIS AGE *209*

CHAPTER 19. EMBRYOLOGY AND BIOCHEMISTRY IN THE DARWINIAN ERA 233

CHAPTER 20. THE AGE OF PASTEUR AND THE DEVELOPMENT OF THE MICROSCOPE *249*

CHAPTER 21. BIOLOGY IN THE TWENTIETH CENTURY *269*

CHAPTER 22. T. H. MORGAN AND THE RISE OF GENETICS *291*

CHAPTER 23. ON THE TRAIL OF DNA *331*

CHAPTER 24. EVOLUTION AND GENETIC ENGINEERING *363*

EPILOGUE *371*

ENDNOTES *373*

INDEX *377*

CHAPTER 1

The Beginnings

It is difficult to say when biology as a separate science, or even science itself, really began. Though they had virtually no knowledge of what anyone would today call physics or chemistry, the men and women of the Paleolithic and Neolithic periods probably knew some primitive medical techniques. The Paleolithic period was a preliterate epoch which lasted from roughly two million to 10,000 B.C., while the Neolithic age was a later preliterate period that lasted from about 9000 B.C. until the first civilizations of Egypt and Mesopotamia. The people of the preliterate eras may have even classified a variety of plants and animals, although in no sense did they have any knowledge of these facts as a "science." However, the shrewdest of the early species of humans of the later Paleolithic era, Cro-Magnon, at the very least, would have soon found out which plants were toxic, which were not, and which were suitable for medicinal purposes. They also probably knew which kinds were appropriate for dyes, poisons, and so forth. Nor did organized biology begin to emerge even with the earliest true civilizations in the West—the ancient empires of Egypt and Mesopotamia, which began roughly 3500 B.C. Despite considerable progress in the arts and culture, generally in Egypt and Mesopotamia and the many civilizations that followed them, neither science *per se* nor organized biology really existed as a separate, organized body of thought. This would remain so at least until the time of Aristotle, in the fourth century B.C. In the case of biology specifically, whatever knowledge did emerge would remain intertwined with medical practice, again, until the time of Aristotle. The biological knowledge the ancients did discover and develop came about primarily because of practical surgical and medical needs between the time of ancient Egypt and Aristotle, frequently out of a necessity to aid soldiers in the battlefield.

EGYPTIAN MEDICINE

Thus, at least the rudiments of biological/medical learning undeniably existed in the Egyptian age. Five thousand years ago, Egyptian priests were already starting to gather a tremendous amount of medical data. We know this from their hieroglyphic stone tablets, many of which linguists have almost completely deciphered. Among other things, they ostensibly had a highly advanced knowledge of plants and their various medical applications. Thus, the real story of Western biology begins in Egypt.

The reason for this is not hard to understand. The study of ancient civilizations has shown that cultural progress cannot take place in periods of rampant misery. Progress in philosophy, abstract science for its own sake, literature, art, and the like is greatest when life is most serene and unburdened with problems. There are innumerable examples: one thinks of Greece in its "golden age," where men like Pericles and Cleisthenes ruled democratically in a nation that offered almost anything one could want. In such an atmosphere, literature and philosophy flourished through people such as Aristophanes, Plato, and Aristotle. It is no surprise that in the work of Aristotle we do in fact see the study of biology for its own sake. Aristotle was interested in medicine, but he also conducted myriad scientific investigations merely for the sake of learning about what he did not understand, unburdened with concerns for feasible "practical" applications.

Unfortunately, in Egypt, there were few extended periods of peace or tranquility and, therefore, little leisure time for the pursuit of abstract science. (Historians generally divide the history of Ancient Egypt into the Old, Middle, and New Kingdoms, with the Old Kingdom beginning at about 2770 B.C., and the New Kingdom ending about 1087 B.C.) Unquestionably, this was true of the Old Kingdom, where one sees such tyrants as Zoser, who spent virtually all of the public money on a grandiose and wasteful series of shrines to himself called the pyramids. Perhaps the closest Egypt came to a "golden age" of unrestricted science and art was during the Middle Kingdom, which lasted from about 2000 to 1785 B.C., and which comprised the eleventh and twelfth dynasties. In this period, the pharaoh had more or less made peace with the nobles and priests who had challenged the might of the earlier Pharaohs. Also, with the advent of fewer egomaniacal Pharaohs, the frenzy of pyramid building had begun to diminish. In turn, there was considerable progress in astronomy, math, and the practice of medicine. Mathematics, including geometry and arithmetic, developed to aid in massive construction projects, and astronomy developed to compute the times of the flooding of the Nile. In medicine, various documents dating from about 1700 B.C. show that both diagnosis and treatment of disease were highly developed. The Egyptians had learned the significance of the pulse and the circulatory system, could treat broken bones, and, as the ancient *Materia Medica* shows, had catalogued quite a list of medicines.

In the late Middle Ages, by contrast, life was one continual, dreary struggle for the next meal. Life was filled with constant plague, and people had little hope for a better life. It is thus no wonder that philosophical, scientific, and literary accomplishments were comparatively few in the later medieval period. There were exceptions of course: the fourteenth-century philosopher William of Occam is, defensibly, one of the exalted thinkers of antiquity. But, in general, this law of history holds true.

One of the earliest "physicians" then of ancient Egypt was Imhotep, who apparently lived sometime during the third dynasty, or first ruling family of the Old Kingdom. A versatile man, historians believe he also orchestrated the construction of the first pyramids under Zoser, one of the earliest of the legendary Pharaohs. There is some reason to believe, based on what has been learned from written records, that early Egyptian physicians like Imhotep acquired a respectable amount of their medical information from the custom of mummification. The first step in this process consisted of carving out the innards of the deceased. This, of course, gradually allowed early physicians to accumulate considerable anatomical data. Then, by experimenting with different plants for preserving the internal organs, they acquired some knowledge of biochemical action. By trial-and-error testing of a variety of chemicals on the body to determine their biological effects, they found which ones preserved the body best. In this group were wine, aromatics, myrrh, cassia (a legume with medical properties), salt, and several other chemicals. They realized too that extracts from a wide selection of plants covering the linen-wrapped body would also inhibit decay.

MYTH AND MEDICINE

Nevertheless, it was nearly impossible to separate magic and superstition from science. In part this was a function of ignorance, in part a function of power. Throughout the Old, Middle, and New Kingdoms, the Pharaohs had to contend with the domination of the priests. Possibly because the masses believed them to have magical powers and influence over the deities of nature, the priests had immense leverage, and no Pharaoh could afford to ignore them. Imhotep was among the most commanding and influential of the priests. Many centuries later, by the time of the New Kingdom, many regarded him as essentially a divine being. The masses worshiped him, offered gifts to bronze images of him, and prayed to him to let their crops grow.

THE MEDICAL PAPYRI

A measure of how far the ancients were from modern science is the fact that they did all of their research under the belief that spirits dwelt in terrestrial objects. The Egyptians believed that some god or other was the steward of every part of the body. However, specific papyri, such as the famous *Edwin Smith Surgical Papyrus*, show that not all of their medicine was perceived as a mere appendage of religion. For example, during the New Kingdom, which was the last of the major epochs in Egyptian history, surgery, according to the papyrus, was free of theology. The Egyptians judged that purely mechanical damage to bones, muscles, and joints had natural, rather than theological, explanations. Even so, much of the Egyptian population still believed that cures came only through the worship of divinities such as Aton, the lord of the universe in the New Kingdom, Amen-Re, the sun god, and others.

The same papyrus also explains that by the period of the New Kingdom, Egyptian physicians understood how to set bones very well. Even their diagnostic procedures resembled contemporary medical practice; in the Edwin Smith papyrus there are various lists of questions, such as "Can he lift his arm?" "How rapid is his heartbeat?" and so forth.[1] Although the papyrus is not entirely explicit on the reasons for these questions, it seems a safe bet to assume that they were diagnostic. The papyrus also discusses injuries to many different parts of the body, including the face, head, and abdomen. It tells how to bandage battle-ax as well as knife wounds, and how to apply tourniquets.

The Egyptians knew that the heart was a pump of some sort, and they may even have had a rudimentary comprehension of the circulation of the blood. Interestingly, the Greek pre-Socratic philosopher Empedocles also studied the heart and circulatory system, though today historians know him best for his philosophical views. Although he made many errors in his descriptions, such as claiming that the human psyche was located in the heart, his first stumbling forays were important stimuli to later research on the circulation of the blood.

The Egyptians acquired all of their erudition through the method of trial and error. We know, for instance, that they first realized the antibacterial properties of penicillin through this method. They had no theory as such; they knew only that certain techniques helped particular ailments. In like fashion, they came to know that onions were valuable in protecting them against epidemics, apparently functioning as a kind of primitive antibiotic, and that the yarrow plant, by constricting blood vessels, functioned as does a modern styptic pencil to control bleeding. Finally, although they had insufficient knowledge of alcohol's properties, they shortly discovered that giving alcohol to patients, especially via mixing beer with other medications, gave the patient a decidedly useful sense of well-being. The Egyptians also acquired some biological knowledge through agriculture. Even in preliterate ages, as early as 8000 B.C., humans were learning about the soil conditions necessary for raising wheat and corn. In doing so, they acquired practical botanical knowledge about how soil conditions such as moisture, acidity, and the like affected the development of plants.

Papyrus Ebers

Another well-known text, the *Papyrus Ebers*, offers descriptions of over 600 medicines, explaining in considerable detail how the Egyptians used each of them. Allegedly, it was over a foot wide, over sixty feet long, and written wholly in hieroglyphics. According to legend, a nomad discovered it near the Nile River sometime during the Middle Kingdom of Egypt, although assorted historians believe it was written in about 1550 B.C., or around the end of the second intermediate period. The sections of it that have survived rest permanently today in Leipzig, Germany. This historic papyrus prescribes exercise, hypnosis, diet, and fasts. More specifically, it discusses the function and dysfunction of the various senses and their sense organs, including long sections on ailments of the ear and eye. It says much about tumors and blocked glands and offers various medicines designed to heal burns, wrinkles, freckles, and nearly every skin condition imaginable. It prescribes medicines for baldness as well. Some sections deal with pediatrics, for medical specialization was common in Egypt.

The Therapeutic Papyrus

The *Therapeutic Papyrus* also suggests well over 600 distinct remedies for a broad spectrum of ailments. The remedies include copper, acacia, castor beans, olive extract, saffron, and ginger. Now and then the "medicines" border on the preposterous. For example, parts of this papyrus imply that moisture from a sow can cure illness. Furthermore, the information in the papyrus was not purely medical. It is evident from the *Therapeutic Papyrus* that Egyptian biologists knew something about how tadpoles changed into frogs and how beetles emerged from eggs.

Many Egyptian cave paintings as well as hieroglyphics show explicitly enough that the Egyptian priests were practicing surgery of all types. They apparently did abdominal surgery, surgery of the groin area, and even eye surgery—much of it successful. It is arguable that there was not a comparable achievement in surgery until the mid-seventeenth century, when the Germanic physician Johann Shultes wrote his *Armamentarium Chirugicum*, or "Tools of the Surgeon," which proposed, among other things, an unpleasant procedure for performing a mastectomy. Despite their superstitious beliefs concerning science, Egyptian approaches at times were strikingly reminiscent of recent medicine. For instance, the Egyptians had a wide variety of medical specialties, including obstetrics, ophthalmology, cardiology, neurology, and even psychiatry. Often psychiatrists would talk to patients, suggesting that they might improve by adopting more "positive" attitudes toward life and the gods. Apparently they also had some sense of social responsibility, for there is some indication that they used contraceptives, possibly utilizing dried animal skins as condoms.

In other nations at the time, biological/medical knowledge was less sophisticated; in Syria, for instance, although medical practitioners had organized themselves as a professional group, they still based a substantial amount of their medical speculation on astrology. Much the same was true of many other countries in the Middle East.

In sum, the biology of Egypt was a very mixed bag. In some ways it was highly advanced, and at the same time it was shrouded in magic, mystery, and superstition. Nor was biology, at this point, really distinguishable from medicine. This union would persist for many centuries.

CHAPTER 2

Mesopotamia

The land known as Mesopotamia was the home of the Babylonians (historians sometimes call the underlying culture *Sumer*). There were significant dissimilarities between Egypt and Mesopotamia. Located between the Tigris and Euphrates rivers, Mesopotamia was much more easily invaded than Egypt. Not surprisingly, the medical arts, and therefore biology, progressed much faster in Mesopotamia than in Egypt, since schooling in medicine was imperative for the treatment of wounded and ailing soldiers in the field. Among their other accomplishments, the Sumerians developed early erudition in, and distinguished some surprisingly modern areas of, biology. Studies show that Sumerian scientists delved into endocrinology (the study of hormones), histology (the analysis of tissues), comparative anatomy (an approach to anatomy that progresses by comparing one animal with another), and many other topics.

MAGIC AS A SYSTEM OF EXPLANATION

Hosts of intrepid researchers also developed a considerable fund of botanical material. As in Egypt, science, magic, and superstition were inextricably mixed. Indeed the hard "science" that the Sumerians did discover was often researched not only through medical inquiry, but through the use of charms, hexes, amulets, and other implements of the occult. As in Japan, China, and other countries of the Orient, Sumerians held that the entrails of essentially any beast could tell them the future and predict possible propitious events or ill-fortune. As a consequence, the priests had just about as much political muscle as the official leaders of the nation, for the masses thought that all of the powers of the universe lay at the command of the priests.

Mesopotamian farmers were also interested in the science of biology. They were learning modern irrigation methods, and by 3000 B.C., they were domesticating donkeys in what is now modern Israel.

The Sumerians assembled much of their medical lore during the reign of the fabled Hammurabi, who ruled from 1792 to 1750 B.C. Although usually thought of as primarily a legal document, the ancient Code of Hammurabi actually contained a surprising amount of medical information. For example, several precepts of the code discussed physicians' fees, often limiting the amount they could charge for either services or medicine. Also included were laws providing penalties for malpractice. While some parts of the code were highly contemporary, others were brutal and illogical; for example, a surgeon who erred during surgery would be operated on and made to suffer the same injury himself—this, presumably, being a logical extension of the Hammurabi idea of "an eye for an eye." As the code stated:

> If the doctor shall treat a gentleman and shall open an abscess with the knife and shall preserve the eye of the patient, he shall receive ten shekels of silver. If the doctor shall open an abscess with a blunt knife and shall kill the patient or shall destroy the sight of the eye, his hands shall be cut off or his eye shall be put out.[2]

This is not the sort of thing that would promote calm nerves and steady hands in a surgeon. Unfortunately, clusters of these eccentric doctrines remained influential for centuries. Succeeding civilizations practically without exception used the Code of Hammurabi as the basis for their own legal systems. Some of these include the Hebrews, who flourished in the Arabian desert around 1900 B.C., the Hittites, the Syro-Hittites, the Phoenicians, the Assyrians, the Akkadians, the

Kassites, and several others. In fact, the influence of the Code reaches even into the bowels of the U.S. legal system.

Physicians and historians of Sumeria saw to it that all legal, literary, scientific, and medical facts survived in the majestic libraries of antiquity. One of these was the library of Ashurbanipal in Sumeria, established in the seventh century B.C., during the time that the Assyrians conquered Sumeria, and named after the seventh-century B.C. leader of Assyria, King Ashurbanipal, who was himself a student of learning and proffered every assistance in maintaining and protecting these records. To date, scholars can boast of over 25,000 clay tablets scribed in the ancient Sumerian writing system known as cuneiform (stemming from the Latin *cuneus,* meaning "wedge-shaped," since the ancient scribes used wedge-shaped tools to write the characters of this language). Today, collections of these tablets are kept in the British Museum. Among other things, these tablets reveal that the ancient Sumerians had a profusion of data about disease, mixed, of course, with liberal portions of superstition. The role of demons in medicine and infirmity is as pivotal in these records as is that of medical potions and surgical procedures.

Nevertheless, much of the medical lore was useful in practice, the "hypothesis" behind it notwithstanding. The cuneiform tablets show that the Sumerians had various uses for roots, olive branches, garlic, cinnamon, all manner of flowers, and many herbs. There is also reason to believe that the Sumerians used other plants as well, since much of their lore remains untranslated because of the difficulty of cuneiform and the wretched state of some of the tablets.

Beyond herbal remedies they used all sorts of animal body parts in treating eye, skin, and dental ailments. Assorted mineral compounds appear on the tablets as well, including copper, iron, mercury, and so forth. Sour milk was a common remedy, as were a variety of salves and ointments.

Historically, there has been considerable confusion about the presence of genuine physicians in Sumeria. It seems evident that, as indicated earlier, there were no physicians in the modern sense of the term, since their real occupation was either religion or sorcery. The Greek historian Herodotus, among others, who lived around the time of Plato, made this latter claim. However, other evidence, for instance, information on cuneiform tablets, shows otherwise. According to the tablets, there were institutions resembling medical schools around the time of the Assyrian invasion of Sumeria. Also, letters survive that a Babylonian physician apparently had written, giving recommendations for stopping a nosebleed, treating blindness, and so forth. Whether the author of these letters was a physician or a sorcerer continues to be a matter of debate.

The Babylonians also kept careful records of their treatments—possibly out of fear of some horrible retribution called for in Hammurabi's Code? Such records show that the physicians, such as they were, did cautiously follow the progress of a disease. They also noticed that one person could pass on a disorder to another. Of course they knew nothing of present germ theory, and suggested that demons may have been culpable. As part of their medical procedure, they would often offer a sacrifice to the demon in the hope that it would simply leave the patient's body. Usually, they put lambs next to the ailing individual to accomplish this. They even had a procedure for "verifying" this idea: If the lamb became ill, they concluded that the demon had passed from the person to the lamb to show its compassion toward the stricken family.

While this may seem laughable in light of today's learning, the "demon" idea really was scientifically sound—in this sense: In the absence of a scientific canon, all ancient civilizations sought to fathom the workings of the universe in some other manner. Very often, they attributed commonplace events to demons, witches, and so forth. Silly? Not really. They were speculating in a theoretical manner, and the demon supposition was at least an attempt to explain the transmission of illness. It is worth remembering that according to scientific methodology, no supposition is "silly" if no better one exists. For most ancient civilizations, the demon assumption was the best they could do under the circumstances.

The Sumerians, the dominant culture in Mesopotamia for most of its history, had already begun something resembling modern dental practices. We know, again from cuneiform tablets, that they were able to drill teeth and fill them (and conceivably even overcharge patients). It is fascinating that the Etruscans, the people of Italy who predated the Romans, also developed a form of false teeth; in fact, they used many of the same dental practices the Sumerians used.

Perhaps parts of the Sumerian culture somehow reached the Etruscans, although historians are not sure how, or even if this is true.

The Sumerians had an amazing sense of the significance of cleanliness. We know that they had fairly sophisticated sewage systems and an intuitive comprehension of the importance of isolating people afflicted with certain ailments.

All in all, given the admixture of fact and myth, the progress of Sumerian medicine resembles Egyptian, except that it progressed more quickly. That this is so is not surprising; the Sumerians, unlike the Egyptians, did not have natural protective geographical boundaries. As a consequence, their history is one of invasion by one aggressor after another. With a civilization almost constantly at war, it is no wonder that the healing arts progressed so rapidly.

CHAPTER 3

Greek Medicine

The closest approximations to a global, theoretical, and scientific worldview in Egyptian times were magic and demonology, although the Egyptians had considerable practical information about mathematics, biology, and even engineering. Even so, it is only in ancient Greece that we find anything resembling true scientists. Although they did not have the sophisticated technology that we have today, men like Thales, Anaximenes, Anaximander, and others were possibly the first to adopt scientific methodology, theory construction, and testing via observation.

This merits some further comment, for even historians often fail to appreciate how similar the "Greek mind" was to the contemporary scientific mind. To grasp this, it is necessary to try to attend to the world from their perspective. They had none of today's technology—no microscopes, no telescopes, and no body of scientific theory. Consequently, they had to conduct their scientific speculations by the "seat of their pants," so to speak.

THALES

Thales, the first of the pre-Socratic philosophers, hypothesized that all material substances in the universe were merely different forms of water. So, for example, fire was "rarefied" water, while stone was "condensed" water. The idea is of course false, although Thales was on the right track. We know, for instance, that water, after all, does cover three-quarters of the earth and that the human body is nearly 100 percent water. In a word, Thales correctly realized that water was exceptionally weighty in the scheme of things—he merely took this concept a bit too far. Others of the pre-Socratic era answered this question in a different way. Anaximenes surmised that the fundamental material was air, and Heraclitus believed it was fire and so forth.

HERACLITUS

Heraclitus is worthy of special mention. He, like Parmenides, had a tremendous influence on Plato. Although Plato did not accept the materialism of Heraclitus or Parmenides—*materialism* being a theory that accepts only the reality of physical things and denies the existence of spiritual entities such as the soul—he did accept the Heraclitean doctrine that change is a pervasive feature of the physical domain.

Heraclitus's biology was well thought out, and to some extent he tried to ground his deliberations on careful investigation. Born in 540 B.C., he came from a well-to-do and influential family, even acquiring a prominent post later in life in the Greek government. However, dissatisfied with political corruption, he abandoned the position to pursue philosophy full time. In direct contrast to the teachings of Parmenides and Zeno, he believed that change was the fundamental feature of the cosmos. Everything everywhere was forever in a state of flux. It is scarcely surprising, therefore, that he hypothesized that the volatile element fire was the most essential element in the universe. He believed further that the universe was a colossal cylinder, with fire entering on one end and emerging out of the other. Thus, fire is the source of all matter and all life. Life arises from fire and, in death, will revert to fire. Fire also comprises man's "soul," though in his use of this term a caution is necessary. Heraclitus's use of terms like "life" and "soul" has misled many historians of science and philosophy into believing that he had at least some kind of rudimentary theological worldview. This is undeniably false, and again, his references to man's "soul" are merely references to physical matter. He had no conception of God, a divine nature, or anything else that even faintly partook of the "spiritual."

Naturally, fire played the essential role in all of his biological speculations. He saw it as the key ingredient in the soul of man, though, again, "soul" has no spiritual connotations. The more "fire" there is in the body, the more "alive" it is. Disease, then, was the departure of fire from the body. Also quite logically, Heraclitus associated liquids in any form with sickness. His most famous saying, certainly, is "The dry soul is wisest and best." Water and any form of alcohol were, therefore, enemies of existence—a surprisingly sophisticated attitude, actually. Unfortunately, most of his biological observations no longer exist. Although numerous modern philosophers believe that he performed what a contemporary biologist would call comparative anatomical studies and physiological research, all of this has been lost. Again, his greatest influence was on Plato. In stressing as he did the volatile and ever-changing nature of the physical cosmos, Plato came to accept this view as part of the ultimate truth.

More prophetic than the answers, however, was the question itself. The Greek thinkers were obsessed with the search for unity in the universe. That is, they were trying desperately to show that all the *apparently* different types of material substances in the universe were really just different forms of one and the same substance. That search, in modified form, has not yet ended.

THE MATHEMATICAL PHILOSOPHY OF PYTHAGORAS

Still another important sage of the pre-Socratic era was Pythagoras, who lived during the sixth century B.C. Incontrovertibly, he was one of the first eminent philosophers, scientists, and mathematicians, as well as being a religious mystic. He was born in Samos near the coast of Asia Minor and subsequently taught there, though political problems and enmities ultimately drove him to flee to the Greek city of Croton on the Italian peninsula.

After finally settling down to serious labor, he began research on political, scientific, and theological reform. It is evident in his writings that Eastern metaphysics guided him. Indeed, Eastern ideas may have affected a great number of ancient Greek thinkers. Aristotle, to give one example, possibly derived his teachings about the "golden mean," or the making of wise intermediate choices, from the similar Buddhist notion found in the *Pali Texts* of that creed. Pythagoras may have taken the notion of the transmigration of the soul, or its passing from one body to another, from Hinduism—a cosmic ideology that stretches back to the Vedic texts of Hinduism written over 2500 years before Christ. Also, much of Pythagoras's mathematics derives from Hindu sources—including the "Pythagorean" theorem so habitually misascribed to Pythagoras.

Still, not everything Pythagoras said or taught derives from Hindu sources; some of it he purloined from his fellow Greeks. Like Heraclitus, Pythagoras proposed that fire was the fundamental substance of the cosmos—the medium that would unify and, therefore, "explain" all of nature. Roughly, his assumption was that the "primordial fire" abided in the center of the universe. From it came the planets, stars, and so forth, which revolved eternally around the central fire.

Perhaps Pythagoras can legitimately lay claim to the concept of planetary orbits. To his credit, he continued to defend the idea, even though several other pundits of the day attacked it rather savagely. Plato finally secured its fate when he merely ignored the doctrine, although Aristotle returned in some sense to Pythagorean thinking when he argued that the universe consisted of a series of concentric circles within circles—the outermost being the *primum mobile,* or prime mover. Later, of course, in the Renaissance, Copernicus dramatically revived and offered solid evidence for the "orbital" picture of the universe.

Mathematics is the discipline most closely linked to Pythagoras's name, as is the proposal that the heavens constantly produce music with mystical properties. That is, he held that the planets continually generate musical sounds that could affect, in mysterious ways, all the goings-on in the heavens. Mathematics, according to Pythagoras, was also a supreme explanatory principle: everything that took place, including the beginning of life, was explicable mathematically. Although his authority was notable during his lifetime, chiefly philosophically, his eminence eventually paled in the radiant light of Aristotle and Plato.

THE EMPIRICISM OF XENOPHANES

Yet another important early Greek philosopher was Xenophanes, who lived roughly a century after Thales. Born in the city of Colophon on the coast of Asia Minor, he initially devoted himself to Anaximander's teachings. Like many other philosophers, political tumult sent him into exile, ultimately to the city of Elea. Doubtless his most prominent accomplishment derives from Anaximander's *On Nature,* only pieces of which survive today. Following his mentor, he taught that the world began when water and "primordial mud" began to condense. Xenophanes was also something of an empiricist (a believer in the worth of knowledge derived through the senses), in that he appears to have actually searched the Elean countryside studying plants and animals. One of his more remarkable conclusions was that because the fossil fragments of ocean creatures were embedded in mountain rock, the mountains must have been under the oceans at some point in the history of the earth—a conviction not all that far from the truth.

Also, Xenophanes was conceivably the first to advance a version of "catastrophism"—the belief that the biosphere, or parts of it, suddenly vanished virtually overnight. Xenophanes believed this of the earth itself for instance. This dogma disappeared for centuries until some of the early Enlightenment and Victorian philosophers revived it. After this, Darwin's authority squashed catastrophism again. But catastrophism seems to be a worldview that is all but indestructible. This tenet would reappear again with the work of Stephen Jay Gould late in the twentieth century.

PARMENIDES AND THE IMMUTABILITY OF THE COSMOS

Arguably the strangest of all emerging scientific/philosophical theories in the West was that of Parmenides, whom Plato describes in great detail in his dialogue of the same name. Parmenides was the polar opposite of Heraclitus; where the latter assumed that all things in the universe were in constant motion, Parmenides postulated that there was quite literally no such thing as motion. Since Parmenides obviously did not place much reliance on the senses, or else he would never have adopted such a view, philosophers classify him as one of the prototypical rationalists, or philosophers who believe that all human wisdom derives from unaided reason rather than from the five senses. His apprentice Zeno would quickly become an ardent defender of the Parmenidean "no-change" teaching.

Curiously, although Parmenides talks of the "soul," this does not prove that he adopted anything resembling a contemporary religious picture. It is plain that for him, the soul is actually another form of matter. Thus he describes the soul as "hot," while the body is "cold,"—a logical conjecture since when the "principle of life" leaves a person, that is, when that person dies, the body does in fact grow cold. As far as the origin of life is concerned, he based his views chiefly on those of Anaximander, contributing little that was innovative.

It is important to note that Parmenides enormously influenced the thought of Plato. Although Plato rejected Parmenides's materialism, he did accept the notion that ultimate reality was fixed and immutable. What Plato professed, however, was that this immutable reality was located in the spiritual, rather than the physical, realm.

EMPEDOCLES AND THE FORCE OF LOVE

So far as biology is concerned, without question the most significant Greek mind after Aristotle was Empedocles. Above all, Empedocles devoted himself to explaining physical transformations in the world. To this degree, he was a direct opponent of Parmenides and Zeno. Cosmic alteration for Empedocles consisted of varying mixtures of the four basic elements—earth, air, fire, and water. At this point, however, Empedocles took a heroic scientific step forward and an equally giant step backward. Like any conscientious contemporary scientist, he realized that change could come about only by the interplay of forces. However, he lapsed into absurdity when he speculated that these forces were "love" and "hate." In short, he anthropomorphized the physical universe. It is worth noting in fairness, however, that this appears to be an instinctive tendency in ancient philosophers and scientists. The ancient Oriental philosophers from Lao-tzu to Confucius also anthropomorphized the universe in this way, as did pedagogues and philosophers throughout the Middle Ages and well into the Renaissance.

Thus, when love is stronger than hate, matter comes together and new physical objects form; when the reverse is the case, matter decays and fragments. But these primordial forces were not always opposed to one another. In fact, more often than not they cooperated. Accordingly, a primordial moist mass of earth and water separated out so as to give rise to the earth and the oceans. Also, in the spirit of recent science, Empedocles conjectured, accurately, that the total of all the elements always remained the same. No additional matter could arise out of nothing, and no matter could ever disappear; matter could only change form.

This outlook then led him to his biological tenets. He appears to have surmised that animals, rather than plants, were the most basic forms of life, and he interpreted all alterations in the biological cosmos according to what he noticed in animals. Correspondingly, he believed that all living things emanated from the earth. Plants came first, and they obtained food and water from pores in the leaves and stem. Animals also arose from the earth, though in a most peculiar way. Empedocles first postulated that limbs of beasts appeared first. Later, through the force of love, the limbs joined to form complete animals.

Interestingly, Empedocles also appears to have had some sort of a rudimentary idea of evolution or, at least, of a "survival of the fittest" sort of doctrine. He believed that the forces of love and hate did not behave in any systematic, planned way. That is, there was no benevolent deity guiding them. Rather, chance was the controlling factor (as it would be afterward in the philosophy of Lucretius, the Roman poet and author of *On the Nature of Things.* In the eighteenth century, similar proposals would find their way into the philosophy of the British empirical philosopher David Hume). Thus, the creatures that resulted from these forces might be extremely hardy and adaptive, or they might be irreparably nonfunctional. The more fit would survive while the less fit would perish. Empedocles explained mankind in much the same way. Since he had no theological views, he had no particular grounds for believing that man was uncommon or in any way dissimilar from the lowest brutes. Accordingly, he believed that human limbs leaped out of the "subterranean fire" from the center of the earth. Again, through the force of love, the limbs combined to form entire human beings. His theory as to the origin of men and women, however, was unique, and he did not apply this part of his philosophy to the lower animals. Men had a naturally "warmer" constitution, so they originated in the Southern Hemisphere. Women, on the other hand, were more cold-blooded, so they surfaced in the North.

Empedocles attempted to explain respiration by theorizing that air enters the mouth and nose as well the pores of the skin. It then mixes with the blood and finds its way to all parts of the body. As he says in the fragments of his writings we do have:

> In this wise do all breathe in and out. All have bloodless tubes of flesh stretched over the surface of the skin pierced with pores closely packed so that the blood is kept in, while an easy way is cut for the air through the openings.[3]

He also generated something of a theory of embryology and reproduction which was rather more modern than other parts of his teachings. He believed that the embryo took part of itself from the mother and part from the father. He conjectured that after birth, the child grew because of the gradual accumulation of heat from the environment. Correspondingly and consistently, he calculated that the inevitable deterioration of the body in old age was traceable to loss of heat from the body.

Likewise, Empedocles attempted to clarify the operation of the five senses by intimating that all objects give off "atoms" which then enter the various sense organs. He supposed that parts of each sense organ were specific for one of the basic elements. Therefore, water in the ear detected the presence of water, earth in the nose detected earth in the environment, and so forth. He occasionally came close to a correct explanation. He said that the ear, for instance, detects tones that develop when objects move through the air. He also maintained that these tones entered the auditory canal; he went awry, however, when he further asserted that the animal, according to his "physiology," then heated them, thus allowing the creature to hear the tones.

The eye, he said, is much like a lamp. When a person sees light, he sees the "fire" of the eye, and when he sees darkness he sees the "water" in the eye.

Perhaps most portentously, he anticipated prevailing psychophysical analyses of mind. He speculated that what we call "mental processes," such as thought, reason, belief, fear, and the like, were actually processes in the brain or even other parts of the body. More precisely, he believed that thinking occurs in the blood. When a person thinks, the four basic elements "seek out" one another. Thus the better the four elements mix and harmonize with one another, the higher the person's intelligence. There is a peculiar anomaly in his metaphysics, however. Like every one of the pre-Socratics, Empedocles was clearly a materialist—one with the view that there are no realities other than matter. That is, he left no room in his philosophy for notions like "soul," "mind," "spirit," and so forth, even though such terminology appears from time to time in his writings. Typically, materialist philosophers, such as Hume, D'Holbach, Marx, and many others, also hold to the principle of empiricism—the doctrine claiming that the true fount of all human knowledge is sense experience. Philosophers who downplay both the value and even the reality of the physical universe, such as Plato and Descartes, believe that pure reason or pure thought is the ultimate source of genuine human knowledge.

Empedocles's approach is an odd amalgam of both. He held unvaryingly to a materialist philosophy, but he also believed that reason, not the senses, was the true source of human knowledge. While the senses are fallible, rationality is not. Perhaps only the eighteenth-century German philosopher Immanuel Kant defended a philosophy even remotely like this. (However, Kant's philosophy was not exactly like Empedocles's either: it is similar only in the general sense that Kant tried to construct a philosophy of human sagacity that included the predominant ideas from both the empiricist and rationalist traditions.)

In sum, Empedocles came reasonably close to a modern scientific approach, at least insofar as he appreciated the importance of observation, even while he distrusted it as an ultimate authority. He did have some influence on later researchers, most notably Aristotle. Even so, though we know more about Empedocles than we do about some other pre-Socratic philosophers, his sway over the later Greek philosophers Plato and Aristotle was less than that of the pre-Socratics Parmenides and Heraclitus.

THE ATOMIC THEORY OF DEMOCRITUS

Democritus is another of the most influential of all the Greek savants, for it was Democritus who advocated the first "atomic" hypothesis of matter, thereby suggesting the atomic hypothesis of today. He was born in the Greek colony of Abdera on the Thracian coast around 470 B.C., and scholars believe him to have lived to a mellow old age, roughly eighty-plus years. He studied with still another pre-Socratic, the philosopher Leucippus, who possibly first mentioned the atomic theory to Democritus. Luckily for him, Democritus came from a wealthy and powerful family and was therefore able to receive the finest instruction and upbringing. His father, always an advocate of schooling and, expressly, scientific knowledge, sent him on regular trips to look at the flora and fauna of other lands.

Still, Democritus was in some respects the stereotypical philosopher. Pathetic at handling money, he lived only through the benevolent graces of family and friends, even well into manhood. His brother supported him for some years, as did some wealthy scientific and philosophically minded patrons in Abdera. Historians believe that, like scores of other philosophers, Democritus was a man of wide interests who wrote numerous essays, although as so frequently happens this far back in time, most of his writings have not survived.

As with Socrates, we learn something about Democritus from Aristotle. It is primarily through Aristotle, as well as through the few shreds of Democritus's writings that have survived, that we know what little we do. Like the other pre-Socratic philosophers, Democritus held tightly to an unadulterated materialistic conception of the cosmos. As further proof that Democritus did not accept any sort of "spiritualism," despite his terminology, he says that not only are obviously physical objects like trees and the human body composed of atoms, but so is the soul. The only disparity is that the soul is composed of atoms that are "finer" than atoms composing other physical objects. One of his endeavors was to try to explain the make-up and launching of the universe. Democritus taught that the cosmos consisted of two and only two key elements—atoms and the void. That is, he posited that the universe consisted of a measureless span of space called "the void." In the void fell an infinite number of particles. These particles had lasted through the infinite past and would so persist into the endless future. And, since space was infinite, the downward "fall" of these particles was also infinite.

He further taught that the atoms were irregularly shaped and that their motion was random. Thus in their downward course, atoms would occasionally bump into one another. From these collisions, aggregates of atoms would form. Some of these would then be more stable than others. The less stable ones would, by definition, quickly fall apart. The more stable ones would remain and enlarge, as more and more atoms smashed into them—sort of a cosmic snowball effect.

Democritus had still more "explanatory" precepts. Since he was a materialist, he did not perceive the cosmos to be under the control of any benevolent or malevolent deity; instead, all happened according to ordinary causal laws. In the latter respect, at least, he veered tantalizingly close to contemporary physics. He also endorsed something similar to the universal law of conservation of energy and matter when he argued that nothing can come from nothing. All change in the universe is merely a rearrangement of atoms. This is similar to the way modern physics says that all physical transformations are from matter to energy and vice versa.

DEMOCRITUS'S BIOLOGICAL WORK

Democritus's specifically biological contributions are numerous. We know he conducted comparative anatomical observations on a wide variety of animals, from the simplest to the most complex. In fact, it was from Democritus that Aristotle derived his distinction between "sanguiferous" animals, or vertebrates, and "bloodless" animals, or invertebrates—animals without a backbone. Democritus also believed that every kind of animal life, no matter how rudimentary, had an assortment of sense organs. Democritus was well aware that no one could detect such sense organs in the most primitive creatures. But in a rationalist, rather than an empiricist, spirit, he believed that logic dictated that they must have such organs to survive. The fact that humans could not detect them was evidence only of the imperfection of the human mind.

Democritus also speculated on embryological development. He believed, for example, that the sense organs developed first, followed later by the organs of the digestive tract. Again, he was often right. He concluded correctly, to offer one such instance, that a spider manufactures its web inside its body. This is an arresting contrast to the erroneous Aristotelian view that a web is simply skin that has dried up. On the other hand, Democritus also surmised that one could clear up the mystery of infertility in mules on the grounds that the uterus had "overcontracted"—not one of his better hypotheses.

So far as humans were concerned, he anticipated the outlook of the seventeenth-century German philosopher Leibnitz that every atom of the human body "mirrored," or was a microcosm of, the rest of the universe. (In much the same way, an individual Fascist might be said to reflect the entire philosophy of Fascism.) Again, he tended to characterize humans much as he did the other animals, in terms of "organs." He believed, for instance, that the heart was the organ of courage, while the brain was the organ of thought and the liver the organ of sensuality. Intriguingly, many of the late medieval biologists and philosophers erred in rebuffing some of this. Following dogmatically the teachings of Aristotle, they claimed that the only purpose of the brain was to "cool" the blood. Democritus further explained sensation in a manner similar to Empedocles when he suggested that infinitesimally fine atoms "emanate" from physical objects and impinge on the eyes.

Democritus's explanation of respiration was more or less on the right track; he argued that inhalation consists of taking in atoms, while exhalation consists of giving them off. Although Democritus had no real grasp of air or its composition, in general this view is right. He also believed, not surprisingly, that when fresh "soul atoms" cease to enter the body, a person perishes. He analyzed sleep also as due to a loss of soul atoms.

Consistent to the last, Democritus applied these ideas to illuminating the various diseases. He held that atoms colliding with the earth brought about epidemics and that inflammation of the nerves caused hydrophobia, an irrational fear of water.

In short, Democritus was a surprisingly modern thinker. He used the scientific method and speculated about a number of biological processes—several of his conjectures being right on target, as when he expressly noted that the brain was the "organ" of deliberation and sentience. He also understood the role of the law of causality in explaining physical phenomena. His atomic theory is, in principle, much like the present atomic theory, although it lacks many of the details of today's version, since Democritus had no facts about electrons, protons, or other subatomic particles and many other refinements. Indeed, modern science has really come up with few truly novel recommendations since Greek times. It is vital to remember, too, that in addition to lacking all of these conceptual tools, Democritus also lacked all of the technology that we take for granted today. Instruments like microscopes, microtomes, telescopes, and the like, would have to wait centuries before finding their way into the scientific armamentarium.

Another Greek, Diocles, who lived in the fourth century B.C., also furthered the discipline of anatomy by actually publishing the very first book on that subject. Although his conjectures were ingenious and his drawings meticulous, historians of science do not consider much of this work to be accurate today. Even so, in his notebooks, he adroitly described the structure of the eye, as well as the ear, and noted the pertinence of the brain for all "higher" functions. However, whether he studied actual bodies is something of a conundrum. Living on the island of Crete, Diogenes of Apollonia (a different philosopher from the Diogenes of Aristotle's day, who founded the philosophical school known as Cynicism) helped found the so-called Ionian philosophic school. Like all of the pre-Socratics, he held that the universe was entirely material in constitution. To a large extent, Anaximenes's philosophy affected him. Like Anaximenes, Diogenes believed that the fundamental material of the universe was air. Out of air, through processes of condensation and rarefaction, all of the other elements in the cosmos were supposed to have formed. He applied these approaches to biology as well, arguing that "life" was a current of "warm air" coursing through the body. He is possibly the earliest Western thinker to have even a rudimentary grasp of the circulatory system, and some of his drawings in this area have survived to the present. From these primitive opinions he developed a hypothesis about the origin of life. He conjectured that the sun brought together the various forms of air in the cosmos to form simple living creatures. He also held that the "heat" of the mother created the embryo out of the father's seed.

PLATO

Building on some of the propositions of the pre-Socratic philosophers, Plato went on, in the fifth century B.C., to develop some of his own conjectures. Although Plato's own assumptions were drastically far from those of recent science or biology—since he departed from the materialism of his predecessors—his disciple Aristotle did carry on the empirical, or observational, scientific tradition. Unlike Plato, who believed that the study of the physical universe could not yield true knowledge, Aristotle believed that the physical world was worthy of serious investigation, and he based his conclusions on actual scrutiny of living and dead plants and animals, carefully noting their anatomical features, and trying to discern the function of the various organs.

ARISTOTLE

A leader of his own school, the Lyceum, he was *the* most informed man of ancient times. He was born on the island of Lesbos in the fourth century B.C. and he passed away in 322 B.C. He wrote books on zoology, astronomy, botany, poetry, drama, metaphysics, physics, ethics, and a scattering of other, varied topics.

Among his most significant titles are *Nicomachean Ethics, Metaphysics, Physics, Poetics, De Animalia, De Caelo,* and several others. To a great extent, the principal precepts of Aristotle stemmed from those of his master, Plato. Thus Aristotle, like Plato, held to the "theory of forms." A "form" is a kind of spiritual pattern or blueprint, such that the things of the physical universe were copies of eternal paradigms, or forms, of those things existing in some spiritual world of their own. Plato believed that in the *spiritual* world there existed eternal and unchanging blueprints or prototypes of all the different types of things that existed in the *physical* world. So, all actual housecats, trees, and people, were copies of the perfect form of the housecat, tree, and person existing in some mysterious spiritual dimension. (The Christian notion that man was created in the image of God is reminiscent of this Platonic theory.) Like Plato, Aristotle believed that these forms were eternal and immutable. Also like Plato, Aristotle held to the tenet of "dualism," the ideology that man was composed of both a physical as well as a spiritual element, or, in more prosaic terms, a body and a soul. Thus Aristotle, like Plato and Socrates, represented a tremendous advance over the pre-Socratic philosophers. While the pre-Socratics, with no exceptions whatsoever, conceded only a "material" reality, Socrates, Plato, and Aristotle believed in both physical and spiritual realities.

Still, in many ways Aristotle differed considerably from Plato. First, Aristotle was an empiricist, at least more so than Plato. Where Plato thoroughly distrusted the sphere of the senses, Aristotle did not. Like Plato, Aristotle conceded the role of the intellect in the acquisition of learning, but he regarded the senses as no less significant. Thus to him, the physical cosmos was as vital as the spiritual and was equally deserving of sober study. Not surprisingly, Aristotle wrote a number of volumes devoted specifically to physics and biology.

The Metaphysics of Aristotle

Aristotle's metaphysical views also differed considerably from Plato's in other ways. Although, as noted above, Aristotle did hold to a version of the theory of forms, his notion was quite unlike Plato's. Indeed, in his volume *Metaphysics,* Aristotle subjects the hypothesis of forms to a number of scathing criticisms. One of the most truculent of these was that Plato had needlessly introduced a gulf between the spiritual forms and physical things. Plato had allowed his eternal and immutable forms to inhabit a domain totally separate from the world of physical things. Believing that this separation failed to explain how the things of the physical plane could "resemble" the forms, even imperfectly, Aristotle dropped this notion. In his revised approach, Aristotle claimed that forms and matter abide together as one entity. Thus each "substance" in the physical world, such as a rock or tree, consisted of both form and matter together, a doctrine he called "hylomorphism."

In line with this reasoning, Aristotle devoted a considerable amount of time and intellectual energy to explaining change—hardly unexpected since he believed in the reality of the physical world and noticed, of course, that a cardinal feature of the world was that it was in constant flux. From this he developed his famous dogma of the "four causes." All change comes about, he argued, through one of four causal agencies—the so-called formal, final, material, and efficient causes. The "final" cause, for example, was the tendency in every living thing to move in the direction of its fully developed state.

The latter idea had enormous implications for biology, implications that would affect the reflections of scientists and philosophers for centuries to come. The "formal" cause of change in any object was the Platonic form existing *within* it. That is, one could think of the form existing in an acorn as a kind of internal "motor" that causes it to develop into a mature tree. The "final" cause of an acorn was the fully or "finally" formed oak tree—the state toward which the acorn is being drawn. The ideas of formal and final causes are evidently rather similar. It was much as if there was an internal "command" in the acorn that, in the natural course of events, would cause it to "want" to become an oak tree. Likewise, in the human or any animal embryo, there was another internal edict that caused it to "want" to mature into a fully formed adult. The medieval thinkers regarded this precept of change, the so-called principle of teleology, as one of the most momentous insights of the ancient world. Philosophers and scientists would employ it widely throughout the Middle Ages and the Renaissance.

Aristotle spoke of the "material" cause of change as well, a relatively uncomplicated concept that was little more than a reference to the material that was undergoing the change. Thus, the material cause of change in an oak tree was merely the wood out of which it was composed, while the material cause of growth in a human embryo was the flesh out of which it was composed.

Finally, the "efficient" cause of change was roughly our ordinary idea of causality. The efficient cause of any kind of change was some event that preceded the change and gave rise to it. Overeating causing a stomach ache, or throwing a rock through a window causing it to break are perfect examples of efficient causality.

There is yet another metaphysical principle that plays a key role in Aristotle's ideas on reproduction. The principle is his notion of "actuality and potentiality." This principle is among the various abstract tools Aristotle used to interpret all motion and change in the universe. Amazingly, it is a suggestion he actually seems to have come up with alone, since it does not exist in any Oriental faith and positively does not come from Plato or any of the pre-Socratics. According to Aristotle, then, all modification involves going from a "potential" to an "actual" state. The hypothesis is easy enough: an acorn is a potential oak tree, while the tree itself is the actual oak tree. Analogously, a fertilized ovum is a potential human being, while a college graduate is an actual human being. In all likelihood, Aristotle would have been mystified by the abortion controversy; he surely would have said that the question of when human existence emerges is obvious. The embryo is a unfolding potential human being, while the child is an actual human being.

Similarly a brick held a foot from the ground has the potential to fall to earth. If released, it actualizes that potential and does in fact fall. Scholars of physics will recognize this Aristotelian precept. It has stayed with us today in the idea of potential energy in physics. Again, a contemporary physicist could say, using Aristotelian philosophy that the brick held over the table has a certain amount of potential energy which, if released, is actualized. Nuclear physicists recognize that the nucleus of the atom has a vast amount of potential energy, which they can release via an atomic explosion to become actual energy.

The Biology of Aristotle

While it was in the field of philosophy that Aristotle made his most heralded contributions, his accomplishments in zoology and botany were formidable as well. In biology specifically, it is arguable that Aristotle held one of the first genuine hypotheses of evolution, that is, a prototype that at least remotely hinted at Darwin's. For one thing, he speculated that higher forms evolved over considerable periods of time from lower forms, and, like Darwin, he based most of his judgments on direct, scientific observation. He also proposed some embryological ideas. Through observation he saw that embryonic forms underwent, via teleological change—or movement toward some fully developed state—a transition from the primitive imperfect state of the embryo to the "perfect" state of the fully formed organism.

His purely biological works include *De Animalia,* or "On the History of Animals," as well as books on both reproduction and the anatomy of animal forms. In these writings he assembled all of the information that anyone had gathered up to that point, including his own and his pupils' as well as that of earlier scholars. It is crucial to note that Aristotle's entire system of writing was precisely that—a system. He did not view the universe piecemeal, as scientists often do today; he did not, that is, represent biology as existing entirely separately from physics, nor physics as entirely separate from even such apparently disparate fields as poetry and art. He was one of the most eminent "system builders" in the history of science. In the manner of the nineteenth-century Germanic philosopher Hegel and others, nothing was disconnected from anything else. All parts of the cosmic order interrelated with other parts. The same principles (such as the teleological principle and the axioms of natural deductive logic, which he invented and which mathematicians and philosophers still use today) that govern the writing of a play also govern the writing of the book of nature, and the animal as well as the plant kingdoms.

Aristotelian Taxonomy

As a logician, Aristotle was quite automatically interested in categorizing all life-forms. Many consider him, appropriately enough, the inaugurator of systematic taxonomy in the style of Linnaeus, the eminent eighteenth-century founder of the modern biological system of classification. Although his theories and specific classifications have altered considerably, several of the fundamental tenets still influence contemporary biology. Important elements in his system of taxonomy were, understandably enough, behavior, anatomy, and anatomical differences. As he put it, "Animals may be characterized according to their way of living, their actions, their habits, and their bodily parts."

Using behavior and native habitat as a guide, he divided all beasts into land animals, animals that live always in the water, and animals that live periodically in the water. Fish, of course, as well as particular mammals like whales, are in the second category, while the third includes those such as otters, alligators, and beavers. The latter, though they spend most of their lives in water, breathe and reproduce on land. Intriguingly, no one had yet even fantasized about a life-form intermediate between amphibians and fish until the discovery of the lungfish in 1836. Aristotle, in his *De Partibus Animalium*, or "Parts of Animals," describes part of the theory behind his system as follows:

> Some writers propose to reach the definitions of the ultimate forms of animal life by bipartite division. But this method is often difficult and often impracticable. Sometimes the final differentia of the subdivision is sufficient by itself. . . . Thus in the series Footed, Two-footed, Cleft-footed, the last term is all-expressive by itself, and to append the higher terms is only an idle iteration.[4]

Even more consequential, however, in Aristotle's classificatory scheme were internal and external anatomical attributes, including sense organs, blood, organs of locomotion, and organs of digestion. As Aristotle avers, "Many animals allow of association into large divisions, such as birds, fishes, and whales." To this list he added shellfish, and crayfish.

In some cases he found pigeonholing arduous; four-legged beasts, for instance, broke down into those that give birth to live young and those that are egg-laying, but he found it unavailing to subdivide animals in those categories any further. Partly as a result of these and similar complications, his classification system, while it resembled the later and more modern ideas of Linnaeus, had, as a matter of fact, only two principal divisions, the *Genos* and the *Eidos*.

Genos signified broad categories of animals, such as a mammal, while *eidos* referred to specific kinds of animals within the *genos*, such as cats, horses, tigers, and so forth.

However, it is likely that his pupil Alexander the Great had sent him much of his material. It is well known that Alexander penetrated as far into the Orient as the Indus River in India. Undoubtedly, he acquired a liberal amount of knowledge in his travels. Since Indian scientists had been accumulating both philosophical and biological truths since 2500 B.C., there was probably an enormous amount that Alexander could have learned, even though much evidence of this is lost today.

At least one reason for thinking that Alexander may have "pirated" maxims from the Orient is the remarkable resemblance of some of Aristotle's precepts to Oriental concepts. The concept of the "golden mean," among others, so long thought to have been conceived by Aristotle, may well have come from Buddhism, since the exact thought appears in the *Pali Texts*, as noted earlier. The principle in both cases says that in all human actions, the best course is to avoid extremes, whether in eating, ambition, or whatever. Other ideas may have come to Aristotle from the historian Herodotus; for example, Aristotle's depiction of the crocodile is virtually identical with that of the acclaimed historian, and Aristotle repeats without hesitation the claim of Herodotus that the upper jaw of that beast is connected to the lower. Given Aristotle's keen power of description and observation, it is extremely unlikely that he would have made such an incorrect affirmation had he seen the crocodile for himself.

In any case, Aristotle obviously had access to information about both native and foreign fishes, crustaceans, molluscs, sponges, and a profusion of other marine creatures.

Aristotelian Reproduction

Most biologists of the time held to an archaic model that argued that the creation of unique animals was similar to the creation of new plants. In other words, the man provided a seed which, by itself, held a self-contained organism. The mother served only to provide fertile grounds for the cultivation of an already-complete individual, or "homunculus." However, it is not clear whether or not Aristotle held this view, although some historians think he did. What he did say was that the sperm of the male was the "form" while the egg was the "matter" of the new organism, in clear reliance on one of his most important metaphysical precepts.

He also distinguished between sexual and asexual reproduction. The latter he believed occurred via "spontaneous generation"—a fiction that would inhibit biological progress for centuries. In line with later thinking on this supposition, Aristotle believed spontaneous generation occurred primarily in more primitive animals, such as fleas and mosquitos, which could arise, spontaneously, from decaying substances.

His account of sexual reproduction, however, is bizarre. Here is the astonishing theory behind this alleged phenomenon. Aristotle, unlike Christian pedagogues of the Middle Ages, believed that all living beings have souls. The difference between man and all other beasts, nonetheless, was that man had a tripartite, or three-part, soul, a legacy from Plato. In man the three parts, or faculties, of the soul were the "rational," "spirited," and "vegetative" parts. All lower animals, however, lacked the rational soul and partook only of the vegetative soul—the part responsible for basic biological desires and processes. Aristotle believed that it was this part that could cause sexual desire and thereby give rise to a unique organism.

Yet even if he did not fully grasp sexual reproduction, Aristotle was the first to realize that the mother and father were of equal weight in the creation of a new organism. The following excerpt from *De Generatione Animalium,* his terminology aside, sounds rather modern and even Lamarckian (the view that acquired qualities can be inherited):

> Now it is thought that all animals are generated out of semen, and that the semen comes from the parents. Wherefore it is part of the same inquiry to ask whether both male and female produce it or only one of them, and to ask whether it comes from the whole of the body or not . . . it can be argued that the semen comes from each and every part of the body. . . . First, the intensity of the pleasure of coition . . . secondly, the alleged fact that mutilations are inherited. . . . And these opinions are plausibly supported by such evidence as that children are born with a likeness to their parents, not only in congenital but also in acquired char-

> acteristics. . . . when the parents have had scars, the children have been born with a mark in the form of the scar in the same place.[5]

Nonetheless, he went awry in claiming later on that the father was the more complete and "warmer" element, while the female was unfinished and "colder." Again, there may well be Oriental influences at work. The ancient Taoists, already well established in their faith by Aristotle's day, held just the same opinion, that man was the "warmer" while woman was the "colder" element, an instance of the grand opposite forces of Yin and Yang that the Taoists assumed pervaded the entire cosmos. The principal difference here is that Aristotle was, manifestly, too much of a scientific empiricist to accept the view that the universe could be viewed as a manifestation of "warm" and "cold" sexual elements. He restricted these concepts to literal men and women.

It is essential to refer to Aristotle's doctrine of potentiality–actuality, discussed above, in order to understand his somewhat aberrant conception of sexual reproduction. While it is true that the modern ideas of potential and actual energy are valid applications of Aristotelian metaphysics, Aristotle nonetheless erred in applying these ideas to sexual reproduction. He contended that the father was the actualizing principle, while the egg of the mother was the potential principle.

Using still another principle discussed earlier, the distinction between "matter" and "form," or the hypothesis of hylomorphism, as it is usually called, Aristotle further declared that the man gives form or shape to the matter provided in the ovum of the mother. He then argued that the seed of the father emerges "fully-cooked" from the blood, which gives its purity and form-making abilities. The female principle is, interestingly, also a sperm "seed," but literally a "half-baked" one.

Physiology and Anatomy

Some of Aristotle's most definitive work is in the field of comparative anatomy. Though this branch of contemporary biology as such did not arise in ancient times, Aristotle accurately states the central precept of comparative anatomy in his *De Animalia*, in which he says that one should compare poorly understood anatomical structures in one animal to better understand corresponding structures in other animals. He proposed too, with Empedocles and Democritus, that all living organisms consist of the four basic elements, earth, air, fire, and water. It seems incontrovertible, however, judging from Aristotle's own words, that he never examined a human cadaver, being content to extrapolate, as others had done before him, from lower animals to man. As a consequence, his summaries of human anatomy are only vaguely right.

Aristotle also worked hard on the circulation of the blood, prefiguring the accomplishments of Fabricius, the great sixteenth-century Italian anatomist, and Fabricius's student, the fabled British biologist William Harvey, who first fully appreciated and described the circulation of the blood. Aristotle concluded that the heart controlled the flow of blood and was also the source of "animal heat," an antique idea similar to the vitalism—the conviction that a undetectable, nonphysical "force" keeps animals alive—which appeared later on. Both are enigmatic, but they roughly incorporate the notion that organisms are more than matter in motion; there is some extra element that gives "life" to the tissue. This ideology also reached back into antiquity. It survives in the writings of classical Hinduism as well as Chinese Taoism. What distinguishes animate from inanimate matter, such as a rock, is precisely the fact that the latter does not have the "vital force," while the former does. Aristotle believed too that the heart was the organ of the soul and the seat of the rational faculties in man—those faculties peculiar to man, which distinguished him from other animals. Oddly, he somehow reached the conclusion that the brain secreted mucus that acted to "cool" the blood. Similarly, Aristotle suspected that "spirits" came from the lungs, while the pulse was, in his way of thinking, a "boiling" that occurred when blood and spirits, the substance containing the vital force, mixed together. Obviously, biology had a long haul ahead.

Aristotle's views on the digestive system are another peculiar amalgam of fact and fiction. Although his reports of the various anatomical parts of the digestive processes are roughly accurate, he seemed utterly confused about the physiology of digestion. His analogy here was apparently with cooking, which, given the purpose of digestion, is more or less understandable. He believed that the intestines "baked" the food and that the heart aided it in this process.

He also tried to explain the brain and nervous system, but had scant success. He believed that the brain was "cold," the spinal marrow "hot," and the nerves and tendons "confused." He does describe the anatomy of the ear reasonably well and makes a valiant effort at describing the functioning of the eye, believing that the "moisture" of the eye acts as the target for visual impressions emanating from the environment—more or less following the lead of Empedocles here.

All in all, Aristotle is clearly first among equals in ancient thought. Though others anticipated the modern empirical methods of science, it was Aristotle who brought this approach to its highest state of development in the ancient world. His contributions to anatomy, physiology, and classification—imperfect and curious as they often are—nonetheless continue to impress modern thinkers.

HIPPOCRATES

Perhaps the most exalted mind of them all was Hippocrates. He, arguably, was the very first to found a "profession" of medicine, although some would quarrel with this view in light of the Egyptians' accomplishments.

He is best known today for the famous Hippocratic oath still taken by about twenty percent of physicians entering the field of medicine. Although it is possible to overemphasize, he apparently did do something to separate out occult and mystical ideas from medical and scientific ones. Historians believe that Hippocrates lived from 460 to about 377 B.C., roughly contemporaneously with Plato and Democritus. He was born on the Island of Cos in Greece to the Asclepiads family. For generations this family had been extremely influential in medicine and science. Indeed, his father, Heracleides, taught a considerable amount of medicine and biology to Hippocrates in his youth. In the years following, Hippocrates went on to study in Athens with the quasi-philosopher Gorgias, whom Socrates mercilessly assaulted in the Platonic dialogue of the same name. Later Hippocrates lived in the Balkans, Asia Minor, and ultimately settled permanently in Thessaly. He supported himself through the practice of medicine in Thessaly and taught numerous generations of prospective physicians.

As with Plato and other ancient writers, there is some dispute as to whether Hippocrates actually wrote many of the publications attributed to him. Despite this, at least some of the writings in the so-called Hippocratic Collection appear to be authentic. His foremost collection is certainly "Airs, Waters and Places," which offers a wealth of insightful observations on both the environment and geological issues. In it he also speculates about both the causes and the treatments for an extremely large variety of human ailments, including the use of cauterization with hot irons for controlling bleeding. There is also a considerable amount of research in physiology, embryology, and anatomy in the collections of his writings, though again, it is not certain that Hippocrates was the author of all of these sections. But whoever did author them was obviously a meticulous researcher, producing detailed descriptions of various parts of the anatomy. Some, albeit, are less cogent than others. One often-heard supposition to explain these lapses is that the author of these writings did not actually perform dissections on humans, but instead relied only on dissections of lower animals, and then extrapolated to humans.

Although there is relatively little direct evidence to support this contention, it is true that the ancients regarded the human body as sacred and would definitely have perceived attempts to cut it open as pure sacrilege—no matter what the motivation. Short of grave-robbing, which was not unheard of either, it would be an ordeal to obtain permission from a dead person's relatives to dissect the body.

Nevertheless, through whatever means, a smattering of scientists did unearth human bodies to scrutinize. Grave-robbing was certainly one method. The Greek philosopher Alcmaeon, a pupil of the mystical philosophy Pythagoreanism, was apparently the first person to examine dead bodies purely to acquire some appreciation of anatomy. And the Greek philosopher Herophilus acquired considerable anatomical wisdom through dissecting corpses.

Anatomy and Physiology in the Hippocratic Collection

From these studies exceptional knowledge of the skeleton emerged early—which is no surprise since bones decay much slower than the rest of the body. In this way early physicians "mapped out" fairly reliably the bones of the skull, forearm, legs, and so forth. The Hippocratic Collection also tells us something about the musculature, accurately describing most of the muscles of the extremities, although, understandably enough, there is little data on the many smaller muscles in the body, such as the muscles of the eye. The reason for this is simple. The large muscles were close to the surface of the body, and if the deceased was lean enough, an astute observer could easily discern the muscles' structure and function at least. Muscles like those of the eye, on the other hand, were not only small, but not visible on external examination. The observer would have had to remove the eye from the socket to see them.

Physicians knew something, though not much, about the various glands and organs in the body, again because it was difficult to look at them in human cadavers given the rate at which the body decays. They had a rough outline of the anatomy of the digestive tract, but knew almost nothing else about it. They believed that the glands acted to extract excess water from the body, a view close to the truth in some cases but wide of the mark in many others. They appeared, nonetheless, to grasp both the structure and function of the lungs and windpipe, as well as the anatomy of the heart. They had, however, no clearheaded understanding of the disparity between veins and arteries. The Hippocratic Collection proposed that the brain acted to "cool" the blood. This is true in the broad sense that mechanisms regulating body temperature are found in the brain, but the theory had little to offer outside of this extraordinarily sweeping remark.

If there is one single problem that interfered with the valiant struggles of these as well as numerous other intrepid pioneers of modern biology, it was their inability to differentiate one class of tissue from another. The science of histology would have to wait years to take on any real form. Thus, with abandon, they confused veins with tendons and nerves with blood vessels.

By contrast, early physicians managed to assemble a reasonably good knowledge of the eye. They appear to have been aware of the optic nerve, though they had only a hazy conception of its function. Although they knew nothing of the function of the lens, they did recognize and describe the iris and the pupil, as well as what anatomists today call the vitreous and aqueous humors. They gained the same level of understanding of the human ear. They knew about and described fairly accurately the auditory canal and the ear drum, but again had only the haziest understanding of their functions.

Returning to the knowledge we know for certain comes from Hippocrates alone and not some other, unknown, author, we see that he engaged in an enormous amount of physiological speculation. Following assorted pre-Socratics, Hippocrates speculated that earth, air, fire, and water make up the human body. Parallel to these forces were four other elements or "juices"—phlegm, yellow bile, black bile, and blood. In Hippocrates's own words in "Nature of Man, Humours, Aphorisms and Regimen":

> If the seasons proceed normally and regularly, they produce diseases that come easily to a crisis. If the summer proves bilious, and if the increased bile be left behind, there will also be diseases of the spleen. So when spring too has had a bilious constitution, there occur cases of jaundice in spring also.[6]

In this regard Hippocrates may have influenced Plato enormously, since Plato alludes to these juices repeatedly in his later dialogues. Hippocrates further guessed as to their origin and function. The spleen manufactured black bile, while the liver manufactured yellow bile. Indeed, although these conclusions are wrong, there was some evidence for them. Hippocrates realized, to report one intriguing case, that blood coagulates; some parts are black, some red, and some were yellow. To Hippocrates, the role of these various elements was distinct enough; the way they intermingled and their relative proportions to one another determined the health of the body.

If Aristotle merits the reputation as the greatest biologist of ancient times, surely Hippocrates deserves to be called the finest physician of ancient times. Because of his dedication and industry, he remains in the highest esteem today.

OTHER GREEK BIOLOGISTS

Giants like Hippocrates and Aristotle did not work in a vacuum. Often there were lesser, but still fine, minds at work in the background. Had Aristotle and Hippocrates not lived, these men might themselves have reached the greatest heights in ancient Greece. One of these was the Greek biologist Theophrastus, who took over the Lyceum after the passing of Aristotle. He had been a close friend of Aristotle's since the time Aristotle had studied with Plato. Though he was already at an advanced age when he took over the Lyceum, he nonetheless lived for another thirty years and remained a respected teacher and scholar throughout. Although he made no innovative contributions to scientific method generally, he did survey and classify several new species of plants and animals, and it is arguable that his tomes on botany were as important in the ancient world as Aristotle's contributions to zoology. He also wrote about physics, and some of the manuscripts have survived to the present day. Though again there is relatively little that is new here, it did serve as a sturdy compendium of all the knowledge in physics accumulated up to his time.

Still another important Greek that any chronicle should mention is Thucydides (born ca. 470 B.C.). Although we know him today primarily as a historian, he describes exhaustively the enigmatic plagues that hit Athens and, to some extent, Sparta during the long wars between these city-states. The affliction may have been similar to the ones that decimated the Roman population in the closing centuries of that majestic empire.

Only slightly less significant in the history of biology was Herophilus. The ancients knew a considerable amount about the digestive system, to some degree because of the labors of Hippocrates and Aristotle, but mostly because of the efforts of Herophilus. He was born in Chalcedon in Asia Minor. He studied early in life in the schools of Cos and Cnidus, later conducting research and teaching biology in the city of Alexandria. Although not much knowledge about his life and work has survived, scholars believe that he lived in approximately 300 B.C., which would make him a contemporary of Aristotle. As with Socrates and many other legends of antiquity, we know of his work principally because of what others said about him.

His forte was anatomy, and historians almost universally regard him as one of the greatest anatomists of antiquity. In fact, prevailing science has named the point at which the four great cranial venous sinuses meet after him, the so-called torcular Herophili. Conceivably his most prominent contribution was his pioneering use of human corpses as well as live subjects, to learn anatomy, despite the very vocal disapproval of many who dreaded retribution by some malevolent deity.

He took special notice of the circulatory system, where he made comparative analyses of the arteries and veins as well as heart rates. In some ways he unwittingly succumbed to obsolete dogma; for instance, he held firmly to the doctrine of the "pneuma"—a kind of mystical "air" that gives the "life essence" to the body—found in so many ancient authors. On the other hand, he for the first time clearly distinguished between nerves and tendons. He also thoroughly probed the nervous system and brain, identifying many parts of that organ, including the ventricles.

He also thoroughly studied the anatomical structure of the eye, explicitly identifying the retina, iris, and other parts. He also studied the digestive system as well as the liver.

In sum, Greece emerges as the very summit of ancient achievement both in science and art. This is not surprising, for Greece was far, far ahead of all previous civilizations politically and socially. The Greeks were the first to introduce democracy, albeit not a representative democracy. Above all, the dominant theme in Greek life was rationalism—the reliance on reason rather than myth and superstition. It is no exaggeration to say that many civilizations after Greece made whatever contributions they made by imitating and building on the work of the Greeks. Certainly this was true of the Hellenistic civilization that came after Greece, and of the empire of Rome.

THE ORIENT

It is at least arguable that many Western ideas really came from the Orient. Thus, even though this is a history of Western biology, some mention of the Orient has to be included. Conceivably, for example, Oriental thinkers came closer than anyone in antiquity to discovering the function of blood circulation. But they too, even more than many Western metaphysicians, could not purge mystical and occult elements from their philosophy—principally the creed of the sect of Taoism. In 304 B.C., the Oriental biologist Hsi Than produced his "Record of Plants and Trees of the Southern Regions," in which he describes, among other things, how to protect a variety of fruits from insects. At about the same period, the Chinese realized that a chemical found in some flowers could be useful as an insecticide.

Manifestly, the foundations of modern biology came from the Greeks. With their emphasis on reason, the leisure for science and quiet study, and the genius of men like Aristotle, it could hardly have been otherwise.

CHAPTER 4

Rome

Rome was never the equal of Greece either in science, philosophy, or mathematics. Indeed, most of the thoughts, speculations, and even myths that we associate with Rome really originated with the Greeks. Still, there were some Roman thinkers worthy of mention. A biologist of moderate caliber named Columella lived during the reign of Marcus Aurelius in the second century A.D., although even he, esteemed though he was, clung to a belief in the existence of supernatural spirits as critical elements in the sober study of medicine. He was born in Spain in the dawning days of the Christian era, though he spent most of his professional career in Rome. Columella was something of an anomaly in ancient biological science in that he was one of the few who devoted considerable time to agriculture, writing a twelve-book study of the subject. Considering the economic makeup of Rome, however, his work on agriculture was eminently understandable. Since the opening decades of the Etruscan civilization, Rome had been primarily an agrarian society. There was negligible overseas trade and little manufacture. Indeed, it was in part because of Rome's very heavy dependence on farming that the empire eventually collapsed. During the second and third centuries A.D., Rome had become grossly overextended. Having conquered practically every square inch of land surrounding the Mediterranean, including Carthage in northern Africa and much of Europe, the Romans barely had the manpower to both farm the land and maintain an army. In the ensuing years, crops began to fail because there was only some comprehension of how to rotate crops to maximize soil fertility. Also, the plagues that hit Rome further decimated and weakened the population. That forced Rome to thin her manpower even more in a desperate attempt to both preserve a strong army and feed the people. Seeing that Rome was on the brink of collapse, the various German hordes such as the Goths, Ostrogoths, and Visigoths saw their chance, and, in the fifth century A.D., they finally vanquished Rome.

That Rome did not collapse even faster was possibly due to Columella's treatises on agriculture. He surveyed and wrote about the best ways of caring for livestock, and even made helpful suggestions about taking care of crops.

PLINY

One of the outstanding biologists of the Roman age was Gaius Plinius Secundus (Pliny). Born in A.D. 23 in the town of Como in what is now northern Italy, he came from a notable and politically active family, and he himself strayed into the political arena for a while. He was well educated and served Rome in both a civilian and military capacity for some years. Yet even while diligently carrying out his various official duties, he managed to carve out time to carry out his scientific investigations, though he was also curious about and researched a mélange of issues in everything from linguistics to military tactics and philosophy.

Like most intellectuals of Rome at this time, he was swayed by philosophies such as Stoicism and Epicureanism. The former was a philosophy that encouraged bravery in the face of adversity and the latter stressed the notion that pleasure, but only in moderation, was the ultimate good. Thus, Pliny displayed open indifference to the bounteous pleasures available to him in ancient Rome. As ever, Aristotle's supremacy lingered and, naturally enough, touched Pliny. He appears to have accepted Aristotle's cosmology of the universe as consisting of concentric spheres, as well as many of Aristotle's views on motion, such as the precepts of actuality–potentiality and the principle of the "four causes" of change—material, final, formal, and efficient causes.

His most notable achievement is his *Historia Naturalis*, or "Natural History," which exerted leverage over biologists and agriculturalists through the Byzantine era—shortly after the fall of Rome—and well into the Middle Ages. This series consisted of some thirty-seven books, covering everything from medicine and economics to physiology and comparative anatomy. In *Historia Naturalis* he devotes a huge amount of space to zoology. He describes, among other things, the behavior and the practical use of certain animals, such as in farming, and gives some information about their anatomy.

Perhaps the chief drawback of his writings is the fact that he incorporates a massive amount of legendary and mythological "data" as if it were scientific fact. He also restates ancient legends about animals of other lands as if they too were factual, attributing a hodgepodge of humanlike properties to the elk, for instance. He says, for example, "Among land animals, the elephant is the biggest; also its intelligence is closest to man's, since it grasps language, can obey orders and has a powerful memory . . . it also worships the heavens, including the stars and the moon . . ." (my translation). He claimed, too, that animals "care deeply" for smaller and weaker creatures, without any facts to support his thesis.

Pliny more accurately described the anatomy of animals he could examine directly, such as farm animals of all sorts, even noting how various organs change, from one species to another. Again, Aristotle obviously impressed him, inasmuch as various of his anatomical descriptions appear as if he had snitched them from Aristotle rather than having come up with them himself. Even so, Pliny made helpful comments about caring for and prolonging the useful life of cattle, and he goes into some detail about the anatomy of molluscs, fish, caterpillars, and so forth. Interestingly, he spent a considerable amount of time studying bees—conscientiously describing their behavior, anatomy, and the like. Thus, although Pliny did do a middling amount of original research, his most famous contribution consisted of assembling together an enormous amount of material appropriated from previous biologists. According to Pliny's own testimony, in writing his "Natural History," he relied on at least 2,000 publications by earlier authors. Despite the peculiar mix of fact and mythology, later Renaissance theorists such as Gesner, and even Da Vinci, found Pliny a valuable source of information. Pliny perished while witnessing the eruption of Mt. Vesuvius in A.D. 79.

LUCRETIUS

Asclepiades was a Bithynian physician who argued that he had cured innumerable patients by merely bathing them plus prescribing a sound diet and moderate exercise. As modern as that sounds, he nonetheless retained mystical and occult elements in his theory, because he argued that "disturbances" within the body caused disease—an idea he had pirated from Plato and, to some extent, the pre-Socratic philosophers.

Aulus Celsus, who lived around the time of Christ, assisted in putting together one of the first medical encyclopedias of the ancient world, titled *De Medicina,* or "On Medicine." The books of this collection consist of assorted helpful and fundamental observations about the anatomy of the eye, tooth, ear, and numerous other parts of the human anatomy. Following him was Alderotti Thaddeus, a wealthy and ingenious Florentine of the first century A.D. who authored the book *De Virtutibus Aquae Vitae,* or "The Virtues of Alcohol." In this, he described innumerable medical applications for alcohol, including reducing fever.

Yet men like Asclepiades and Celsus pale by comparison with the most illustrious Roman thinker—Titus Lucretius Carus. Born in B.C. 99, he came from a wealthy Roman family and may well have known intimately such statesmen as Caesar. His preeminent writing was his long poem, *De Rerum Natura,* or "On the Nature of Things." Taking the lead of his predecessors, such as Leucippus and Democritus, he too was an "atomist" and believed that atoms and the "void" were the two basic constituents of the cosmos. A brilliant and resourceful thinker, Lucretius introduced a modification to earlier atomistic doctrine. A problem that bothered him was exactly how atoms collided to build the aggregates of matter that eventually resulted in the planets, stars, and earth. To address this problem, he introduced what the history of science has come to know as the theory of the "swerve." According to Lucretius, a periodic and entirely random swerve from their downward path punctuated the fall of atoms. Such variations in the downward path would lead to the collisions Democritus spoke of, forming ever larger masses of matter. In Lucretius's own words:

> This point too herein we wish you to apprehend: when bodies are borne downward sheer through void by their own weights, at quite uncertain times and uncertain spots, they push themselves a little from their course; . . . If they were not used to swerve, they would all fall down like drops of rain. . .[7]

Like Aristotle, Lucretius believed that one could explain the human soul with the laws of physics. He therefore attempted an "atomistic" analysis of it. He accepted the tripartite division of the soul of Plato and Aristotle mentioned earlier, albeit with some modifications. He supposed that the three principal parts of the soul were the *anima, mens,* and *animus,* that is, "life-force," "reason," and "spirit." Like the rest of the body or indeed any material thing, he believed that the soul was composed of atoms, albeit atoms that are much tinier than the atoms of the body. He further subdivided these atoms into "warm," "air," and "aura." There was also a fourth element; he never actually labeled it, but he did insinuate that it constituted human consciousness. He believed he was thus able to explain all human emotions, such as anger, fear, and the like, in terms of the various ways in which the highly mobile and volatile atoms of the soul intermingled. Fear, to offer an example, consisted of "cold atoms."

Although Lucretius used the term "soul," his meaning of the word was unrelated to that of the Christian philosophers of the medieval period or even Plato and Aristotle, who believed that it was absolutely crucial to the teachings about the soul that it be immortal. While the body might perish, the soul could not. Understandably then, Lucretius vigorously assaulted philosophers like Plato, since Lucretius rejected the Platonic "spiritual" constitution of the soul. In fact, Lucretius's ideas about the "soul" constitute a rudimentary version of what philosophers and scientists have come to call the "psychophysical identity" theory. In this view, what we normally call "mental states," or "states of the soul," such as memory, thought, fear, and so forth, are merely regions in the brain. (This way of looking at the human mind emerged in the 1950s with philosophers like J. J. C. Smart and U. T. Place. Some, such as David Armstrong of the University of Adelaide, still cling to it, although succeeding theoreticians such as Norman Malcolm of Cornell University have argued strongly that the central precepts are indefensible.)

Still, Lucretius made the atomistic approach appear plausible enough. He was of the persuasion that with it he could analyze such phenomena as sense perception, arguing that physical objects give off extremely "light atoms," much as Empedocles had argued, and that these light atoms then impinge on the eye, giving rise to vision. He furnished similar explanations of the other senses. Conceivably, one could even interpret dreams with atomistic tenets. According to this view, the images in dreams are merely the remnants of images that have come from physical objects and "imprinted" themselves on the eyes and ultimately the soul during waking life.

It is surely worth remarking that like Plato, Lucretius was not only a brilliant philosopher, but a writer of immense power. He was one of the few philosophers of yesterday who managed to put his reflections down in poetic form and still communicate substantive scientific and philosophical thoughts. In his writings he expresses his assurance that the strength of human reason is practically unlimited; given sufficient time it would surely unlock the secrets of the universe. Thus it is scarcely a shock to learn that he had scant use for dogma or religion of any form, believing that since it bypassed human rationality, it could lead only to ignorance and darkness.

So compelling was the presentation of his ideas that they survived well into the Middle Ages and the Renaissance, despite relentless efforts by medieval theologians to suppress them. He influenced such thinkers as Giordano Bruno and Dante, and more recent atomic scientists, such as Dalton and Berzelius—who lived and worked many centuries later—owe their key ideas to Lucretius and his predecessors. Indeed, the atomic hypothesis is, arguably, the main conceptual underpinning of all modern scientific theory. The applicability of this theory to biology has been demonstrated over and over again.

Science has long since proved beyond any doubt that all biological molecules are composed of atoms.

GALEN

If Lucretius was a significant influence on modern physics, the biologist Galen was an equally significant influence on future biology. He was born in 131 in the city of Pergamum in Asia Minor. Although born a Greek, he later moved to Rome, fell in love with Roman culture, and even altered his name to the more Roman Claudius Galenus. His father was an architect, and according to legend, had dreamt that destiny called for his son to become a healer. Like so many, his education, even in Rome, consisted in large part of Greek philosophy, such as instruction in Plato, Aristotle, Stoicism, Cynicism, Epicureanism, and so forth. (The explanation of this common anomaly is apparent enough. Rome, as noted earlier, had appropriated much of its culture from Greece. The Roman gods were really just the Greek deities with different names—Poseidon became Neptune, Herakles became Hercules, and so forth. In philosophy, Rome contributed little that was new. Even some of the most eminent of the Roman philosophers, such as Lucretius and Epicurus, relied heavily on Greek thought. There is no philosopher of Rome who even remotely approached the importance of Plato and Aristotle. Even by the Middle Ages, the philosophers of the Church were relying on the Greeks rather than the Romans for the intellectual foundations of their theology. St. Thomas Aquinas relied virtually exclusively on Aristotle, for example, much as St. Augustine relied heavily on Plato.)

After his philosophical training, Galen began to study medicine, first in Corinth and later in the Hellenistic city of Alexandria. He returned to his native city in 158, becoming a court physician at the city's school for gladiators. Tiring of this after a few years, he headed for Rome, where he began teaching science. Before long, his reputation in medicine and biology had become formidable. Finally, he became the personal physician to the most majestic of all Roman emperors, the philosopher-statesman Marcus Aurelius.

Galen's Physiology

Galen used the pulse to diagnose illness, although he had no understanding of the circulatory system as such. Even so, he did put together something resembling an encyclopedia of medical knowledge. Like so many others, Galen relied heavily on Aristotle, but he did offer a few ingenious speculations of his own regarding, for example, the circulation of the blood, and he described the aorta and the primary veins of dogs, pigs, sheep, and other beasts. His discussions of the digestive system are an interesting blend of fact and fiction. He explained digestion by a mysterious force he called the "transformation power," which existed in the stomach. First the food entered the stomach (so far so good), after which it underwent "coction," or boiling. After breaking down the food, the stomach sent these more basic elements to the liver for conversion into blood. From the liver, the blood went to the heart and then to the lungs. The waste products were converted to "black bile," which the body excreted via the bowels. Finally, a mixture of blood and pneuma, or "vital spirit," headed into the aorta and traveled to all parts of the body to provide nourishment. This odd belief, as it turns out, would be in use until the age of St. Thomas Aquinas.

Galen's Anatomical Ideas

To pursue his anatomical interests, Galen examined creatures of nearly every imaginable species. He studied them both alive and dead, being unafraid to pursue vivisection, except on human beings. His descriptions of the circulatory system were not bad; he was the first to reject the tattered mythology claiming that the heart, especially the left chamber, contained air. Although he had no grasp of the importance of oxygen in physiology, he did in a way anticipate this view when he defended the tenet that there was some hitherto undiscovered element in the blood which was the basis of life. Oxygen, of course, was that element.

Despite his superb contributions, Galen did make mistakes. And several of his disciples, not yet fully appreciating the scientific method, treated Galen's word as gospel, much as others had treated Aristotle's. In his depiction of the anatomy of the heart, for example, he mistakenly maintained that both veins and arteries carried blood away from the heart, and that the wall between the left and right ventricles was porous, allowing blood to pass between them. He also claimed that the liver was the "seat" of the venous system.

On the other hand, his explanation of the blood vessels and genitalia is tolerably accurate. He also described both the foot and hand in exceptional detail, along with the brain, nervous system, and spinal cord. He appears to have been aware of the fact that the latter two structures carry nerve impulses to the brain and vice versa. He showed that nerve branches innervate every part of the body. He also devised a detailed "map" of the brain, even describing with fair accuracy the paths of a cluster of nerves that contemporary neurologists call the cranial nerves. Again, however, he clung to the antiquated philosophy that the "soul pneuma" was lodged in the brain and circulated throughout the body. Galen's scientific output, though not totally reliable, was impressive for its sheer bulk alone.

SUMMARY

The progress of biology ground ahead only haltingly during the Roman era. While there was some good work in comparative anatomy, taxonomy, and physiology, much of Roman medicine was borrowed from the Greeks, just as most other Roman accomplishments were. And Galen's work notwithstanding, arguably the greatest contributions to science were not to medicine or biology, but rather to physics, with the work of Democritus.

CHAPTER 5

The Middle Ages

After the fall of Rome in 476 A.D., there was an immensely long period of relatively little progress. Of course the confusion had already begun in the third century A.D., when economic hardship and political confusion was growing. History teaches us that, in such circumstances, most people have little time for abstract speculation and scientific research. Instead, as is usually the case, the people of Europe turned toward religion, to the new cults of Christianity, to Serpis, and to Mithraism. To add to this confusion, there were at least three civilizations ready to fill in the power gap caused by Rome's fall: the Byzantine civilization, western Christianity, and the new religion of Islam.

There are again men of great genius in both science and philosophy, by the time of the high Middle Ages, from 1050 to 1300, including Avicenna, Averroes, and St. Thomas Aquinas. Here, historians have recently rethought the old dogmas. While western Christianity, Islam, and the Byzantine empire have traditionally been regarded as backward and unprogressive (especially western Christianity), historians have begun only recently to appreciate the full measure of this age's accomplishments, e.g., Islamic advances in science and philosophy, especially those of Averroes (though he did not appear until the 12th century). Indeed, the first to do any really significant work were the Muslim *faylasufs,* who essentially served as the scientists in Muslim society. In addition to significant work in chemistry, such as the discovery of silver, nitric acid, alum, and many other substances, faylasufs did some important work in optics, including the refraction of light and the construction of lenses.

In biology, Avicenna showed that diseases such as tuberculosis, pleurisy, and others, were contagious. In fact his *Canon of Medicine* was the most highly regarded writing in medicine until the time of Descartes in the seventeenth century. Other Islamic physicians studied cancer and could treat a variety of eye diseases.

The reasons for such an intellectual rebirth in the high Middle Ages are complex; this much can be said, however: everything changed for the better with the great spread of schooling in the high Middle Ages. Schools began, with the support of the papacy, to appear in the larger cities. Then, as economic conditions improved, the more well-to-do families were able to send their children to these schools, and the schools no longer served just to provide an educated clergy. In fact, by the fourteenth century, most schools were no longer even controlled by the church. The universities began to appear, the first being the University of Bologna in the 12th century.

In this climate, science finally began again to prosper. So strong was the impetus towards education and science that even during the late Middle Ages, despite disease and poverty being again rampant, great thinkers such as William of Occam nontheless emerged.

During the Middle Ages, numerous Arabic scholars recovered the treatises of Aristotle, and his writings permeated all subsequent thought, right through the Renaissance and even well into the Enlightenment. Not only did Europe see considerable biological and scientific progress during the medieval period, but the Middle East did as well. In that part of the world there were systems of hospitals and a magnificently organized medical profession that in some ways was reminiscent of ancient Egypt.

ARISTOTLE IN THE MIDDLE AGES

Without reservation, the biological authority during the Middle Ages was Aristotle. Nearly all of the intellectuals of this period regarded the exalted Greek philosopher as the ultimate authority on all matters philosophical, physical, biological, astronomical, and even poetical. So towering a figure was he that even St. Thomas Aquinas referred to him simply as "The Philosopher."

As was the case in the past, even scholarly men hardly ever valued biological learning for its own sake, but rather because of its medical applications. Among the most distinguished biologists was Albertus Magnus, astute medieval sage and also the teacher of St. Thomas Aquinas. Born around 1193 A.D., Magnus was a devout Catholic, and despite coming from a rich and powerful family, early in life he entered the austere Dominican order. Eventually he joined the faculty of the University of Paris and later became Bishop of Regensburg. But soon afterward he returned to the monastic order and steeped himself in his scientific and philosophical meditations.

It is perhaps a tenable position to assert that Magnus was one of the first real chemists of history. Although many before him had ventured into alchemy, he was clearly the first to appreciate the distinction between the scientific and the occult branches of chemical studies. He was the first to synthesize arsenic, and he began to realize that certain chemicals had a tendency to combine spontaneously with certain others. It is probable that Magnus influenced Aquinas' great skepticism of alchemy, as Aquinas relates in the following excerpt:

> Even in these art can produce a similitude, as when alchemists produce something similar to gold as to exterior accidents. But it is still not true gold, since the substantial form of gold is not [induced] by the heat of fire—which alchemists use. . . . Hence such [alchemical] gold does not operate according to the species [of real gold], and the same is true for the other things that they make.[8]

Magnus believed, as did his pupil St. Thomas Aquinas, that the Almighty had called him to base Church teachings on the philosophy of Aristotle. Like Aristotle, he too fell into absurdity in proclaiming that the brain is "cold" and that the arteries contain "air," and so forth. Nonetheless, he also disagreed with Aristotle, which was a hazardous undertaking at the time. He disavowed Aristotle's belief, for instance, that the soul was located in the heart. Instead, anticipating modern views, he suspected that the soul was in the brain. He traveled throughout Europe, relentlessly observing and cataloguing plants and animals of every conceivable type. An expansive thinker, he studied geology and physics as well as biology and classified a number of minerals into discrete categories. Perhaps Albertus Magnus's greatest contributions to biology lay in his work describing and cataloguing the plants of Europe, although biologists now know that much of this work is incorrect.

FACT AND FICTION IN THE MIDDLE AGES

Yet despite the rationality of St. Thomas and the paradigm of the scientific method that Aristotle had tendered, magic and superstition persisted in playing a dominant role in medieval biology and science generally. This is scarcely surprising, because the science of the day was necessarily empirical, that is, lacking in any solid theoretical basis. On the one hand, physicians found that individual medicines cured certain maladies and so they continued to use them. More theoretically minded thinkers, on the other hand, were not content with the mere fact that a medicine or technique worked; they wanted to know why. Having no modern scientific knowledge, they turned to magic and superstition to provide the answers.

A conspicuous feature of ancient and medieval science was the abundant use of superstition, astrology and magic as "theories" intended to explain why certain things happened in the universe. Medievalists often appealed to magic, for instance, to show why a given medicine performed the way it did. Although such unsound "theory" did not help the progress of medical biology, it did not really hurt it either. There was little chance the ancients would have been able to develop anything resembling contemporary scientific doctrine with the meager knowledge they had about the way the cosmos worked. So they continued to proceed empirically, cataloguing the effects of various plants and minerals (for example, they knew that opium and mercury could function as anesthetics). With the passage of decades, this list grew ever longer. Later, in the seventeenth century, the English physician Thomas Sydenham and several others would augment the uses of these resources as well as introduce others.

It could hardly be said, then, that the Middle Ages was a period of any unusual contribution to or progress in biology or medicine. The authority of Aristotle was simply too great, holding back original thinking on the part of many. Probably the greatest contribution of a medieval thinker to science was the work of Albertus Magnus in chemistry.

CHAPTER 6

The Renaissance

The Renaissance was a time of great interest in education, in comparison with the late Middle Ages. Progress accelerated in science, art, literature, and most other branches of human knowledge. One exception to this progress was philosophy, for reasons no one has ever really grasped, save perhaps for the fact that people were plainly more interested in comprehending the physical world around them than in abstractions. Aristotle remained a dominant power, and Galen continued to influence anatomical research.

LEONARDO DA VINCI

One of the giants of the Renaissance was Leonardo Da Vinci. Not only were his biological labors of the first importance, but he achieved immeasurable fame in art as well as other branches of science. The artists of the Renaissance realized that magnificent art could emerge only from a thorough grasp of human anatomy, so Leonardo's, as well as other artists', varied interests were eminently understandable.

Leonardo was born in the village of Vinci, near Florence, in 1452. As a child he spent innumerable hours watching the fauna of the Florentine hills. He made countless drawings of trees and other plants growing in his hometown. Recognizing his enormous ability, his father sent him to begin his tutelage with Verrochio, one of the prominent artists of the period. From Verrochio he mastered the realist style of the Renaissance, which urged the artist to paint and draw things as they were—a fortunate coincidence for the biological artist. Although no one is sure, many historians surmise that Leonardo actually dissected as many as thirty corpses in his youth. He also recorded his commentary in the form of notes and drawings that fill 129 notebooks. In fact, there are about twenty notebooks and bound volumes of his writings reaching 4,000 pages still extant. Leonardo's finely detailed drawings helped others make headway in understanding the muscular and skeletal systems, and, most significantly for biology, the relationships between them.

Somewhat subsequent to his studies with Verrochio, he obtained permission to begin his education in anatomy at the hospital of Santa Maria Nuova in Florence. His adult life was turbulent and unsettled, and he wandered endlessly throughout Europe, ultimately landing in France. He was an empiricist through and through, always basing his studies, whether in physics, art, or anatomy, upon close observation. Fortunately, some of his drawings and other artwork have survived until the present day.

He was, from the beginning, an Aristotelian in that he followed the empirical method to the letter, though he did not hesitate to disagree with an occasional conclusion of Aristotle's. He was a thorough adversary of Scholasticism, the medieval doctrine espousing that human reason alone could comprehend theological mysteries. As he said in one of his notebooks, "Wrongly do men cry out against innocent experience, accusing her often of deceit and lying demonstrations." He challenged this doctrine not so much because the Scholastics relied so heavily on Aristotle, but because he perceived them as imperious and unwilling to confront the evidence of their senses. In this way he disparaged all who followed a similar path. Fortunately for the progress of both art and science in the Renaissance, he inspired essentially all of the important scholars of that era.

Following Da Vinci's lead, other Renaissance authorities began to read the medieval metaphysicians with an extremely critical eye. In doing so, they chastened them for everything from their tortured Latin prose to their dogmatic and intolerant inferences. In a word, Da Vinci and his followers believed that the Scholastics erred in following only the conclusions, and not the methods, of Aristotle.

Da Vinci did not limit his pursuits to human anatomy; he also studied the physiology of sense perception. His interests to some extent even extended to geology, and he carried on a number of wary surveys of geological strata, even arguing the altogether modern position, against Aristotle, that fossils were the hardened remnants of ancient animals.

CHEMISTRY DURING THE RENAISSANCE

At this point in history, scientists were starting to understand the role of chemistry in biology. The key name in this field during the Renaissance is Paracelsus. He was christened Theophrastus Hohenheim in 1493. Although miscellaneous details of his life are not known, he apparently was born at a monastery in Maria-Einsiedeln in Switzerland. Numerous historians believe that his father was the illegitimate child of a knight in the noble family of Bombast von Hohenheim and a peasant woman. Doubtless because of his illegitimacy he saw few benefits from his royal lineage and grew up in abject poverty instead. Even so, his father and some parish priests saw to it that he received a solid education at Basel University. Still, he inevitably grew weary of the dogmatic and rigid Scholasticism he found at Basel and withdrew to learn alchemy under an abbot who had established some semblance of a laboratory in his monastery. In the ensuing years he broadened his studies to include metallurgy.

Eager to broaden his scientific vision still more, he began a trek through the European states as a sort of *scholare vagrante,* as many called people who did this kind of traveling. At various times, he surfaced in France, Italy, Spain, Germany and Sweden. During these travels he became fascinated by the occult and explored witchcraft, demon worship, and other assorted systems. Even so he retained enough of a foothold in biology and real science to become the town physician in Basel as well as a lecturer at the university.

Hardly known for his tact, he usually got his way chiefly through his persistent and overbearing methods. An oddity even in the classroom, he would start each semester by burning the tomes of Aristotle, Galen, and others with extreme ceremony, to parade his contempt for dogma, no matter how cherished the creator.

After still more years as a *scholare vagrante,* he finally moved to Salzburg, at the invitation of Archbishop Ernst, around the middle of the sixteenth century. Paracelsus took a step toward modern chemistry by emphasizing the biological effects of different kinds of substances, although he remained in the grip of alchemy and, like Copernicus and Kepler, concluded that the forces controlling the cosmos were spiritual rather than physical. He authored the *Paragranum,* in which he contended that medicine was little more than a branch of physics. He argued further that the laws of biology were really not independent laws at all, but merely specific cases or applications of the laws of physics, a view still extensively debated today by philosophers of science. Beyond this he did not automatically bend to the authority of the old philosophers like Aristotle. Instead he traveled much, always in search of better medicines as well as more knowledge about how the environment affected the body. Most significantly, his appreciation of the relationship between chemistry and medicine anticipated the science of pharmacology.

BOTANY IN THE RENAISSANCE

Men like Brunfels and Fuchs compiled sizable amounts of information during the Renaissance on the subject of botany. Otto Brunfels, born in 1489, is often called one of the "fathers of botany." He had at first intended to serve the Church, but later abandoned this plan to begin his apprenticeship in medicine. A bookseller in Strasbourg commissioned Brunfels to write a medical book based on plants and herbs that were very commonly used in medicines. From this emerged his *Herbarum Vivae Eicones,* or "Living Pictures of Herbs." Unfortunately, Brunfels presented nothing that was really new here, relying mostly on classical biologists such as Discorides and, of course, Aristotle. Also, various of the drawings in the book, done by the artist Hans Weiditz, were plainly faulty.

Paracelsus (1493–1541). (Courtesy of the Library of Congress.)

Scarcely less important were the contributions of Leonard Fuchs. Born in 1501, he initially began his schooling under the umbrella of Scholasticism. Nevertheless, after reading Luther and Calvin, he converted to Protestantism. As was customary at the time, he entered the field of biology via a medical degree. Like Brunfels he devoted most of his hours to taxonomy, even though his official occupation was professor of medicine at the University of Tübingen. He was a practicing physician as well.

After years of research, Fuchs produced his classic *De Historia Stirpium* in 1542. In this book he systematically described and arranged over 400 plants. The book, not surprisingly, derives much of its biological theory from Aristotle and Plato. Doubtless influenced by Plato's complicated classificatory system in his dialogue the *Sophist*, Fuchs divides the analysis of a plant into several categories, such as "character" and "form." For each species of plant he studied, he noted its structure, at what time of year the plant flourished, and where it was found.

Oddly, he adds two bizarre categories called "temperament" and "powers." In all likelihood, these categories stem from the Aristotelian idea that vegetative forces subsist in all living things causing them to grow and flourish. Doubtless the book's main positive attribute, however, is that it was far better researched and documented than anything earlier botanists had composed.

VAN HELMONT AND SPONTANEOUS GENERATION

In 1500, Jakob Nufer of Switzerland, for example, actually carried out a successful Cesarean operation—probably the first documented case in history. There was not to be another comparable achievement in surgery until the mid-seventeenth century, when Johann Shultes wrote his *Armamentarium Chirugicum*, or *Tools of the Surgeon*, which proposed a not very pleasant procedure for performing a mastectomy.

Learned men everywhere have always wondered about the beginnings of life on earth. Many explanations were, of course, spiritual. Others, not satisfied with religious theories, sought natural ones.

One of the earliest attempts to explain the origin of life was through the notorious doctrine of "spontaneous generation." This concept touched a mystical nerve. According to this esoteric teaching, life could arise from nonliving matter Aristotle, among others, gave the idea a long shelf life when he claimed that he had witnessed minute organisms arising "spontaneously" in ponds and rivers. The proposal of spontaneous generation was, therefore, a theory about life-forms whose origins the scientist could not directly observe. In the sixteenth century, men like the Belgian biologist J. B. Van Helmont also supported this assumption, even suggesting that larger animals like rats arose spontaneously. In van Helmont's view, if one placed moist, dirty clothes and bran inside a vessel in a dark corner and surrounded them with food, mice would appear spontaneously. Van Helmont also believed that he had produced gold from quicksilver. So much for "scientific method."

To be fair to Van Helmont, in other endeavors he did do better than this. In plant physiology, for instance, his techniques were much more rigorous. Van Helmont was born in Brussels in 1577 to a wealthy family. His father died when he was young, but he seemed to recover swiftly from the loss and by the age of seventeen he had already completed his college degree in natural philosophy. He later entered an obscure Jesuit school to begin his instruction in theology. Having something of a nervous and mystical temper since childhood, he found the philosophies of both Middle Platonism and Neoplatonism, with their rococo and arcane splashes, enormously intriguing. Like the Neoplatonists, he believed for much of his career—foreseeable in part because of inordinate personal anxiety about his own fate and his nervous disposition—that there was some mystical realm of absolute perfection to which the soul could rise through meditation. He accepted too the "emanationist" premises of Neoplatonism claiming that the created universe "emanates" or "spills" out of the spirit of God in a series of stages. Now and then he would stare for hours into bright lights in an effort to produce mystical trances. Not surprisingly, the mystical reflections of Paracelsus affected him deeply as well.

Nevertheless, the more scientific part of his personality continued to bedevil him, and he soon relaxed his pursuit of mystical philosophy to master medicine and biology. He received the doctor of medicine degree at age twenty-two, and like so many others, he became one of the *scholares vagrantes,* roaming through Europe studying here and there wherever he could spot a library or someone practiced in science who was willing to speak with him.

He studied, among other things, the process of fermentation. He showed that it produces an odd sort of air, which was the same as that produced when he burned charcoal. Although he understood few of its details or that it was carbon dioxide, he named this peculiar air "gas," a term that has found a permanent place in scientific nomenclature. He linked these results to digestion, contending that the body processed food through some kind of fermentation. Although many of his conclusions were way off the mark, some were not. He found, among other things, that an acid in the stomach played a critical role in digestion and that the gallbladder would neutralize too much acid.

Because of his innate philosophical bent, however, Van Helmont sought to probe even further into the puzzle of life itself, beyond the mechanical processes of the body. Here he again lapsed into bizarre mysticism. He advanced the idea that there was an organ called the "archeus," which was located near the stomach and which controlled all physiological processes. Corresponding to this was a spiritual archeus, which he called the "intellectus," a concept scarcely more than a modification of the crude Aristotelian precept of the soul. According to Van Helmont, this intellectus would have made the soul "blessed" had man not fallen in the Garden of Eden. The idea is cryptic and all but meaningless, save that it seems to be roughly equivalent to man's "perfect goodness," which he lost by disobeying God in the Garden of Eden. With that event, man received an inferior type of soul Van Helmont called a "ration," which fettered the soul to earth in a material body.

The pre-Socratic philosopher Thales may have influenced him, since Van Helmont claims that water was the most basic constituent of the cosmos. To prove this, he contrived one of his better experiments. It consisted of planting a willow tree in 200 pounds of dry soil and measuring the increasing amounts of water in it as it grew. He then compared the weight of the fully mature tree with the weight of water he had put in the soil and concluded that the only thing the tree required for growth was water. Although his conclusion is not correct in that he overlooked the role of nutrients as well as air and sunlight, his method of tackling the problem was basically sound.

HOMOLOGY

A major concept in modern biology that surfaced at about this time was the concept of "homology"—two unlike structures that have, nonetheless, similar embryonic and evolutionary origins. Possibly the first academic to grasp the existence and import of homologies was the French biologist Pierre Belon. Born in 1517, near Le Mans in France, he grew up in an impoverished household. Recognizing his immense abilities, however, a local bishop saw to it that he received the finest education imaginable. Ultimately, he became a physician. He continued his studies in Germany and, again through the assistance of a number of wealthy sponsors, he undertook an extensive world tour. He pursued his biological examinations in such far-off lands as Syria, Persia, northern Africa, and the Hellenic world. When he returned, King Henry II decided to support his scientific investigations. Unfortunately for science, a gang of thieves murdered him in 1564.

His short lifetime notwithstanding, Belon carried out extensive comparative inquiries. He published three volumes, *History of Birds, Natural History of Strange Marine Animals,* and *The Nature and Diversity of Animals.* Much of his analysis consisted of some rather erratic groupings; he mistakenly identified as "fishes" such disparate species as seals, crustaceans, sea anemones, beavers, otters, and even hippopotami and certain species of lizards. Still, one must concede that he did introduce some helpful distinctions, based on internal as well as external anatomical features. He distinguished between cartilaginous and bony fish—still a significant distinction.

In his reflections, Belon first found homologies between particular bones in fish and certain mammals. Although the homologies he claimed were habitually inexact, he did succeed in proving that there were fundamental skeletal similarities between apparently disparate species. One of his most convincing performances was his comparison of the skeleton of a man with the skeletons of birds, where he pointed out distinct similarities in the skeletal and musculature systems. Because he traveled widely, Belon paid considerable attention not only to Western animals, but to many species native to the Orient.

SUMMARY

It is unarguable that scientific progress went ahead rapidly during the Renaissance, especially under the eyes of men like Leonardo Da Vinci. With the concept of homology, comparative anatomy took a large step toward its present form. At the same time, Renaissance scientists hardly appear to have made a complete break with old ideas. Aristotelianism lingered, as did alchemy, even as Paracelsus was beginning the long road to modern chemistry.

CHAPTER 7

Pestilence in the Renaissance

Precisely as it had been during the Middle Ages, disease was the scourge of the Renaissance. In the camp of the French general Charles VIII for example, during his offensive on the Iberian peninsula, the Spaniards he was attacking infected his troops with syphilis. The French army introduced the infection upon returning home, and the disease spread throughout most of Europe. The subsequent loss of life was catastrophic.

SMALLPOX

Smallpox seemed to be everywhere. It ravaged, for example, the tiny island of Hispaniola as well as scores of the Aztec Indians, allowing Sir Hernando Cortés and his army, who had been exposed to smallpox in childhood and were therefore immune, to effectively capture the illness-ravaged Indians. A similar situation occurred in Peru, where the disease killed the Inca ruler Capac as well as thousands of his people. Centuries later, the infamous Great London Plague would annihilate nearly eighty thousand people. Smallpox had been described as early as the Middle Ages by Arab scientist *Al-Razi*. Science would not be able to take even the most meager steps toward controlling smallpox until 1701, when the Italian physician Giacomo Pylarini single-handedly invented the field of immunology by inoculating several children in Constantinople with a primitive vaccine against smallpox; historians differ about the true success of his project.

By 1717, the wealthy English letter writer and amateur naturalist Lady Mary Wortley Montagu had introduced the practice of inoculation to England from Turkey, even venturing to have her own children vaccinated. She describes her progressive worldview and her equally crude "inoculation" methods in a letter written in 1717 in Adrianople:

> Apropos of distempers, I am going to tell you a thing that will make you wish yourself here. The small-pox, so fatal, and so general amongst us, is here entirely harmless by the invention of ingrafting, which is the term they give to it. . . . People send to one another to know if any of their family has a mind to have the small-pox; . . . the old woman comes with a nut-shell full of the matter of the best sort of small-pox, and asks what vein you please to have opened. She immediately rips open that you offer to her with a large needle . . . and puts into the vein as much matter as can lye upon the head of her needle, and after that binds up the little wound with a hollow bit of shell . . . There is no example of anyone that has died in it; and you may believe I am well satisfied of the safety of this experiment, since I intend to try it on my dear little son.[9]

The vaccine she introduced probably consisted of weakened strains of the virus causing the disease; Montagu's child did not in fact contract the sickness, although historians are not sure whether this was due to the vaccination, mere good fortune, or even false a diagnosis. There is a large element of uncertainty about the extent of smallpox in history, since physicians of earlier eras often confused it with chicken pox or even syphilis. In any event, a few years later, the American physician Zabdiel Boylston introduced a version of Pylarini's vaccine to America while the ravaging contagion was tearing into the city of Boston. So far as historians can tell, the results appear to have been favorable.

MALARIA AND YELLOW FEVER

Subsequent to this, the science of immunology progressed still further, when the eminent Italian biologist Giovanni Lancisi, in his

"On the Noxious Effluvia of Marches," argued that mosquitos could transmit malaria and yellow fever. While the Western nations, at least, did not suffer from these plagues during the Renaissance to the extent that populations did in the late Middle Ages, they were seldom immune either. One of the worst episodes was a massive attack of yellow fever which decimated the island of Barbados in 1647. Not until 1898 did physicians such as Walter Reed draw public attention and medical funding to this dreaded malady. Even so, it would not be until 1892 that the renowned Italian biologist Camillo Golgi would explain that various forms of malaria stem from parasite infection. And only in 1897 did the English doctor Ronald Ross find that the *Anopheles* mosquito carried the malaria parasite. Later, in 1904, Ross would publish the memoirs of his battles with the anopheles mosquito in his book *Researches on Malaria.* In 1881, Alphonse Leveran found the protozoon causing malaria.

GLISSON'S CONTRIBUTIONS

Yet more scourges lurked. Francis Glisson of England published in 1650, his *De Rachtide* (translated in 1668 by Nicholas Culpper) in which formulated a detailed depiction of rickets, a childhood disease characterized by soft and weak bones caused by insufficient sunlight or vitamin D in the diet.

Born in 1597, Glisson came from a family of moderate income. He began his university education at Cambridge in the study of medicine. Directly after that he began teaching medicine and biology, though the English Civil War forced him to flee to London. There, without delay, he began a medical practice which brought him a considerable reputation—so much so that the scientific elite of the day would make him one of the first scientists to be invited to join the Royal Society. Glisson conducted careful studies of both the liver and the intestinal tract which anatomists still hold in high regard even today. In honor of this effort, science has named a tissue layer in the liver the Glisson's capsule.

Oddly, even at this late date, Scholasticism and old-fashioned Aristotelianism still blighted the struggles of some thinkers, and Glisson was no exception. He seems to have accepted the doctrine of hylomorphism, or the idea that all things are composed of matter and form, and saw the cosmos as evidence of the heavenly Father's handiwork. As a physician, he was interested in a liberal assortment of ailments, and it was this concern that led him to an acute exploration of rickets. Unfortunately, though he understood much about rickets, he had precious little to say that was of any value in treating it.

WILLIS AND TYPHOID FEVER

Scientists eventually conjectured that slaves coming to England from Africa had brought yellow fever with them around the 1640s. By 1659, Thomas Willis of Great Bedwyn, England, gave civilization invaluable information about typhoid fever. He first described the symptoms and progress of this disease in his book *De Febribus*. In his honor, biologists named that part of the circulatory system where the carotid and vertebral arteries meet the circle of Willis.

Willis was born in 1621 and took his bachelor's degree at Oxford, while simultaneously fighting parliamentary troops as a soldier in the Royal Army. Because of his loyalty to the crown in these battles, the King made sure he received a professorship at Oxford. Discontented with university life, however, he soon emigrated to London to begin a medical practice. He, too, shortly became a member of the prestigious and still embryonic Royal Society. He was apparently extraordinarily conservative in his politics, and always loyal to the crown. Writers of his time describe him as being "of the highest moral character."

Doubtless his outstanding contribution, as alluded to above, was his comparative anatomical investigation of the brain and the nervous system of humans and many other vertebrates. He described with a remarkable degree of accuracy both the brain and the nerves emanating from it. He then supplemented a selection of these analyses with accurate drawings. His obliging colleague, the eminent architect Sir Christopher Wren, created the drawings for him in recognition of both their camaraderie and the precision of his investigations.

In his conception of life, the seventeenth-century French philosopher Descartes exerted considerable leverage on him. Like Descartes, Willis embraced the concept of "dualism," the belief that both a body and a soul comprise a human being, with the soul linked to the body via the pineal gland. In spite of this, he differed from Descartes in locating some "mental states" literally in the brain. He thought, for instance, that the cerebral cortex stores memory (anticipating similar recommendations in the twentieth century by such theorists as Wilder Penfield of the Montreal Neurological Institute and Eliot Valenstein of Antioch College). Only slightly more interesting were his philosophical speculations. Following Aristotle rather than Descartes, he thought that any animal whatsoever has at least a "vegetative" soul, if not the "rational" soul of humans.

CAUSES AND CURES OF DISEASE

Other scientists devoted themselves to the study of disease. Syphilis was rampant, and it was no wonder that a number of scientists sought to understand this dreaded plague. One of these was Girolamo Fracastoro, who wrote *Syphilis sive Morbus Gallicus,* which gives a tolerably authentic description of the symptoms of the disorder and also champions some treatments, now known to be ineffective. In 1546, Fracastoro put forth a revolutionary suggestion when he proposed something resembling the modern germ theory of disease, though his ideas would require centuries to be revived. Even earlier, in 1498, Francisco de Villalobos studied this scourge.

Valerius Cordus soon established himself as another of the "fathers of botany." Born in 1515, he became a pharmacist as well as a botanist. He received most of his education, unlike other botanists of the day, chiefly from his father, Euricius. In fact, his father was also a botanist of some ability, having penned a number of books, although current historians consider none to be a major contribution. By age twenty, Valerius Cordus had put together the *Dispensatorium,* a commendable encyclopedia of drugs and their possible medical applications. More momentous was his next publication the *Historia Plantarum,* which described well over 400 distinct species of plants. Not only did he describe the external anatomy of plants, but he described a considerable part of the internal anatomy as well, including the flowers and the fruit. Feeling that he had concluded his studies of native plants, he moved to Italy to investigate still more species, though he passed away not long afterwards, at the age of twenty-nine.

At this point in the history of biology, although some scientists had made contributions to the treatment of disease, they had accomplished next to nothing in the way of understanding the underlying physiological mechanisms of illness. Indeed, the overly imaginative van Helmont proposed, in 1626, that beings from another world caused contagion and pestilence in humans as well as animals. Continuing this bizarre fantasy, he believed that periodic invasions of these beings was the correct explanation of the famous plagues of the past! (They were "invaders," of course, but hardly from another world.)

It would not be until the time of Jacob Sylvius that, under his authority, physicians would abandon the antiquated superstitions. Born in Paris in 1478, Sylvius followed Leonardo Da Vinci in urging the rejection of dogmatic Scholasticism, and supporting further critical instruction in Greek and Roman masterpieces. In his youth, Sylvius himself studied the classics as well as the classical languages of Roman, Greek, and Hebrew. Later on he devoted himself to the medical sciences and began lecturing extensively on classical medical texts.

Here, however, he fell into the same pattern of muddled thinking that he had so admirably criticized in medieval logicians. While disclaiming the dogmatism of Scholasticism, on the one hand, he uncritically followed such authors as Galen—even intimating that Galen was "divinely inspired" and, therefore, infallible. Since the age of Plato's *Timaeus,* biologists had condoned without question the proposal that an "imbalance" in the "four humors"—phlegm, black bile, blood, and yellow bile—causes disease. Even so, with the help of Galen's research, knowledge of human maladies, at least, would advance rapidly. By the mid-seventeenth century, for example, the naturalist Daniel Whistler would proffer the first medical characterization of rickets as part of his doctoral work at the University of Leiden.

SUMMARY

The conquest of contagion was a dominant theme in the Renaissance. With Lady Montagu's introduction of the practice of inoculation, Glisson's study of rickets, Willis's work on yellow fever, and Fracastoro's work on syphilis, the world was on the trail to conquering these dreaded maladies.

CHAPTER 8

The Age of Vesalius

The most exalted biologist of the Renaissance was Vesalius. He was born in 1514 in Brussels to a family that had a long tradition of scholarly involvement and dedication to medicine. Not wishing to part with tradition, Vesalius too studied medicine. Early in his schooling he was interested principally in art, philosophy, mathematics, rhetoric, and so forth, rather than science *per se*. So it was on his own initiative that he began to probe the riddles of biology. He pored over classical texts when and wherever he could get hold of them. He also began dissecting every kind of animal he could seize, from moles and mice to cats, dogs, and even weasels. As a teenager, he was sent by his father to the University of Louvain, expected to carry on the long family heritage of practicing physicians.

Unfortunately, Fate threw him upon his own resources. The leading authority in Paris was Jacob Sylvius, and, realizing the dogmatism of Sylvius, only after going to Paris Vesalius concluded that he would have to attempt his biological studies alone. Thus thwarted by the intellectual atmosphere at Louvain, he moved to the University of

Paris while still in his teens, where he studied anatomy and medicine for three years.

DE HUMANI CORPORIS FABRICA

At age twenty-nine and after many years of investigations, Vesalius published his book *De Humani Corporis Fabrica* ("The Constitution of the Human Body"), in 1543, the same year that Copernicus's transcendent masterpiece "The Revolutions of the Heavenly Spheres" appeared. Vesalius's book was a masterpiece of Renaissance art and science together in one treatise. It offered, for the era, reasonably satisfactory depictions of nerves, organs, tissues, muscles, and the brain. Beyond doubt, *De Humani Corporis Fabrica* was one of the first, if not the first, comprehensive textbook of human anatomy, remarkable as much for its accuracy and detail as for the beauty of its illustrations. It is a large volume of some 700 pages, with a concordance of 30 pages. Vesalius's epochal work is not a mere assemblage of anatomical charts. Anticipating the spirit of present-day biology, he attempted to show how disparate anatomical parts blended with one another as well as depict how they looked. He first discusses the skeletal system, followed by the muscles, circulatory system, nerves, and internal organs. The exposition of the book followed roughly the order of his public lectures.

On the negative side, he postulated the existence of "pores" though which blood was supposed to flow from one side of the heart to the other. This was arguably the only really grave slip in this majestic volume (though "pores" *are* seen in congenital malformations and in many vertebrates).

The illustrations were the handiwork of Vesalius's fellow Belgian Jan van Calcar. Jan van Calcar was born in 1499 in Belgium and had served under one of the exalted artistic virtuosos—Titian. His celebrated lifetime achievement was, in fact, his illustration of *De Humani Corporis Fabrica*. Certainly van Calcar depended on the outstanding volume of anatomical information provided to him not only by Vesalius but by the Renaissance in general.

Assuredly, this book, besides the labor of Copernicus, was a consummate attainment of the Renaissance—perhaps the ultimate merger of art, theory, and practice.

Throughout his life, Vesalius maintained his old habits of collecting and reflecting on creatures of all ilks. He pressed on and made new discoveries and prepared more accurate reports of nerves and the circulatory, musculature, and skeletal systems. Not only did he offer meticulous descriptions of internal human anatomy, but he actually managed to assemble an intact human skeleton out of random bones that others had swiped from recently executed felons.

His reputation spread briskly by word of mouth, and physicians began asking him to publicly lecture and dissect human cadavers. His fame became so immense that the University of Padua tendered him a professorship when he was only twenty-two years of age. With that security, and the use of ample scientific facilities—then as now, the University of Padua was among the choicest medical schools—he could teach and persevere in his work in relative harmony and security. Typically he would lecture to hundreds of interested followers, faculty, and curious laypersons. His lectures followed a regular pattern; he held his public dissections in the winter months, when corpses rotted more slowly. He would begin by demonstrating the various bones of the skeleton, explaining both their structure and function. He would then proceed deeper into the corpse, exposing and describing blood vessels, nerves, and internal organs.

INFLUENCES ON VESALIUS

Initially, Vesalius had followed Galen's teaching but soon noted error after error in Galen. He found that these blunders marred Galen's inquiries, parts of which were vague and often illogical. Thus by the time he began his acclaimed treatise, he had apparently begun to free himself from the spurious authority of Galen, Aristotle, or anyone else, whether a medieval or classical author. He scarcely hesitated to declare the "authorities" wrong. Nevertheless, his criticism of Galen was by no means global. His veneration of him endured, and Vesalius shared Galen's theological vision that the Almighty was the ultimate architect of the universe and that the creations of the cosmos reflected His Wisdom and Benevolence. He also followed Galen on many aspects of the circulatory system and shared Galen's belief that there was a vital connection between the circulatory system and the liver. He did not even totally abandon Aristotle and, like him, believed that the stomach "cooks" food and that the point of breathing is to "cool" the blood. Even on the subject of reproduction he followed Aristotle's thoughts in suggesting that the seed of man plus menstrual blood of the female gives rise to the embryo.

PUBLIC SCORN

By going against the teachings of Aristotle, Galen, and others, Vesalius infuriated many of his followers. Several began speaking and writing against him. Medicine and science still harbored the twin nemeses of ignorance and superstition. Not only did many claim that Vesalius was dead wrong, but they even launched personal attacks against him. Some maintained that he was of "degenerate" character, while others warned that demons possessed him and that he regularly practiced vivisection on humans. This latter charge probably stemmed from the fact that to properly apply the empirical method, Vesalius had to dissect corpses, typically stripping away the skin covering so that he might better study the anatomical details. Appalled at the thought of "desecrating" the human body, numbers of the clergy, lay public, and even other physicians accused him of grave-robbing. These accusations effectively ended Vesalius's research career. Nonetheless, history recognizes Vesalius as one of the very greatest anatomists of the Renaissance.

Vesalius was a pioneer. Although it took another century before his influence really began to exert itself, numerous others would follow and begin questioning the authority of Galen. In the mid 17th century, Richard Lower, for instance, carried on the tradition when he pointed out that Galen was also wrong in maintaining that phlegm began in the brain.

THE ORGANS OF SENSE

Another great anatomist of the Renaissance was the Italian naturalist Bartolommeo Eustachio, who gave some of first descriptions of the anatomy of various sense organs. He was born in 1520, but little is known about his youth, although we do know that he had a medical practice in Rome and that he was a professor at the Papal Medical Academy. His greatest tract is his *Opuscula Anatomica*, which carries his magnificent achievements on the human ear as well as reliable discussions of the circulatory system of the embryo. In 1601, Julius Casserius augmented this book with his *De Vocis Auditusque Organis Historia Anatomyic*, a highly accurate textbook describing both the structure and function of the larynx and ear.

It is fair to note that Eustachio had actually written another larger dissertation on anatomy which was set for publication. But Eustachio died abruptly in 1574, and the book did not appear in print until 1714. Though it was dated by then, the book again reveals Eustachio's vast abilities at anatomical description. Among other things, he argued that the biology of humans is no more important than the biology of the lower beasts—an obvious shift away from Church dogma characteristic of the Renaissance. Scientists subsequently named the Eustachian tube of the ear in his honor.

The Italian biologist Realdo Colombo also made a special study of sensation, particularly the structure and function of the human ear. His results appear primarily in his book *De Re Anatomica*, or "On Things Anatomical." Still, the book makes it eminently clear that regardless of his superficially scientific approach, he, unlike Eustachio, had departed only a hairbreadth from the ways of many of the dogmatic Scholastics of the Middle Ages. When he died in 1559, he had scarcely influenced anatomical thinking, chiefly because of his unpleasant disposition, being self-assured to the point of arrogance. Science recognized his contributions for their real importance only after many decades.

RENAISSANCE ZOOLOGISTS STUDY TAXONOMY

One of the earliest true zoology books of the day was the *Historia Animalium* by the Swiss biologist Konrad von Gesner. Gesner was born in Switzerland in 1516, the son of a soldier and ardent patriot who ultimately lost his life in the Battle of Kappel in 1531. The battle was an epochal point in history in that the followers of Zwingly, one of the formidable Protestant Reformers, lost decisively to the Catholic forces of the Counter-Reformation. Dispirited at the loss of his father and afraid to stay in his native land, Gesner headed for such romantic ports as Paris and Montpellier. In these cities he began his schooling in both classical and Oriental tongues as well in as medicine, mathematics, astronomy, and science. During his life, he became professor of Greek at Lausanne and the official physician of the city of Zurich, ultimately dying of the plague in 1665.

The publication of his *Historia Animalium* helped launch contemporary zoology as a separate science. Over 3,500 pages long, it occupied four giant volumes. In it he arranged fauna using the classificatory rules of Aristotle—in no sense a forgotten figure, even in the Renaissance. Accordingly, he classified animals into viviparous—those that brought forth young without eggs—and oviparous, or egg-laying animals. However, Gesner had the foresight to omit some of Aristotle's rather less scientific attitudes, such as in the following passage from Aristotle's *Historia Animalium:*

> Hence woman is more compassionate than man, more easily moved to tears, at the same time is more jealous, more querulous, more apt to scold and strike. She . . . more prone to despondency and less hopeful than the man, more void of sham or self-respect, more false of speech, more deceptive, and of more retentive memory. She . . . also requires a smaller quantity of nutriment.[10]

Having sidestepped such pitfalls, Gesner went on to catalogue fishes, insects, birds, and reptiles.

Not the least notable feature of this series was its design. Gesner described animals under eight categories, including the animal's name in eight languages, its regional habitat, typical anatomical features, and the makeup of its "soul." (Recall that Aristotle believed all living things had a soul.) He speculated about its usefulness for human beings, including edibility and medical uses, and carried out vague philosophical reflections about its role in the cosmos. Although odd mythological inferences in the manner of Pliny sully the accuracy of this work, there is definitely far less mythology in Gesner than in Pliny.

ANATOMY

A significant anatomical discovery occurred in the last third of the seventeenth century, when Richard Lower showed that the heart was as much a muscle as any other in the body, and even offered a fairly reliable anatomical portrait of this organ. Sadly, his heretical theological outlook, suggesting that the Almighty was less than all powerful, had the effect of convincing the Protestant crusader John Calvin to burn him at the stake.

THEOLOGY AND THE CIRCULATION OF THE BLOOD

Another illustrious biologist was Michael Servetus. His real name was Miguel Servet Y Reves, and he was born in the village of Villanueva in Spain, probably around 1511. His developmental years were indulgent ones, filled with wealth and noble parentage, though we know little else about them. Almost from childhood he became mesmerized by the wonders of the biological realm. Like Leonardo Da Vinci and others, he traveled widely in search of biological knowledge, including trips to Germany and Italy. He spent a substantial portion of his career in Strasbourg, and while there he wrote his *De Trinitatis Erroribus*, or "On the Errors of the Trinity." As if there were not enough of a war between Catholic theologians and Protestants, Servetus provoked still another conflict when he engineered an offense to both by defending the heretical doctrine of Arianism—the idea that Christ was less than God. In the aftermath of this, he had to leave Strasbourg in fear for his life.

Servetus dared offend ruling tenets further by implying that the correct route of the blood, contrary to previous authorities including Galen, was from the heart to the lungs, back to the heart, and lastly to all parts of the body. Furthermore, he taught that the "vital spirit" of the Lord abides in the circulatory system and the liver adds material to this spirit, formed by the combination of the choicest elements of air and blood. Ultimately and most significantly for science, his discussion of the pulmonary circulation of the blood, despite its bizarre mystical overtones, was at least partially reliable. Despite his courage, what he did not realize was that the blood passed through an entirely separate system—the veins. The world would have to wait for Harvey in the opening days of the seventeenth century for that information.

Evidently, there is a robust religious dimension to this work. To understand religion and the relationship of the physical to the spiritual, it is mandatory, according to Servetus, to understand that the body has three vital elements—the "spiritus vitalis" in the heart and arteries, the "spiritus animalis"—a ray of light in the brain and nervous system—and the blood in the liver. Furthermore, he fervently argued that God lives in all of these and that the Almighty's spirit travels from the heart to the liver. He thus veered perilously close to the anti-Christian tenet of pantheism, the doctrine stating that the Creator dwells in all terrestrial objects.

After leaving Strasbourg, he finally relocated in Lyons, moving in with a physician who hid him and recommended he use a different name and begin to learn medicine. Eventually, he dared to move around more and went to Paris to apprentice under Vesalius. But his zeal for controversy had by no means left him. In the city of Vienna he resumed his analyses of religion with his *Christianismi Restitutio*, or "The Restoration of Christianity." In it he contended that there was a critical link between knowledge of the Holy Spirit and knowledge of human anatomy and physiology. In short, he advocated the aged precept that all of God's creations reflect His wisdom. He also mercilessly assaulted John Calvin himself, one of the architects of the Protestant Reformation. When he learned of this, Calvin, in the style of a pharisaic inquisitor, tried Servetus, throwing him immediately into prison. Eventually, Servetus was able to escape and continue his assaults on Protestantism by allying himself with an anti-Calvinist party in Switzerland. Calvin, however, managed to recapture him and condemned him to be burned at the stake. The Protestant Reformers upheld their "Christian" crusade by actually carrying out the sentence three days before All Hallows' Eve in 1553. Catholic theologians afterwards honored his valiant fight against the enigmatic cult of Protestantism by building statues of him in Paris and Madrid.

Although his major interest was theology, Servetus in fact was a professional physician. But the two disciplines were, for Servetus, connected as noted above. His discussion of the route of the blood was actually a bid to authenticate his theological views, although he hardly succeeds in showing that one has anything to do with the other. Not surprisingly, Servetus's ideas had an immense impact, whether positive or negative, on the thought of his time. Specifically, we find his biological contributions reappearing in the writings of many authors of the sixteenth and even the seventeenth centuries.

One should not overlook still another of the contributions to an understanding of the Circulatory system of the Italian scientist Realdo Colombo, who followed Vesalius at the University of Paris. Colombo had initially been a disciple of Galen as well, but through his own efforts succeeded in proving that Servetus was right with respect to the true route of the blood. Galen had assumed that blood passed between two chambers of the heart, but Colombo and Servetus revealed that this was not so; in reality, blood leaves one chamber of the heart, goes to the lungs, returns, and only then enters the other side of the heart.

Others continued to study the circulatory system. In 1537, Hieronymus Fabricius entered the world in the village of Aquapendente, Italy. He was possibly, after Vesalius, the most admired teacher and anatomical scholar of this period. He both conducted research and taught at the University of Padua for over sixty years. In his studies, he examined a wide variety of species, comparing both their physiological and anatomical attributes. In 1603, he penned his *De Venarium Ostiolis*, or "On the Valves in the Veins." This book explained the presence of valves in the venous system—the network of veins in the body—and offered countless insights on the circulation of the blood generally. Fabricius was among the first to realize that the blood flowed continuously. Furthering the tradition of accuracy as well as thoroughness, he unveiled virtually all of the anatomical details of the venous system, though he understood little about their actual function in routing the blood. William Harvey later built his renowned legacy on Fabricius's ideas; indeed, Harvey was his pupil.

REPRODUCTIVE BIOLOGY

Another vital publication by Fabricius was his "On the Formation of the Fetus," which almost single-handedly established embryology as a separate branch of zoology. In this book, he furnished humankind with a comprehensive overview of blood circulation in the umbilical cord, placenta, and other parts of the reproductive system.

Hieronymus Fabricius's reflections on fetal development were clearly a milestone in the still immature science of embryology.

In his *De Formatione Ovi et Pulli*, or "On the Development of Eggs and Chickens," Fabricius made some of the finest early sketches of the various stages of a developing chick embryo. The later scientist Regnier de Graaf extended Fabricius's work and would make a major breakthrough in understanding female reproduction in vertebrates when he recognized the role of the follicle, a small sac in the ovary that contains the maturing egg. In 1672, de Graaf wrote that he had managed to follow the egg uninterruptedly from the ovary, through the Fallopian tubes, and ultimately into the uterus. This entire idea, especially the assertion that eggs came from the ovary, was a revolutionary thought at the time. In his analysis, he also found spaces filled with fluid between follicles. Biologists later renamed the ovarian follicle the Graafian follicle in his honor. De Graaf also gave a reasonably reliable depiction of the process of ovulation. Clearly, by the mid-1500s, portentous advances were occurring in the still-new field of reproductive biology.

In 1561, Gabriel Falloppio gave an unerring description of the female reproductive organs. Born in 1523, Falloppio grew up in an impoverished household, but still managed, through the assistance of wealthy patrons, to study medicine and anatomy at the University of Padua. Since Vesalius was still at Padua at the time, it is likely that they knew one another, especially since Falloppio adopted many of Vesalius's techniques and approaches to medicine. Falloppio's brilliance became evident early on. By age twenty-four, the university at Ferrar had appointed him a professor of anatomy, and it is here that we observe some of the influence of Vesalius, since Falloppio's lectures followed the same broad pattern as those of the master. Falloppio's total written output was relatively paltry, but nonetheless central in the story of biology. His one book was *Observationes Anatomicae*, in which the influence of Vesalius appears and to whom he openly concedes his debt. Not only did this treatise contain superb contributions to the growing body of knowledge about the sex organs, but it also offered fresh and acute insights into the skeletal system and the ear. Sadly, Falloppio's life was short, in part because he never overcame the malnutrition suffered during his poverty-ridden boyhood. He passed away in 1562. Science later named the Fallopian tubes in his honor.

Regnier de Graaf (1641–1673). (Courtesy of the Library of Congress.)

Later both the Danish biologist Steno and de Graaf proved, to the incredulity of the scientific community, the theory that sharks of the class Chondricthyes have eggs and ovaries. They generalized this concept and reached the conclusion that all mammals give birth to young via the production of eggs. In 1667, Steno used the term "ovary" for the first time in discussing the female reproductive tract. This was an innovation of the first magnitude in reproductive biology, since every biologist before him had believed that only oviparous animals had ovaries.

PALEONTOLOGY

At about the same period as de Graaf, Nicolaus Steno imprinted his seal on the history of biology. Steno was born in Copenhagen in 1638 into a family of goldsmiths. As with numerous other scholars, his talent for biology surfaced quickly in the form of countless questions to teachers and parents about the biological kingdom. He eventually studied medicine under the Danish biologist Caspar Bartholin and soon began his own inquiries, gradually attracting the attention of the European scientific community. Although Bartholin had instructed him, his relations with his mentor were somewhat fragile, given Bartholin's temperament. After failing to procure a teaching position at the University of Copenhagen, doubtless owing to tensions with the administration, Steno finally headed for Paris to continue his studies of anatomy. Fortunately, he procured the favor of the Grand Duke of Tuscany, who extended to him a type of "fellowship" to underwrite his education.

In his declining years, he gradually succumbed to various religious delusions, believing that it was his duty to save humankind from sin. He lapsed into an extreme form of ascetic self-denial, but his health could no longer afford the rigors of an ascetic existence, and he succumbed at the age of forty-eight.

Among his contributions, he participated in laying the foundations of the science of paleontology, or the study of fossils. This interest apparently stemmed from his stay in Florence, where had the chance to scrutinize the fossilized remnants of shark teeth, called "glossopetri," or "tongues of stone," by local peasants. He also anticipated modern geology when he inspected the rock strata in Florence. From a variety of clues, including the fossil relics of an incredibly diverse assemblage of flora and fauna, he concluded that strata, having the appearance they had, must once have been under water. Still, owing to the fact that he saw potential contradictions about the age of the earth when he compared his strata findings with Church teaching, he did not, unfortunately, pursue this inquiry further.

PHYSIOLOGY

The Serbian biologist Sanctorius of the University of Padua, a few decades before Steno's work on ovaries in 1667, constructed a device to measure pulse by means of a free-swinging pendulum. Its virtues notwithstanding, it never became widely used. More influential was his *Commentaria in Artem Medicinalem Galeni,* which told physicians how to build a primitive thermometer, though he credits Galileo for actually inventing it. Another contribution Sanctorius made to the world was his early discussion of metabolism. The study was careful and fairly thorough, recording his careful measurements of pulse, temperature, weight, rates of breathing, pupil activity, and other physiological variables. He is perhaps best known for his idea that "insensible perspiration" exceeds all other bodily excretions.

BOTANY

Scientific interest in this field increased in 1580, when a Venetian named Prospero Alpini made an epochal and revolutionary find that would forever alter botany. Formerly, scientists had thought that sexual reproduction was a phenomenon restricted to the animal kingdom. Alpini showed, however, in his *De Plantis Aegyptii* that plants as well as animals can have two sexes. As the sixteenth century was drawing to a close, interest in botany appears to have accelerated, doubtless due to Alpini's efforts.

SUMMARY

This period witnessed the branching and maturation of biology in all directions. There was progress in physiology, paleontology, and botany. Perhaps for the first time also, real scientific work in embryology and reproduction emerged with the work of Fabricius and Fallopio, as well as the work of Eustachio and Colombo on the human ear. Also, taxonomy took a huge step ahead with the research of Gesner. Arguably one of the most courageous scientists in history worked during this era—Servetus, who hade the audacity to contradict church teachings about the circulation of the blood. From another vantage point, with van Calcar doing the illustrations for Vesalius's *De Humani Corporis Fabrica,* the world had perhaps the first true masterpiece of both art and science.

CHAPTER 9

The Harvey Era

In 1578, one of the truly legendary figures in the story of biology was born. The fabled William Harvey, an apostle of Fabricius, was born in the quiet village of Folkestone, on the south coast of England. He was the son of Thomas Harvey, a local and highly regarded businessman. Of the elder Harvey's seven sons, five traded in turkeys, another was a physician and still another was employed by the Royal Court of James I. William Harvey's early education took place at Kings School in Canterbury. By sixteen, the elders deemed him ready to enter Cambridge University. This he did, performing brilliantly and receiving the B.A. degree in three years. From there he went to northern Italy, enrolling at the University of Padua for further tutelage. Fabricius was there and was starting to attract a sizable following, so it was the perfect atmosphere to nourish Harvey's biological studies. Fabricius, in turn, had been tutored by Falloppio, so Harvey faced the task of preserving a distinguished biological heritage—and he did. Cambridge and Padua had had close relations since 1539, when the found-

William Harvey (1578–1657). (Courtesy of the Library of Congress.)

er of Caius College of Cambridge University, John Caius, had apprenticed under Vesalius at the University of Padua.

In 1602 Harvey received a doctorate in physics from Padua. He returned to Cambridge for another year and received a doctorate in medicine. Two years after that he entered the prestigious College of Physicians in London. In 1607 they honored him further by electing him a fellow. He ultimately became so renowned that both Charles I and James I appointed him as their official royal physician during their reigns. His fortunes nonetheless deteriorated when the English Civil War broke out and he had to flee London with his family. He lost his house as well as valuable artifacts and research notes he had collected over many years. As he was nearing seventy years of age and low on funds, such a situation could have been a great catastrophe. Luckily, Oxford decided to appoint him professor. In part this was due to who had appreciated Harvey's inestimable contributions and put some pressure on Oxford to appoint him to the faculty. He thus quickly rebuilt his resources and, with some assistance from confidants and relatives, was able to live to a peaceful old age.

He had become, through his many years of research, the first scientist to truly understand the circulation of the blood, including the venous system. Harvey inherited his famous distrust of authority from Vesalius and Servetus. But he went a step further than they or anyone else had done by practicing vivisection. He first lectured on the circulation of the blood before the Royal College of Physicians in England in 1616. Scarcely more than a decade after this, Harvey published his *Exercitatio Anatomica de Motu Cordis et Sanguinis*, or "Anatomical Treatise on the Movement of the Heart and Blood," in both Main and Frankfurt in 1628. This was it: the classic of barely more than seventy-two pages, which altered the direction of biology and medicine forever. Some of his lecture notes from this period have weathered the centuries and show clearly the strong concentration Harvey brought to all of his studies. His task was not without difficulties. As he says in his *De Motu Cordis et Sanguinis:*

> When I first gave my mind to vivisections, as a means of discovering the motions and uses of the heart, and sought to discover these from actual inspection, and not from the writings of others, I found the task to be truly arduous, . . . that I was almost tempted to think, with Fracastorius, that the motion of the heart was only to be comprehended by God.[11]

The assumption Harvey rejected held that there was no muscle in the heart and that the heart "dilates" to circulate the blood, carrying it from the veins into the heart. Through his own steady and resourceful inquiries, Harvey showed that not only is there muscle in the heart but that, in fact, the heart is itself a muscle. Harvey also showed that the heart circulates the blood not by dilation, but by regular and periodic contractions. He clarified the direction of blood flow, revealing that the heart's contractions drive blood into the arteries and that only the ventricles and their vestibules participate in this action. He then demonstrated that the blood passes from the left to the right side of the heart via the lungs.

He contributed to unmasking the absurdity of the old ideology regarding the blood's function. According to the traditional way of thinking, food changed into blood in the liver. It then passed through the veins to the heart in order to absorb the enigmatic *élan vital,* or "vital spirit." Harvey intuitively suspected that this was wrong and saw immediately that the notion of *élan vital* really had no scientific meaning or basis whatsoever.

He simply believed that it was unreasonable to suppose that the amount of blood in the human body could come merely from the food one ate; nor was it likely that one ate the steady and regular amount of food necessary to keep blood quantity as constant as he knew it was. He realized that even when a person was starving, the volume of blood in the body did not diminish. Harvey relied on close scrutiny to prove his intuitions. Among other things, he observed the pulse in both healthy people and people with various diseases of the circulatory system. He dissected live creatures, tying up first the vena cava and the aorta to see how these actions affected blood flow. He would sever an artery, and then a vein running parallel to the artery, again to see how it affected the flow of blood. In the end, he concluded unhesitatingly that the blood leaves the heart in the arteries and returns via the veins, though he never did manage to explain just how blood left the arterial system and entered the venous system. Of course, since microscopes did not then exist, he knew nothing of capillaries. It would be half a century before Richard Lower, in his *Tractatus De Corde,* would demonstrate that the passage of the blood throughout the body correlates well with definite changes in the blood such as its change of color in the lungs. None of this implies, however, any substantial medical applications; it was not until the inquiries of Giovanni Lancisi of Rome, and his book "On Sudden Death," that physicians would even start to understand what happens when the heart muscle begins to degenerate.

Several decades after his momentous work on blood circulation, Harvey would contribute to the burgeoning science of embryology by describing the successive stages of the developing embryo. He accomplished this in his *Exercitationes de Generatione Animalium*, published in 1651, which speculated that all living creatures come from an egg.

Harvey is unequivocally an entry of towering eminence in the register of biology. His experimentation was no less than revolutionary and courageous, being unafraid either of condemnation or of falling into error. With his contributions, biology began to make the transition from the old part-superstition and part-fact theories of men like Aristotle and Galen, to a modern approach to science.

SWAMMERDAM AND HIS CONTEMPORARIES

In a slightly dissimilar vein, it is worth relating that although a number of scientists had devoted a considerable amount of time and effort to grasping the movement of the blood through the body as well as its relationship to sickness, no one had done very much to explore the properties of blood itself. Since cytology, or the science of cells, still lay in the future, no one even suspected that blood had a cellular makeup. For these reasons, Jan Swammerdam quite deservedly presented himself as a creditable candidate for inclusion in the ranks of the immortals of biology when, in 1658, he theorized that blood was indeed composed of cells. That same year, he proved it by actually describing red blood cells.

Jan Swammerdam was born in 1637 in Amsterdam, the son of a pharmacist. Fortunately, his father was also an enthusiastic amateur biologist and had built a substantial home museum of preserved animals of many kinds. Many of the younger Swammerdam's contemporaries found him difficult to deal with and given to fits of emotion. In fact, there is substantial evidence that he was not altogether stable mentally. Thus handicapped, he was never able to earn a living and depended on handouts from others and support from his father, which was withdrawn when Swammerdam did not return to the practice of medicine. When his father passed away, however, he left him a nominal inheritance.

Undaunted by adversity, Swammerdam attended the University of Leiden to acquire a medical degree, continuing his education in Paris. Finally, the University of Paris awarded him the doctorate in medicine in 1667. Instead of starting a medical practice, however, he returned to the study of biology, dissecting and making comparative inquiries into innumerable animals. A major turning point in his career came when he befriended the King's librarian Melchisedec Thevenot, a cofounder of the French Academy of Science. Thevenot was so taken with Swammerdam's abilities that he not only introduced him to the influential members of the French Academy, but assisted him financially for the rest of his life. With Thevenot's assistance, Swammerdam, by 1669, had concluded a rather thorough examination of insect metamorphosis. Using this concept, he classified insects into three types: those that did not metamorphose at all, such as silverfish, those that underwent a partial transformation, such as grasshoppers (these have no wings at birth, but develop them later), and, finally, those that underwent a radical alteration, such as flies, bees, and butterflies. In all of the latter, Swammerdam identified the now well-known larval, pupal, and adult phases of development.

Beyond this, Swammerdam prepared drawings based on the developing stages of a frog, which led him to adopt the erroneous doctrine of "preformation." Ferreting out what he believed was similar evidence in insect development, he came to the conclusion that all living things start life as an entire entity and their development consists simply in further growth. Indeed, he went even further than this in submitting that the entire human race already preexisted in the loins of Adam and Eve. When that supply of humans was exhausted the race would face extinction.

In the ensuing years he authored a general book on insects. Although biologists of today do not consider this one of his most significant projects, he did give a detailed description of the reproductive system of many groups of insects. Beyond that he examined thoroughly the metamorphosis of insects. Still, he had no theory to explain why this occurred. In the last years of his life, Swammerdam abandoned science completely to devote himself to esoteric religious practices and meditation. He died at forty-three years of age.

The science of entomology would, somewhat later, move forward again with the publication of *Traité d'Insectologie*, or "Treatise on Insects," by Charles Bonnet of Geneva, Switzerland. In this book Bonnet outlines both his methods of observation and their outcomes, clarifying parthenogenic reproduction—embryonic development without fertilization by a male—in insects known as aphid mayflies. His studies also extended to worms and coelenterates.

At about the same time Swammerdam was struggling away in his cramped laboratory in Amsterdam, Johann Glauber was writing up his chemical theories and observations in his *Opera Omnia Chymica*, a manual that scientists and students would read in universities well into the early days of the nineteenth century. His most publicized, if not his weightiest, contribution was the invention of the eighty-degree thermometer scale, which the biological community named after him.

RÉAUMUR AND THOMPSON AND METAMORPHOSIS

In 1734, René de Réaumur would push forward with Swammerdam's investigations of insects. At the same time the biologist and army physician John Thompson would explore, in the West Indies, metamorphosis in marine animals such as crabs. Indeed, it was Thompson who first realized that crabs, as well as insects, underwent a full-scale metamorphosis. Darwin himself would elaborate on Thompson's work, publishing a booklet on the remarkable metamorphosis of barnacles.

Thompson would go beyond even this in his classic studies of parasitism, the phenomenon whereby one organism lives off of another. Thompson found that certain shore crabs occasionally carry what appears to be a sac on their abdomens. But the "sacs" are just degenerate organisms that science today calls parasites, organisms which have little else besides a mouth and sexual organs: no digestive system, eyes, and so forth.

So far as entomology (the science of insects) is concerned, science today considers René de Réaumur's investigations to be among the seminal contributions to the field.

Réaumur's finest effort was the giant six-volume *Memoires pour Servir a L'Histoire des Insects,* or "Studies on the History of Insects," when collected together the most outstanding amount of material about insect anatomy of any book to date. Even today, entomologists hold it in the highest regard for its accuracy and thoroughness. Among other things, Réaumur observed the social disposition of the honeybee as well as Hymenoptera—insects with two pairs of membranelike wings and a complete metamorphosis, including bees and wasps. The series also offers detailed discussions of the various stages of insect development.

Though he had no theory to interpret insect metamorphosis, Réaumur did at least make a start toward such a hypothesis when he suggested that heat might accelerate it. René Antoine Ferchault de Réaumur came from a wealthy family of noble lineage. He was born in 1683 and was educated first at a Jesuit school. He then studied law at the University of Paris. The world of natural science, however, soon drew him away from jurisprudence. Like Linnaeus, he did most of his research in private. The only organization he ever joined was the French Academy of Sciences. He passed away in 1757.

Réaumur had, possibly, the widest interests and background in science of any of his contemporaries. Along with his insect studies, Réaumur also delved into metallurgy, smelting, and the nature and composition of gases as well as the properties of heat.

PLANT CLASSIFICATION

Up until this time, scientists had not done much to place taxonomy of plants and animals on a firm foundation. Indeed, whatever theory there was in this area was only a holdover from the primitive Aristotelian scheme. Considerable confusion existed not only about what features of an organism a biologist should use to classify it, but even about whether particular organisms were plants or animals. It was not until the opening decades of the eighteenth century, for instance, that the Italian biologist Luigi Marsigli proved that corals were really animals—not plants as everyone had always believed.

His influence was barely more than moderate, however, doubtless because he was a shy and retiring man and did little to "promote" himself—a far cry from many scientists of today.

In around 1580, Kaspar Bauhin began making contributions to the art of classification. He began by looking at plants native to Padua, Italy, while he was at the University of Padua. He persisted in classifying plants after returning home to Basel. By 1623, he had finished his *Pinax Theatri Botanici,* a summary of all the species of plants anyone had described up to that point. He was also one of the pioneers in drying plants for preservation in herbaria—collections of dry plants, or places housing such plants. As the numbers of plants surveyed swelled, Bauhin realized that more systematic groupings were necessary.

He started using a "binomial" system, adopting one name as the species name and another as the genus name. He based most of his classification on the external form of a plant, grouping together those that resembled one another the most. His manner of exposition was to begin with what he thought were the most primitive plants, discussing more advanced plants second, in the belief that one could not adequately appreciate the higher species without a grasp of more primitive ones. Thus he began with plants he believed were simple, like the Poaceae and Liliaceae, following this task with some ruminations on what he thought were necessarily more advanced plants, such as trees. Of course his view is riddled with errors. The *Poaceae* and *Liliaceae* are advanced plants, for example.

Another important scholar of the day was the British noble Francis Willughby, who authored still another volume on taxonomy. Though crude, it too placed another stone on the foundation from which the fabled Karl von Linne, better known as Linnaeus, would construct the system which is still used today, albeit with some modifications.

The Italian botanist Andrea Cesalpino improved Bauhin's taxonomic system. Both a botanist and a physician, he was born in Tuscany in 1519 and began his training in medicine at Pisa. He received his medical degree in 1549 and shortly after joined the faculty of the University of Pisa as professor of pharmacology. It was during his tenure in Pisa that he began his most intense period of study in botany. Eventually he became the chief physician of the Vatican, where he remained until his death. Most of his contributions to botany appear in his *De Plantis*. Published in 1585, it used the characteristics of both the flower and the fruit as a basis for classification.

This astute Italian biologist also tried to find a common base for the physiological activity of plants and animals. That is, he tried to conceive of a "circulatory" system of plants, consisting of the usual elements such as a heart, blood vessels, and so forth. He quickly concluded that the "heart" of the plant inhabits the collar of the root, or the place where the root system and stem joins. He believed too that the veins of animals originated in the heart and not in the liver as men like Galen had assumed, and he proposed that the nerves had altogether the same foundation. He did dare to break with Aristotle by repudiating the Greek philosopher's idea that food lingered preformed, so to speak, in the ground. He believed and came closer to the truth in arguing that all that existed in the soil were the basic ingredients which a plant could use to manufacture food. He further tried to explain, using the laws of physics, how water enters the body of a plant. He held that there was an organ in the plant body that acted something like a sponge, to soak up water.

A man of panoramic interests, he also did research in metallurgy, anatomy, and even chemistry, though his most exalted efforts were in botany. Like so many naturalists of the era, he had a philosophical vision and was in fact as inquisitive about philosophy as he was about botany. Although he added nothing new to philosophy, his worldview was nonetheless interesting. It followed the general lines Aristotle had set down. Thus, Cesalpino believed in a "prime mover" as the Transcendent Architect of the universe and its First Cause. He accepted also, against Plato, Aristotle's emphasis on the empirical method and the pertinence of change. Similarly, he believed that the heart was the first part of the body to come to life and the last to die. One of his arguments, for this was that men felt emotions first in the heart. The twentieth-century philosopher Wittgenstein has repeatedly reminded that one must not get too carried away with pictures and metaphors; Cesalpino obviously did. Despite convictions by his contemporaries to the contrary, he continued to accept the Aristotelian conviction that there were "pores" in the wall of the heart. He also rejected the sound notion that the function of the veins was to carry blood away from, and not to, the heart. Even so, he got some things right, as in pointing out that the heart was the center of blood circulation.

Later, others probed even deeper into these ideas about the hydraulics of circulation. In 1676, for instance, Edme Mariotte, the seventeenth-century French physicist-turned-botanist, read a paper on plants to the French Academy of Sciences. In it ("On the Vegetation of Plants") he argued that since sap in trees traveled upward under considerable pressure, there had to be some mechanism allowing fluid to enter plants, but not letting it escape. He was strikingly up-to-date in his belief that chemical reactions explained the fact that plants could take in basic materials like water and create food out of it. In all his work, he rejected Aristotle almost completely.

THE PHYSIOLOGY OF THE LYMPHATIC SYSTEM

While some thinkers had already taken steps to advance such fields as embryology, epidemiology (the science of the spread of illness), reproduction, and so forth, they had done little to grasp other systems in the body. The lymphatic system began to yield its secrets in 1652, when Thomas Bartholin of Denmark published the first comprehensive explanation of that system in *De Lacteis Thoracicis*. Thomas Bartholin was born in Copenhagen in 1616, the child of an anatomy professor. After exhausting the scholastic resources in Denmark, he became something of a *scholares vagrante,* or wandering scholar, and traveled throughout Europe in search of wisdom, a trek lasting nine years. He studied at the University of Leiden for three years, paying special attention to Harvey's findings. He then conducted research on anatomy for two years at the University of Padua and then moved on to Naples. He returned to Denmark upon receiving an offer of a professorship in anatomy at Copenhagen University and quickly acquired a sturdy reputation as a fine experimental scientist and an enthusiastic teacher. With his presence, the once obscure university quickly became world renowned as disciple after disciple emigrated to work under Bartholin.

His finest legacy, as indicated above, was certainly his investigation of the lymphatic system. Several before him had known of its existence, but they knew little else about it. Bartholin began his own study with the lacteal vessels, arguing that they were connected to the liver and contained "chyle," a milky substance consisting of emulsified fats and other by-products of digestion. He soon realized that he was wrong, however, and learned that these vessels were linked to an odd vascular system that appeared to run through the entire animal body. He also noted that this system contained a diaphanous, waterlike substance. In writing *De Lacteis Thoracicis,* he reversed his previous attitude, admitting that the chyle vessels were not, in fact, connected to the liver. He had discovered what is today known as the lymphatic system. Although some of the data in *De Lacteis Thoracicis* were either wrong or misleading, it was nonetheless a pioneering treatise.

Bartholin's struggles, quite naturally, sparked further interest. An intimate and colleague of Bartholin's, Olof Rudbeck of Sweden, described the lymphatic system to the reigning Swedish monarch, Queen Christiana.

RUDBECK'S WORK ON THE LYMPHATIC SYSTEM

Olof Rudbeck was born in Vasteras in 1630, one of eleven children. His father, Bishop Johannes Rudbeckius, was a perceptive man, and he noted his son's intellect early and encouraged him to pursue his scientific interests. In fact, he persuaded his child to enter a school that he had founded himself and which appears to have had a rather sturdy reputation during Rudbeck's era. At this school Rudbeck learned science as well as philosophy, and soon took up the formal study of medicine.

He spent many of his afternoons dissecting animals in the university laboratory as well as at home. It was during this period, while still in his teens, that he became engrossed in the lymphatic system and in Bartholin's studies. In a treatise of 1652, he sketched a more or less faithful description of the circulation of the blood, also denying that the liver manufactured it. In 1656, he displayed an even greater understanding of the lymphatic system in a essay in which he outlined the route of the "vasa serosa," or lacteals—the tiny vessels that carry chyle throughout most of the human body. He recognized also that lymphatic fluid tastes salty and tends to coagulate when heated.

Unluckily, a mild rivalry erupted between Bartholin and Rudbeck as to who discovered the lymphatic system first (analogous to the ghastly feud that broke out between Newton and Leibnitz over who was the real discoverer of that branch of mathematics known as the calculus). Each protagonist made speeches and published pamphlets attacking the other, generally in not-too-polite terms. Historians now tend to agree that the query is a conundrum.

Whoever was first, the analysis of the lymphatic system was a turning point in biology. Coupled with Harvey's investigation of the circulation of the blood, science was now starting to understand human physiology in general and the digestive processes in particular. Years later, Queen Christina noticed the merit of Rudbeck's findings, and, in 1652, she gave him a kind of scholarship to carry on his research in other nations. He then headed for Leiden for a three-year stay to further his own inquiries.

Upon returning home, Rudbeck became professor of anatomy and also became entangled in the "political" side of medicine. He was contemptuous of the way his country's physicians emphasized speculation and neglected observation. So it was that he built his own laboratory "theater," which still exists today. He managed to carry on his own scientific examinations, even while engrossed in this sort of work, albeit at a vastly reduced pace.

Understandably, historians do not consider the attainments of this period to be extraordinarily memorable, consisting chiefly of some botanical engravings. Still, the Queen's interest in the lymphatic system soon led to a feverish rise in interest in that overlooked dimension of human physiology. In 1656, for example, Thomas Wharton of England gave the first description of the submaxillary gland, which biologists now know to be an integral part of the lymphatic system.

In 1672, Francis Glisson, a graduate of Cambridge, brought the embryonic science of physiology forward still another step when in addition to his work on the liver he proved that living tissues of all classes will react to their environment. Previous scientists appear to have believed that only whole organisms did so. Later, in the Enlightenment period, the Scottish biologist Robert Whytt would take up the banner for this idea, arguing that a broad assortment of stimuli could irritate living tissue.

SUMMARY

The themes of this period, then, were entomology, plant taxonomy, and physiology. Entomology came closer to the modern science with the work of Glauber, Réaumur and Cesalpino. With his *Pinax Theatri Botanici*, Kaspar Bauhin at one stroke summarized and reorganized the sum total of all information on plants up to that time. Finally, in physiology, Thomas Bartholin of Denmark published the first comprehensive explanation of that system, in *De Lacteis Thoracicis*, while Swammerdam entered the modern era by correctly theorizing that blood was indeed composed of cells.

CHAPTER 10

The Age of Newton

During this era different areas of science and natural philosophy would take giant steps forward. The exalted French philosopher and mathematician René Descartes would publish his *Discourse on Method*, which complemented his *Meditations* and which made use of his famous "method of doubt" to try to prove that there actually was such a thing as absolute certainty. This concept would dominate philosophy and science until the present day. Galileo had, just a few years beforehand, published his classic *Dialogue Concerning the Two Chief World Systems*. This book had proposed the novel and correct idea that it was the sun and not the earth that was the center of the solar system.

But, above all else, the seventeenth century saw the publication of Newton's stately *Principia*, which begat the science that has come to be called "classical mechanics." Newton brought all this about when he stated in his *Principia* a number of important principles, the most powerful of these being the law of universal gravitation. In devising it, Newton was trying to answer two fundamental questions: why do objects fall to earth, and what keeps the earth moving?

The concept of a universal attraction between bodies had actually emerged earlier with the Renaissance astronomer Johannes Kepler, though his efforts were hampered by a reliance on old mystical and superstitious ideas. In Newton's new version of mutual attraction, *matter* was the key ingredient and the source of the attraction. Part of Newton's theory of gravitation also was the "inverse square law," according to which every piece of matter in the universe attracts every other piece with a force that varies directly as the product of their masses and inversely as the square of the distance between them. In a word, the larger the objects, the greater their gravitational fields and the greater the distance between them, the weaker is the gravitational force. Because this idea validated previous astronomical ideas that Copernicus, Galileo, and Kepler had suggested, particularly in confirming the validity of the Copernican sun-centered universe versus the antiquated Ptolemaic idea of the earth as the center of the universe, physics was now on its way to becoming a modern experimental science.

More practically, the power and validity of the law emerged most obviously when all observations of the motions of heavenly bodies confirmed it. Indeed, scientists almost immediately found that Newton's concept could even predict tides. For the first time, science had a universal explanation of how all objects in the universe interact with one another.

Newton even contributed to biology when he probed and partially solved the lens problem of "chromatic aberration," or the tendency of convex lenses to split light into the colors of the rainbow, thereby interfering with viewing.

THE ORGANIZATION OF SCIENCE IN NEWTON'S DAY

All progress aside, there remained a formidable obstacle to further scientific progress, not only in biology but in all fields. The problem was the great difficulty of communication within the fields. The contemporary scientist who is accustomed to comfortable travel grants, affordable journals, fax machines, large societies with annual and regional meetings, and so forth may not have enjoyed working during Newton's era, which had none of the above. There were essentially no societies or journals, no telephones, and travel was expensive and arduous. It was only after the passing of Descartes that this began, no matter how sluggishly, to ease somewhat.

Perhaps the most consequential of the early steps to introduce some kind of organization into science was the formation of the prestigious Royal Society. Throughout Great Britain there had been, for many years, enough heed paid to the sciences that many practitioners began meeting together informally to discuss the newest findings. These meetings, routinely termed "curiosity" cabinets, proliferated primarily in the big cities, such as London and Dublin, in the seventeenth century. The first proposal to form a genuine scientific body appeared in 1616, thanks to the enthusiasm of the well-known scholar Edmund Bolton. During the reign of James I, Bolton succeeded in arousing the interest of the King and miscellaneous members of the royal family, as well as members of parliament, to midwife the society. Then came a setback. When all seemed about to blossom, James unexpectedly died. The momentum vanished and all plans for the society disintegrated.

By 1645, nonetheless, the botanist John Wilkis, Jonathan Goddard, and others had approached Charles I, who also showed some interest in creating a scientific society, though he was apparently much less captivated by the idea than James. Undaunted, Wilkis and other scholars in London, still grappling with the problems of scientific communication, and still desperate to have some sort of society for motivated people to trade thoughts, began meeting at Gresham College in London. Alternatively, they would assemble at the home of the Gresham professor of astronomy at Gresham College, Samuel Foster. By 1648, these meetings had flourished—so much so that a splinter group formed from the prototype in Oxford. Wilkis headed the most fundamental of these. Under his sway, the Oxford group, which had previously discussed issues in physics almost exclusively, now began to venture into both zoology and botany. By 1660, Charles II finally gave official approval to the society.

In 1795, the Royal Society would inaugurate the conferring of its prestigious Rumford Medals to scientists whose lifetime accomplishments met the highest standards of experimental and theoretical investigation. That same year would see the creation of the American Association for the Advancement of Science as well. The science of physics also witnessed the founding of the *Annalen der Physick und Chemie*, or "Annals of Physics and Chemistry," one of the preeminent scientific journals in the world. Physics had begun its overdue rise to prominence and would eventually become the full equal of biology and chemistry in prestige that it is today.

By the opening days of the nineteenth century, amateur geologists would found *The Geological Society of London*—the paradigm for all subsequent geological organizations. After that, geology rapidly became an established science as well. Not long afterwards the society would begin publishing its *Transactions of the Geological Society of London*, which still exists today. This period, therefore, was one of the most significant in the archives of science. Perhaps only the appearance of the quantum philosophy in 1925 and the special and general theories of relativity in 1905 and 1915, respectively, rivaled the Newtonian period in revolutionary impact.

THE EVOLUTION OF SCIENTIFIC METHOD

The exploits of this period were not limited to social organization and actual discoveries. Another significant development during this era was a portentous evolution in scientific *method*. In 1668, the Italian Francesco Redi disproved still another classical Greek teaching, the old notion of van Helmont and others that life can arise spontaneously. Redi initially studied medicine at the University of Pisa, later becoming the official physician to the Medici family. Yet as weighty as this work were his contributions to scientific method. Intrigued by the doctrine of "spontaneous generation," as well as skeptical about it, he decided to run what turned out to be some very modern and well-designed experimental trials. He first killed several snakes and put them outside to let them decay in the sunlight. As maggots appeared and feed on the carcasses, he watched vigilantly, day after day, taking copious notes as he went along. He soon found that after a while the maggots became strangely inactive. Ultimately they "awakened" as flies. He performed similar experiments over and over again, using meat from an enormous variety of animals, including fish, geese, chickens, rabbits, cats, ducks, and numerous others. The pattern was always the same; maggots appeared, which quickly turned into flies.

He went further. He noticed that most of the adult flies dropped minuscule objects onto the rotting meat. Redi speculated that these objects might be the precursors of still more maggots. He thus devised another trial to test this hypothesis; he placed dead fish in bottles. Some of the bottles he sealed, others he left open to see what difference exposure to the air might make. In a word, the open bottle acted as a "control." A controlled trial involves testing the efficacy of an agent or a procedure—a new medicine, for instance—on a group of organisms by comparing the effects to another group which did not receive the agent or procedure. Flies quickly appeared in the open bottles, and maggots surfaced directly after that. But inside the sealed flask Redi never saw a single fly or maggot. To Redi this was conclusive proof that life could not arise "spontaneously" from rotting meat, no matter what the conditions. Of course, again in the spirit of a contemporary scientist, he repeated these probings with endless variations over and over again. Despite his progress, Redi stayed with his belief in spontaneous generation for intestinal worms and gallflies.

ELECTRICITY AND MUSCLE ACTION IN LIVING ORGANISMS

Redi's series of tests constitute a precursor of today's "controlled" experiments. Though this research alone would have secured Redi's place among the immortals of biology, he made further inroads into the still-new field of comparative anatomy by dissecting and describing the electricity-generating organ of the Torpedo fish.

Toward the close of the seventeenth century, the Italian physiologist Giovanni Borelli would probe the phenomenon of electrical generation in living tissue much further in his *De Motu Animalium*. Born in Naples in 1608, Borelli was the son of an official in the Spanish navy. His talent in science and mathematics was evident as a child. His family sent him to the University of Pisa at the earliest opportunity, primarily because the venerated Galileo himself was in the nearby city of Florence as court astronomer, having served on the faculty of Pisa for a number of years. Although the philosophy of Descartes exerted considerable leverage over his biological views, Borelli was scientifically very much a product of his teacher, Galileo.

Thus he did not restrict his inquiries to biology, but conducted extensive investigations into physics, astronomy, and meteorology, though biology remains the domain of his greatest contributions.

In his book *De Motu Animalium,* published in the very year he expired, he shows more of the influence of Descartes. He accurately surmises that the Torpedo produced the electrical shocks for which it is famous by rapid, consecutive muscle contractions of the "electric organ." He tried to explain this muscular activity using mathematical principles of the seventeenth-century French philosopher René Descartes. He even attempted to construct models of bird flight and the swimming of fish to further illustrate and clarify muscular action. As he says in the above-cited book,

> As is generally done in other physical–mathematical sciences, we shall endeavor, with phenomena as our foundation, to expound this science of the movements of animals; and seeing that muscles are the principal organs of animal motion, we must first examine their structure, parts, and visible action.[12]

Thus, under the sway of Galileo and Descartes, Borelli approached biology with the ideas of physics. The organization in this book is engrossing. Somewhat in the axiomatic style of Spinoza's *Ethics,* he articulates his views with a series of axioms and propositions. He first explains the most elementary parts of the animal musculature and then proceeds to more complex descriptions of the entire organism. Among others, he inspected fish, birds, insects, and humans, to whom he devotes the most space. One captivating insight that he first adopted enthusiastically, though he shortly dismissed it, was the intuition that a muscle actually shortens when it contracts. He complements this with his interpretation of Descartes's suggestion that there are fluid "currents" flowing through the nervous system that cause the muscle to shorten in this way. The shortening occurs because of "fermentation"—caused by the mixing of blood with the Cartesian fluids, a theory that science has long known is incorrect.

Borelli's ideas on the mechanics of muscular motion are much more able. His notion that a muscle shortens during contraction was definitely on the right track. Also, the fact that he at least tried to conceive of a physiological explanation for muscular contraction was also in the right spirit and amply demonstrates Borelli's genius. He goes to considerable lengths to analyze the mechanics of flight in birds using his principles of "lifting," walking," "jumping," and so forth. He also tries to clarify the movements of humans with the same concepts. Ultimately, he tries to analyze swimming, pointing to several affinities between the motions of human beings while swimming and the movements of fish.

Borelli attacked muscle from still other points of view. He tried to ascertain what muscle tissue itself was like, though it is scarcely plausible to claim that he succeeded, saying little more than that it is composed of "flesh," which is unarguable. Although he speculated on several branches of biology, the consensus is that the most perspicacious of his discussions are the accounts of muscular action.

All told, Borelli unreservedly ranks along with Harvey as one of the founders of contemporary biology. The attempt to explain muscular activity with the concepts of physics is now universally recognized to be correct according to modern biology. In this way, Borelli left a permanent imprint on biology.

SUMMARY

Above all else then, this was the era of Newton and his universal law of gravitation, the three laws of motion, and many other contributions. Perhaps under the inspiration of Newton, science turned inward. Many realized that to accelerate scientific progress even more, it would be necessary to introduce some degree of organization. From that thought came the American Association for the Advancement of Science, the founding of the journal *Annals of Physics and Chemistry,* and the Geological Society of London. The "modernization" of science continued with Redi's debunking of the theory of spontaneous generation and his introduction of the concept of a controlled experiment. Yet this was the period of important philosophical work as well—the era of the so-called Continental rationalists, Descartes, Leibnitz, and Spinoza, all of whom influenced the science of the period. Descartes, in particular, tremendously influenced Borelli's speculations on the generation of electricity in animals.

CHAPTER 11

The Microscope and Leeuwenhoek

A practical problem that impeded the progress of the biological sciences was the lack of instrumentation with which to pursue detailed investigations. The Dutch scientist Hans Lippershey had invented, in 1609, at least a prototype for the microscope, but the early models were expensive, had minuscule power, and were hard to come by. By the time of the German philosopher and mathematician Leibnitz, British philosopher David Hume, and Newton, this was fortunately starting to change. With the advent of improved microscopes, scientists disclosed new entities such as spermatozoa and eggs and began to appreciate their importance in reproduction. Of course, nothing resembling contemporary genetics was in existence at this time. The biologists of this period had no understanding of genes, chromosomes, DNA, and so forth. Instead, numerous scientists came up with a mishmash of queer guesses, which they based almost invariably on ancient Aristotelian ideology. Swammerdam, to mention one example, followed the "seed" assumption, which argued that all of the seeds of organisms living at the time had been created directly at the moment of the heavenly Father's creation—a metaphysics traceable back to Plato's *Timaeus*.

ANTON VAN LEEUWENHOEK

Beyond any question, the towering biologist of the late 1600s was Anton van Leeuwenhoek. He was born in Delft, Holland, in 1632. While still in his teenage years, he moved to Amsterdam to become an apprentice to a cloth manufacturer. After the period of apprenticeship ended, the company promoted him to head bookkeeper.

After some years he tired of the cloth business, married, and returned to Delft, where he stayed until the end of his days. In this period of his career he had acquired a taste for city politics. Against his wishes, however, economic circumstances forced him to open a cloth shop, though he tutored himself in surveying on the side. Nevertheless, by the age of twenty-eight he became the chamberlain to the Sheriff of Delft with the charge of caring for the City Hall. During the next decade or so he still had not begun any real biological ventures, although his letters show that he had been studying and grinding lenses—ample preparation for his subsequent contributions to microscopy and biology. He lived to ninety years of age, dying in Delft in 1723. It is worth noting that, unlike most of his contemporaries, he did not receive a formal education in science or anything else. He also knew no Latin so could not read most of the classical texts.

Eventually, the Dutch biologist Regnier de Graaf recognized Leeuwenhoek's abilities and wrote the Royal Society of London on April 28, 1673, telling them of Leeuwenhoek's work. The secretary of the society, Henry Oldenburg, immediately contacted Leeuwenhoek to ask about his studies. Leeuwenhoek was, of course, delighted and responded with two lengthy letters discussing his microscopic work on mold growth on flesh, as well as his analyses of the mouthparts and stingers in lice and bees. In the ensuing years, Leeuwenhoek published these observations in the *Philosophical Transactions of the Royal Society of London*.

His reputation soon spread to other lands. Allegedly, Peter the

Great of Russia invited him to give a demonstration of his microscopic investigations, as did the Queen of England. Still, outside of a few dignitaries, Leeuwenhoek jealously guarded his privacy and his endeavors, so much so that he actually became something of a recluse, and villagers began to whisper that he had gone mad.

Leeuwenhoek's Work on the Microscope

Among scores of other accomplishments, Leeuwenhoek improved the microscope. He encased several of his lenses between two thin plates of brass, as a holder. According to occasional commentators in his era, although the magnification of these lenses was not significantly greater than others already in existence, they were much clearer.

Leeuwenhoek went much further than simply building ingenious magnifying glasses. Indeed, he tinkered with all manner of technology which, in turn, helped midwife the birth of the microscope. In addition to glass as a magnifier, for instance, he used crystals, diamonds, and other materials. Assuredly his most splendid accomplishment was the creation of a lens with a magnifying strength of around 300 times. After he passed on, historians found about 400 microscopes in his home laboratory, most of which he had willed to the Royal Society.

Leeuwenhoek on Reproduction

We can scarcely ascribe Leeuwenhoek's success exclusively to a better microscope. He had an immensely heedful and patient temperament, carefully noting and measuring everything he saw. He also kept meticulous notes, most of which he sent to the Royal Society for publication. He was the first to notice that the eye of a fly consists of over a thousand parts. He scrutinized sperm, observing them for the first time in 1677. Leeuwenhoek's investigations of sperm were pivotal, yet scholars sometimes misunderstand them. Although a medical apprentice named Louis Hamm had studied them previously, in around 1675, Leeuwenhoek plainly grasped the significance of sperm, whereas Hamm had falsely believed that they caused disease. Leewenhoek verified that sperm were in fact necessary in reproduction. Leewenhoek went on to study the structure and activity of parasitic types of protozoa in rabbits, dogs, fish, and insects. He also focused more thoroughly than Hamm on the process of fertilization in a large mix of animals, especially frogs and several varieties of fish. He first saw the critical link between sperm and eggs in frogs.

He erred, on the other hand, in again lapsing into a kind of distorted Aristotelianism when he claimed that the entire and complete "preformed" organism actually exists in the sperm, while the female provides the needed environment or nourishment for the sperm to thrive. (Though he had not read the classics, Aristotelianism was still very much "in the air," and Leeuwenhoek therefore became familiar with Greek ideas secondhand.) He thought he proved this idea with the results he got from crossing rabbits. He found that if he crossed a brown male with a white female, all the young were gray, making it appear that the male was the controlling influence. Having no knowledge of genetics, his hypothesis was, to some extent, understandable. Thus, throughout this period most biologists accepted the dogma of preformation. The only particular that biologists did dispute was whether the preformed organism was an egg or a sperm. Those believing the former were "ovists," while the latter believers were "spermatists." Leeuwenhoek, as noted above, was a good example of an "spermatist." Eminent biologist though he was, Leeuwenhoek was somewhat less than overpoweringly consistent in his published writings. To mention just a single example, at one point he argues that insect larvae are not actually insects and, similarly, that spermatozoa are not "humans." Instead, he contended that parts of the embryo appear gradually as the organism develops. Yet the principle of preformation would seem to forbid these sorts of beliefs. Still, all things considered, the bulk of his writings make it clear that he was a preformationist.

Leeuwenhoek examined the circulation of blood in the capillaries of a rabbit ear, the foot of a frog, and later in humans, showing that blood corpuscles, the unattached cells that flow through the body in the blood, were a cardinal component of blood. The seventeenth-century Italian anatomist Marcello Malpighi had probably seen corpuscles, but he had hardly begun to appreciate their constitution or function, referring to them only as "fat globules." In this way, Leeuwenhoek further advanced the capillary studies that Malpighi had begun. He also proved beyond any doubt that the veins and arteries join directly via the capillaries and that the flow of blood continues uninterrupted from the arteries to the veins.

Leeuwenhoek and Classification

Leeuwenhoek was the first to realize that, besides sperm and eggs, there is an enormous diversity of living organisms too small for the unaided eye to see. Among the most consequential of these hitherto undiscovered organisms were bacteria. Although neither Leeuwenhoek nor anyone else would fully grasp bacteria's significance in animal physiology until the age of Pasteur, men such as Leeuwenhoek and the self-taught Danish scientist Otto Muller at least managed to extend the science of classification into the microscopic world.

In 1696, Leeuwenhoek placed bacteria in an array of categories, depending on their superficial appearance under the microscope. Leeuwenhoek recorded this in his *Arcana Naturae*, or "Mysteries of Nature." This was the book that first introduced what Leeuwenhoek called "animalculae," organisms hidden from discovery without the microscope. Plausibly enough, subsequent generations of biologists called him an "animalculist." The exploration of "animalculae" was, beyond doubt, among his most significant contributions to biology. He had first noticed these "wretched beasties," as he liked to call them, in water that had collected in rainfall-measuring tubes, and he noted that the dominion of microscopic life-forms was great. He witnessed, among other things, rotifers, free-living protozoans, amoebae, and intestinal protozoans such as *Giardia* in the digestive system of the shrimp. He found the primitive organism *Rotaria* and Infusoria in streams and watched reproduction in ants. (The term "infusoria" has altered over the years; originally it referred to algae, bacteria and even small worms, though later it came to designate only single-celled organisms.) He realized that what biologists had always called ant "eggs" were in reality pupae from the ant. Unlike several biologists of the day, Leeuwenhoek was far-sighted enough to have strong doubts about the doctrine of spontaneous generation via "putrefaction." Although he did not have a crystalline understanding of reproduction, he did realize that even the most primitive animal forms do reproduce, demonstrating this particularly with aphids and fleas.

Over his lifetime he scrutinized microscopically muscle fiber from whales, ox eyes, sheep hair and hair from a huge number of other animals. Through these investigations, he also contributed to launching histology, or the science of tissues, as a separate branch of the zoological sciences. He was plausibly the first to notice striations in muscle fibers, and he also noted the structure of the teeth. He realized, too, that there was muscle tissue in the iris of the eye and was among the first to unveil nerves emanating from the brain.

He even made contributions to botany by first drawing the fundamental distinction between monocotyledons, embryos with a single seed leaf, and dicotyledons, embryos with a pair of seed leaves.

MÜLLER'S CONTRIBUTIONS

Such an exalted scientist as Leeuwenhoek had his followers. One of these was the biologist Otto Müller. Born in Copenhagen in 1730, Müller was perhaps unique in having been the only active biologist of that period whose father was a musician. Understandably enough, he spent his boyhood in poverty. Eventually, through the munificence of friends and relatives, he succeeded in studying theology and ultimately law.

His interest in biology began while he was tutoring in a wealthy family. On their grounds he began to survey the large numbers of insects that he found, soon publishing a short tract on entomology. Linnaeus greatly influenced him, and he imitated the master's methods throughout his career. In one of his surviving discourses, he tries to describe systematically the anatomical details of and to classify the Infusoria, especially the subgrouping Ciliata, single-celled organisms that were customarily associated with rotting meat in Müller's time.

While one cannot ignore the contributions of Müller, the name that was beginning to dominate biology was Leeuwenhoek. His contributions to the technology of microscopy, his extension of biological inquiry into the microscopic world, his study of reproduction—all these mark him as one of the great geniuses in the history of biology.

CHAPTER 12

The Meeting of Biology and Chemistry

A notorious idea appeared in 1697 which would ultimately hold up the progress of chemistry for decades. And since biology and chemistry were growing ever-increasingly intertwined, this infamous fantasy—the "phlogiston" theory—would retard the development of biology as well.

PHLOGISTON

Now long discredited, scientists once supposed phlogiston to be a medium that caused oxidation, as, for example, in the rusting of iron. The theory originated from an intractable problem. Science could not understand what, exactly, ensued during combustion. From this apparent conundrum, various exotic conjectures appeared. The German chemist Johann Joachim Becher, author of *Physica Subterranea,* speculated, for instance, that there existed a compound (which he dubbed "oil earth") that matter released while it was burning. This odd concept so possessed his fellow German chemist George Stahl that Stahl felt justified in renaming the compound "phlogiston." Stahl summarized Becher's ideas in his *Specimen Becherianum.*

Stahl was born to a devout Protestant family in the town of Ansbach, Bavaria, in 1660. He began his schooling at Jena, ultimately becoming a medical practitioner. In the succeeding years the government at Weimar appointed him official court physician. After a few years he moved to Halle, remaining there some twenty years. He came to know the minor, but politically powerful, German physician Friedrich Hoffmann quite well. But Hoffmann's arrogance coupled with Stahl's self-assuredness eventually ended their cordial relations. Stahl realized that with Hoffmann no longer in his corner, there would be few promising prospects at Halle. He therefore left to become the official court physician in Berlin, remaining there until his demise in 1734.

Part of the explanation of Stahl's exceptional status in science is traceable to the fact that, although scarcely anyone else distinguished between the real science of chemistry and the occult pseudoscience of alchemy, Stahl did make this distinction. He initially and rather routinely went along with the mysticism of alchemy and refers to it with some considerable respect in various of his early published writings. Gradually, nonetheless, principally via his own research and reflection, he came to appreciate the fact that alchemy had, in truth, a negligible scientific basis. Thus he began to apply a true scientific approach to the study of combustion. He did this in a kind of "comparative" way, comparing ordinary burning with the process of calcination—the production of an oxygen-ridden residue that remains after combustion. In doing so, he hoped, again in the spirit of genuine science, to unveil some common, underlying principle that would explain both processes. He thought he had found this in phlogiston.

As he saw it, when ordinary burning occurred, the substance would release phlogiston in the form of smoke. Thus if he brought lead and charcoal together and heated the mass, phlogiston would leave the charcoal and enter the lead compound. Some elements, Stahl proposed, contained more phlogiston than others. He assumed, for instance, that coal was almost pure phlogiston.

It was not long, however, before he realized that the precepts of the phlogiston theory generated a massive contradiction. Presumably, if phlogiston left matter during burning, the matter would be lighter. Yet it turned out to be heavier. To "save" the phlogiston hypothesis, Stahl made an assumption that some scientists easily ridicule today, but which is actually perfectly in line with theory formation in science: he merely assumed that the phlogiston had "negative" weight. True, this was a peculiar supposition, but the history of science is replete with weird suggestions, fashioned from desperate efforts to save an idea. The twentieth-century physicist Wolfgang Pauli, for example, proposed the "neutrino," a vanishingly small, uncharged particle, to save the law of conservation of energy—the precept which says that the total energy coming out of a reaction must equal the energy going in. He did this in the face of observed discrepancies. Although some scientists laughed, Pauli turned out to be correct. There were indeed neutrinos. It is true that the comparable move with the phlogiston tenet did not turn out so well. Still, one has to remember that although chemistry finally had to abandon the phlogiston idea, the scientists who supported it hardly merit condemnation. It was an attempt to formulate a workable thesis to understand a previously inexplicable process. Thus, the holders of the phlogiston postulate were scientists in the normal sense of the word, most especially Stahl.

More complications came when Joseph Priestley conducted some experiments with heated mercuric oxide (a "calx"). He found that mercury emitted a peculiar "cast" of air. If he placed a mouse into a bottle containing this odd air, it lived longer than a mouse in a bottle filled with ordinary air. Since Priestley supported the phlogiston supposition, he dubbed this air "dephlogistonated" air. Of course, the mouse lived longer because the "dephlogistonated" air was merely air that was richer in oxygen than common air, since the heating had caused oxygen to leave the mercury.

LAVOISIER AND THE END OF PHLOGISTON

The phlogiston fable ended in 1772 with the noted French chemist Antoine Lavoisier, whom science often regards as the finest scientist of the Enlightenment as well as a man of immense courage. He was one of the many who lost their lives during the French Revolution. Besides his well-known analysis of oxidation, he named oxygen and hydrogen. He also proved that diamond is merely a remarkable form of rudimentary carbon. He even saw the link between chemistry and biology, pronouncing confidently that nature itself was little more than a melange of chemical changes.

Good scientist that he was, Lavoisier became extremely suspicious of the outcomes of some of his own trials. He saw that if he heated phosphorus and sulfur, they appeared to absorb, rather than give off anything. He extended these experiments, showing that it was possible to burn diamond. He later realized that when sulfur burns, it combines with the atmosphere and becomes heavier. Lavoisier realized, of course, that some other substance must have combined with the sulfur to produce such weight gains, and he theorized that it was oxygen. His own subsequent experiments proved that this hypothesis was valid. Lavoisier now realized that both respiration and combustion, or familiar burning, involved oxidation, and he thereby formulated a new conception of burning. The end of the phlogiston myth was at hand. Lavoisier publicly rejected the phlogiston hypothesis when he became fully persuaded that it was incorrect. The "dephlogistonated" air that had enthralled Priestley was a hitherto unknown gas, which Lavoisier named "oxygen."

Lavoisier captured and summarized all of his discoveries in his *Elementary Treatise on Chemistry* of 1789, in which he argued convincingly for his oxygen explanation of combustion. It was oxygen rather than phlogiston that caused routine combustion. Lavoisier followed this research with further investigative trials confirming that metals would gain weight while burning because they absorbed oxygen. By the time of the French Revolution, Lavoisier had identified over thirty chemical elements. He also considered heat and light to be elements, a classification which scientists of today know is incorrect. Lavoisier's preeminent contribution was his discovery of the law of conservation of mass. He judged that while a chemical reaction may modify the appearance and even the qualities of some materials, the total quantity of the substance stayed the same before and after the reaction.

In 1771, Priestley was in midst of completing his *Observations on Different Kinds of Air* when, through a cleverly designed series of experimental trials, he showed that plants give off oxygen. He had made this discovery when he noticed that whenever fire denuded air of its oxygen, placing live plants in that environment would replace the oxygen. Ultimately, Priestley, in 1780, would become the first to synthesize water by combining hydrogen and oxygen (he would also reveal that one could make seltzer by mixing carbon dioxide and water).

However, the Priestley saga has an unhappy ending. Like untold numbers of innocents of the age, the French Revolution destroyed him. A revolutionary himself, he had been celebrating the second anniversary of the fall of the Bastille in Birmingham when a crowd of antirevolutionaries descended on his church house and laboratory. Within minutes they had ransacked it and smashed every piece of laboratory equipment and furniture in sight, willfully torching it all in the process. Priestley himself escaped with his family, although all of his scientific papers vanished in the holocaust. Later, unable to fully recover his warm feelings for his fellow Englishmen, he emigrated to the United States.

MALPIGHI'S WORK ON ANATOMY AND PHYSIOLOGY

Physiology as well as chemistry began to make new strides forward when the Italian genius Marcello Malpighi made some of the earliest authentic studies of the nervous system as well as the physiology of invertebrates.

He was born in 1628 in the village of Cavalcuore, the scion of a wealthy landowner. As soon as he was of age, he began his education in biology at the University of Bologna. At the outset of his studies he became fascinated with the philosophy of Aristotle, and the eminent Greek thinker would strongly influence his future ideas about biology. After a series of family crises, he finished his medical degree in 1653. Within a few years he had established enough of a reputation that the trustees of the University of Bologna appointed him to a Distinguished Professorship. In spite of this, feeling that research facilities were better at Pisa, he accepted an appointment at that university instead. Early on in his career at Pisa he came under the spell of the Italian physiologist Borelli, which blossomed into a friendship that would persist for the rest of his life. By 1691, the Vatican had appointed him the personal physician to the Pope. He succumbed to apoplexy at the Vatican in 1694.

Somewhat out of the usual pattern, Malpighi published his results in a format more resembling a modern journal article than a longer book. He would periodically write up the results of his inquiries and send them to the Royal Society of London for publication.

Some of his preliminary conclusions survive in letters to Borelli, in which he describes his probings of the lung. According to these letters, Malpighi deduced that the lungs were "fleshy." He made this deduction by first expelling the blood from the lung of a recently butchered animal with prodigious quantities of water. He then inflated the lung and dried it. His conjectures about the lung's function were, nonetheless, quite wide of the mark. In his view, the lungs acted to prevent blood from coagulating, so he was unaware of their function in respiration. In another letter to Borelli, he reports his discovery of the capillary system connecting the arteries and veins in a frog's lung, using procedures similar to those described above.

Still more discoveries came from his laboratory. In about 1670, Malpighi and Leeuwenhoek confirmed Harvey's work on circulation by watching directly the flow of blood through the vast grid of capillaries connecting the arteries and veins. With further study, Malpighi also managed to offer a scrupulous description of the capillary system in the liver. Although Malpighi had no sophisticated microscopes at his disposal, he was far ahead of his generation in that he was one of the first eminent naturalists to appreciate the magnifying power of glass. He used a type of homemade "magnifying glass" in his toils, one which was strong enough to allow him to see the capillary system. Most of his conclusions about the capillary system are, in rough outline at least, correct.

He next turned his attention to the brain and nervous system, examining the minutiae of the cerebral cortex. Here he discovered an array of pyramid-shaped cells, which he conjectured were the source of "fluidum," the liquid quintessence that caused muscles to contract. He carried on lucid and scrupulous investigations of the blood circulation in the brain. Much of this work dominates his classic tome, *De Cerebro*, or "On the Cerebrum." In this work, he proves beyond doubt both that the spinal cord is an aggregate of bound fibers and that it connects directly to the brain. He also examined invertebrate anatomy, describing the anatomy of the silkworm, which led him to his famous discovery of the organs of excretion now known as "Malipighian tubules."

Not satisfied with this victory, he went on to write *Silkworms,* giving the world the first really precise anatomical "map" of the insides of any invertebrate. In 1673, he produced what is doubtless his most significant treatise, the *De Formatione Pulli,* or "On the Formation of the Chick in the Egg," which completed the previous efforts of biologists like de Graaf by disclosing the developmental path of the egg, or ovum, in the female of many species of animals. Still another excellent book along similar lines was his *Observations on the Incubated Egg,* published in 1689. Thus, he added to and verified similar inquiries that Harvey and Fabricius had begun.

MALPIGHI AND PREFORMATION

Still, Malpighi erred in lapsing into preformationism—believing that there was a wholly formed embryo inside an unfertilized egg. More specifically, he believed that the heart existed fully formed from the moment of creation of the organism. One conjecture as to why Malpighi held to this, despite the fact that he was obviously more rigorously scientific in his approach than his predecessors, was that he simply could not observe the beginning stages of chick embryo development—the first twenty-four hours or so. Therefore, he supported the erroneous doctrine of preformation. In fact, the historical ledger of much of biology during and even after this period is filled with accounts of the battle between preformation and the competing philosophy of epigenesis—the supposition that the various tissues appear from an initial undifferentiated embryo. Preformation goes back to the Greeks, as do so many concepts in biology. Plato is one of the most conspicuous supporters of the idea, though it survives in the writings of innumerable theologians of the High and Late Middle Ages. It is not until the eighteenth century that the German anatomist and physiologist Caspar Wolff of St. Petersburg would repudiate any version of preformation and substitute instead the current doctrine of epigenesis.

MALPIGHI AND BOTANY

In 1675, he triumphed once more with his *Anatomy Plantarum,* which some historians of science count as the first indisputably significant treatise on plant, rather than animal, anatomy. This volume was a significant step forward for that science since botany had, for centuries, been lagging behind zoology—an odd kind of anthropocentrism, given the vast significance of plants to the treatment of various infirmities. This essay described his ten years of comparative studies of floral anatomy, including both ligneous (woody) plants and herbs. He gave exceptionally conscientious descriptions of the barks of trees, the buds and leaves of flowers, and so forth.

Probing deeper, he began to realize that plants were composed of cells, or "utriculi," as he called them. These he could see with his magnifying glass. Soon he realized that all such cells connected with one another to form a layer of tissue called the "cuticle" or bark. By looking carefully at an atypical pattern of spiral vessels he found in different plants, he hit on the rather ingenious idea of relating them to the trachea of insects. That, in turn, led him to formulate a global hypothesis of respiration, a theory that he believed to be applicable to both plants and animals. One might call this the first "unified field theory" of biology, that is, via an analogy to physics, it was a bid to see all forms of life as intimately related to one another—at least so far as respiration was concerned. Just as animals inhale air through the trachea, plants usher in air through these spiral vessels.

It is perhaps unfortunate for Malpighi that names like Lavoisier and Priestly all but overshadowed him. Though Malpighi contributed much to botany, to our grasp of the nervous system, as well as to our understanding of the nervous system of invertebrates, numerous errors dim his luster. Malpighi certainly merits a place of honor in the biology of the seventeenth century. Yet, perhaps his reputation would have been even greater, for example, had he not stubbornly clung to the antiquated notion of preformation.

The great work of Lavoisier on ending the phlogiston myth in the generation following Malpighi's makes progress in chemistry one of most dramatic and fruitful areas of investigation of the latter part of the eighteenth century.

CHAPTER 13

Ray and the Emergence of Cell Theory

Another intrepid naturalist, Robert Hooke, entered the world in 1635. Among his many distinctions was his role in helping to shape the destiny of the Royal Society of London. He too contributed both to the technology and the popularity of the microscope. In his published discourses he constantly pressed readers to make extensive use of it. He first devised the concept of the cell, even giving it that name. This he accomplished in his *Micrographia,* which he wrote when he was not yet thirty years of age. In it he gives the scientific community the very first accurate description of cells in cork.

There has been some controversy, historically, about what exactly Hooke contributed to cell theory. According some authors, he contributed more than is commonly acknowledged. Robert Downs, for example, in his book *Landmarks in Science,* claims this, in quoting from the American biologist Edwin Conklin:

> Cells were first seen, named, described by Robert Hooke 170 years before the work of Schleiden and Schwann. Hooke described among many other things the little chambers or cells which he had seen with his simple microscope in sections of cork.[13]

A substantial part of Hooke's *Micrographia* was a survey of past studies with the microscope as well as a survey of the various technological improvements in it up to his own time. He described the compound microscope in tremendous detail, even suggesting ways of improving the illumination of the subject under investigation. One of his most significant innovations was to use a globe of water against a backdrop of oiled paper. The globe acted as a kind of "magnifying glass" to focus sunlight more efficiently.

In his study of biology proper, Hooke made several indispensable recommendations. He investigated and partially described the diffraction of sunlight into its component colors. He inspected the pores in thin slices of cork and first used the term "cell" in 1664 to refer to these pores. He gave a tolerably accurate description of animal respiration and even ventured into what science today calls paleontology, with his descriptions of fossils.

Less theoretical but no less helpful parts of this monumental tract included descriptions and innumerable drawings of the structure and function of both bird and insect wings and the fibers of silk.

Yet technological development alone was insufficient to streamline the study of biology. So far, no really adequate theory of plant and animal classification existed. So it was that others, like the British biologist John Ray, began to make a fresh assault on the old problem of classification that had puzzled biologists since Aristotle. Indisputably, Ray ranks with Linnaeus in the chronicles of the science of taxonomy. He was born in 1627 in the village of Black Notley, in Essex, England. The son of a thriving blacksmith, Ray entered Catherine Hall at Cambridge University when he was only sixteen. A little over a year afterward he transferred to Trinity College to learn both theology as well as the classics with the distinguished Greek classicist James Duport. His final calling, of course, was not the classics but science, although he did not turn to this until comparatively late in his career. As was often the case in this epoch, political turmoil abruptly interrupted the career of this exalted scientist. The government of Charles I had demanded that all clergy acquiesce to a so-called Act of Uniformity, to eliminate novel opinions and rebellion. Since Ray was a clergyman, he could not simply ignore the Act. But as a man of considerable bravery, he abandoned his position rather than cave in to what he thought was an unconscionable directive that would, if successful, squash free reflection everywhere.

From then on until the end of his days, he did all of his meticulous research in the privacy of his own home laboratory. Fortunately, a wealthy amateur scientist and a former pupil of Ray's, Francis Willughby, supported Ray's scholarship and remained a loyal friend throughout his life and even in death; Willughby named Ray executor of his will and provided for a yearly income of sixty pounds for Ray until his death. So supported, Ray continued his work for many productive years. He died in 1705, leaving three daughters and an immense amount of new biological knowledge to the world.

RAY'S CONTRIBUTIONS TO ZOOLOGICAL TAXONOMY

In the sphere of animal life, he wrote his *Synopsis Animalium Quadrupedem et Serpentini*, or "A Survey of Quadrupeds and Reptiles," in 1693. In it he tried to revive the old Aristotelian agenda of dividing animals into the "blooded" and "bloodless." Nevertheless, in other respects he abandoned Aristotle. He discarded his view that a scientist should classify animals into those with numerous toes, those with cleft hooves, and those with uncleft hooves. In place of this he substituted his own system, placing animals into only two classes—those with horn-covered toes and those with toenails. Although biologists preoccupied with classification pay meager attention to this concept today, some of Ray's pronouncements in the same book, such as the claim that whales were mammals, turned out to be valid.

RAY'S CONTRIBUTIONS TO BOTANICAL TAXONOMY

Ray's most astute studies were in botany rather than zoology. By 1660 he had written a complete tract on the plants of Cambridge. One of his most crucial ideas was the notion that the "species" was the fundamental unit of classification. A pious man, he believed that the Supreme Being had created all species fixed and incorruptible. Ray speculated further that ranking could not be arbitrary and that, in fact, God had already classified every species of plant or animal. The task of the scientist then was merely to discover the Divine Systematic Classification. In pursuit of the Creator's plan, he argued that herbs and trees constituted the two major branches of the plant kingdom. We know today, however; that the characteristics of herbs and trees may be so similar as to make them an illogical basis for two separate orderings.

His classic treatise is the *Historia Generalis Planatarum,* or "General History of Plants," published in three installations in 1686, 1688, and 1704. Here Ray collected together extensive information on anatomy, physiology, color, and even geographical distribution of plants. All in all he described nearly 19,000 separate plants—essentially all that were known at this time. In his *New Method of Plants,* he split plants into two groups—the monocotyledons and dicotyledons, based on the number of seed leaves in each group—classificatory ideas still extant today. This book complemented his *Historia Generalis Plantarum,* alluded to above, providing another story in the huge classificatory edifice that so many others had already begun to construct. Among innumerable others, the renowned Linnaeus himself spoke highly of his labors. Perhaps the most important feature of the *Historia Generalis Plantarum* was Ray's success in defining the biological notion of a "species" via the doctrine of "common descent," which Ray introduced into biology for the first time. Although other scientists eventually introduced alternative theoretical assumptions on which to build a system of taxonomy, this endures as a significant idea even today. (Note: the word 'taxonomy' did not exist until the comparatively minor Swiss biologist Augustin de Candolle introduced it in 1813 in a twenty-one volume encyclopedia of plant life. This was a campaign to describe and codify all known plants, and though he did not finish it during his lifetime, his son did so after his passing.)

CAMERARIUS'S WORK ON BOTANICAL CLASSIFICATION

Still another advance in the science of plant taxonomy came with the publication of the German physician Rudolph Camerarius's book

De Sexu Plantarum Epistolam, or "Letter on the Sex of Plants," which appeared toward the close of the seventeenth century, and exactly twelve years after the publication of Nehemiah Grew's classic, *The Anatomy of Plants*.

Rudolph Camerarius was born in 1665. He came from a family of scholars, a smattering of whom had achieved some distinction both in his own and previous generations. Subsequently he became professor of medicine at Tübingen. Early on, the work of Grew influenced him and he tried to extend it further. He designed and conducted several experiments on plant reproduction. He severed the anthers from specimens of the castor-oil plant, *Ricinus*, discovering that the plant then yielded only degenerate seeds which failed to develop. From this result he inferred that he had removed some of the plants' sex organs. Further research persuaded him that these plants were hermaphrodites—plants with both male and female reproductive parts on the same plant.

He reported many of these findings in a letter of 1694 to the professor of botany at Glesen, Gabriel Valentin. It was these letters that entered the biological literature as the *De Sexu Plantarum Epistulam* noted above. Here he first describes pollen as "male" and ovary as "female." Historians of science also credit Camerarius with producing the first artificial plant hybrid, a cross between hemp and hops plants. His leverage extended well into the eighteenth century, when numerous botanists were using pollen from one species to fertilize flowers from another species of plant to create hybrids.

THE WORK OF GREW ON PLANT ANATOMY

Of course, scientists saw God's mastery in plants too. So it was that another major advance in botany came with the studies of Nehemiah Grew. Born in 1628, he was the son of a British preacher who had aligned himself against the Crown in the terrible British Civil War. He began his schooling at Cambridge University and then transferred to the University of Leiden in Holland.

Like Swammerdam and many other leading biologists, he first studied to be a physician, receiving his medical degree in 1671 with a thesis on the nervous system. Also like Swammerdam, he did try to organize a medical practice, but because his teachings on botany had already swept the land, he realized that his hopes for renown lay in that field rather than in the practice of medicine. Thus he abandoned medicine to pursue research in botany in London. He passed away in 1712.

His approach to botany was generally the same as Malpighi's, with the exception that he seldom practiced comparative anatomy between animals and plants, preferring to focus his attention exclusively on plants. In 1672, he completed his marvelous *Philosophical History of Plants*, and in 1682, he published *The Anatomy of Plants*, which catalogued, distinctly and clearly, the enormous variety of cells making up the various parts of a plant. Beyond this he gave the biological community accurate descriptions of pistils, or seed-manufacturing parts of a flower, as well as stamens, or the pollen-manufacturing part. In this essay, he proposed that the plant pistil was roughly analogous to the female sex organ. He claimed also that the stamen was analogous to the male organ, since it contained the seeds or pollen of new plant life. He further noted that, in a few types of plants, both of these organs dwelled in one and the same flower, and he and other botanists quite plausibly theorized that the phenomenon of "hermaphrodism," or the presence of both sexes in a single individual organism, might not be limited to the animal kingdom. He conducted extensive studies of the vascular system in plants as well. Undoubtedly, his discoveries here led ultimately to the realization that plants were fashioned entirely from cells.

Peculiarly, despite the rather marked distinction of his deeds, Grew, like Malpighi, had surprisingly little effect on his contemporaries. Many theories to explain this have appeared, but the neglect may have been due merely to the obscurity of his writings.

THE WORK OF HALES

Neither the continuing progress in plant anatomy nor the incipient progress in animal physiology had shed much light on plant physiology. For a thorough appreciation of that specialized discipline, the world had to wait for the work of the British biologist and clergyman Stephen Hales, who took the first really decisive steps in understanding the metabolism of plants. Born in 1677 in Beckesbury in southern England, Hales studied science and mathematics at Cambridge. Later he read theology at Cambridge and even took holy orders in the Church of England. Thus he did not begin real scientific investigation immediately. He became the Vicar of Teddington, a parish in Middlesex, England, where he ultimately perished in 1761. Throughout his days, his God-fearing zeal stayed with him. He would intermittently dedicate his energies to philanthropy and preached constantly about the decay of public morals in Europe, especially in England.

His dedication to science always ran parallel to his interest in theology. The sciences he focused on at Cambridge were primarily physics, chemistry, and botany. It would be quite some time before he would take any notice of zoology, and his contributions to the latter field are comparatively minor. Still, he did do some interesting zoological work. In a series of experiments, he scrutinized the circulation of the blood and even managed to measure blood pressure as well as the velocity of the blood in the arteries and veins.

He took the incontrovertibly modern view that alcohol could be unduly damaging to an organism, particularly to the circulatory system and the walls of the blood vessels. Not unexpectedly, he became a devout adherent of the Temperance movement in England. In this period, of course, the canon of Newton was dominant, and scientists all over Europe were trying to explain the natural order via the principles of physics. Hales was no exception. He firmly embraced the notion that Newtonian postulates could clarify the birth, maturation, and reproduction of plants and animals. However, Ray's conclusions also inspired him tremendously, especially the descriptions of plants around the university.

HALES AND PLANT PHYSIOLOGY

Following the methodology of Descartes and others, Hales tried to explain the workings of the plant in terms of the laws of physics. In his *Statistical Essays on Nutrition of Plants and Plant Physiology,* he used surprisingly innovative experimental procedures. He was the first to realize that liquids flow through a plant's "circulatory system" and that plants continually emit gases from their outer layers of tissue, though Hales knew little about these gases. In 1733, he published his *Statical Essays*, which, among other things, added to our comprehension of blood flow in animals and the flow of sap through trees. In this treatise, he described how he had measured the volume of water plants soaked up from the ground, while in other experiments he calculated the ratio of the water left in the earth to water wrested from the ground by the plant's roots. He further measured the rate at which roots absorb water. A courageous and perhaps somewhat arrogant thinker, he did not hesitate to criticize others, particularly Grew and Malpighi. In his *Vegetable Statics* he says:

> Had they fortuned to have fallen into this statical way of enquiry, persons of their great application and sagacity had doubtless made considerable advances in the knowledge of the nature of plants. This is the only sure way to measure the several quantities of nourishment which plants imbibe and perspire . . .[14]

Few momentous advances appeared in plant physiology after this until 1754, when Charles Bonnet, best known for discovering parthenogenesis and for his preformation theory, wrote his *Recherches sur l'Usage des Feuilles des Plantes*, or "Investigation of the Function of the Leaves of Plants." Less than a decade after that, the eighteenth-century French botanist and chemist Henri Duhamel, inspired by Bonnet's findings, revitalized and extended his work on plant physiology to trees. Indeed, he devoted himself to learning the function of trees, their structure, and how they are related to one another. In 1779, the Dutch physician and botanist Jan Ingenhousz found that respiration occurred in two ways in plants; during the day, a plant will absorb carbon dioxide and give off oxygen, while in the evening the reverse occurs. In his "Experiments on Vegetables, Discovering Their Great Power of Purifying the Common Air in the Sunshine and of Injuring It in the Shade and at Night" of 1779 (long titles were common, though this is a bit much), he revealed that an essential element in the emission of oxygen by plants is sunlight. Such processes, he surmised, accounted for the gradual accumulation of starch in plants. Finally, he argued that the plant uses the carbon dioxide assimilated during daylight as an ingredient in food, while the carbon dioxide the plant gives off at night is a by-product of plant respiration. He further proved that the plant requires sunlight to carry out photosynthesis—the process wherein green plants use sunlight to manufacture carbohydrates from carbon dioxide and water.

CHEMISTRY AND ROBERT BOYLE

In the early days of the nineteenth century, the Swiss geologist and botanist Nicholas de Saussure, in his *Chemical Research on Vegetation,* broke fresh ground when he found that plants took up carbon dioxide, not from the ground as botanists had believed all along, but from the air. What the soil gave them was nitrogen. However, he erred in claiming that the green color of leaves was irrelevant to tree health. As indicated previously, some of the older scientists had already begun to appreciate the importance of chemistry in biology, most notably Paracelsus. Since his time, however, science had not learned very much until the labors of Jan Baptiste van Helmont and the Irish scientist Robert Boyle in the Newtonian era. The experimental method had only barely crept into chemical "research"; little was known about how substances combine chemically, and the mysticism of alchemy continued to infect scientific methodology. Assuredly, it was Boyle, a graduate of Eton, born in 1627 at Lismore castle, who in fact transformed the superstition-ridden nonscience of alchemy into the laboratory science that chemistry is today. Today's tenets about bases, acids, pH (a measure of the acidity of matter), and gases and their behavior all trace their evolution to Boyle's efforts. The above concepts, along with his notion of "corpuscles," which hinted at future ideas on chemical bonding, appear clearly for the first time in the story of science in Boyle's epochal book *Skeptical Chymist*. In his *New Experiments Physico-Mechanical Touching the Spring of Air,* he demonstrated that evacuating a container of all of its air would prevent respiration in living organisms, undeniably a major step forward in the science of physiology. Eventually he discovered what we know today as "Boyle's law," the rule that states that in a given volume of gas under ideal circumstances such as at a constant temperature, the volume varies inversely with the pressure applied to the gas.

Ultimately, Edmé Mariotte, a physicist of the French Academy of Sciences, would provide a more comprehensive statement of this powerful physical law. John Dalton also made a major contribution when he formulated the gas law that says that the pressure exerted on the sides of a container by a mixture of gases is equal to the sum of the partial pressures of the individual gases.

It is critical to mention that during this era, medicine and biology were starting to go their separate ways. Possibly it was the invention of the microscope, drawing the attention of biologists to a world of new biological phenomena, that caused this split, although it is hard to be sure. In 1672, Boyle added to his luster when he proposed the idea that if one heated certain metals, a gas (afterwards identified as hydrogen) would leave the metal's surface and catch fire. Along other lines, Boyle documented, in a theater demonstration before the Royal Society of London, that he could keep an animal alive with a mechanism of his own design for sustaining breathing—an anticipation of the modern artificial "respirator."

Much later, in 1729, the French chemist Louis Bourget became the first scientist to adequately discriminate organic from inorganic growth. This distinction he laid out in his imposing *Philosophical Letters on the Formation of Salts and Crystals and on Generation and Organic Mechanisms.* As it turned out, this contribution would have vital significance for the study of viruses in the twentieth century, when science realized that there was no sharp line between living and nonliving organisms. Indeed, historians of science realized that the older generations of scientists had wasted many hours trying to understand life when they were, in fact, simply witnessing crystal growth—the formation of stalactites and stalagmites in caves being a good example.

Not all scientific work was aimed at academic researchers pursuing abstruse and arcane topics. Less technical "popularization" of science was starting to appear in this era; in 1771 the Swiss mathematician Leonhard Euler would write a weighty, yet stylish, volume on an assortment of scientific topics, including chromatic aberration. He suggested that one could avoid this unwanted splitting of light into its component colors by constructing a lens consisting of several lenses fused together—"triplicate" lenses, for instance.

SYDENHAM'S STUDIES OF DISEASE

Another authoritative figure of this period was the British physician Thomas Sydenham. Sydenham eventually acquired a most distinguished reputation. Like Pasteur, who lived generations later, Sydenham believed that infections could cause disease, but it was not until long after his demise that the scientific community began to appreciate his eminence as a scientist. He came from a relatively well-to-do family and began his schooling at Oxford. At the onset of the British Civil War, nevertheless, he allied himself with Parliament, even extending his services as a physician. With his virtuosity as a physician and encyclopedic knowledge of biology, he traveled in the highest circles, befriending such eminent thinkers as John Locke and the chemist Robert Boyle.

It was not uncomplicated to be a physician in England during the seventeenth century. Sanitation and medical facilities were primitive, and contagion was rampant, but it gave the intrepid physician a wonderful real-life "laboratory" in which to do his research. As Sydenham said,

> That botanist would have but little conscience who contented himself with the general description of a thistle and overlooked the special and peculiar characteristics in each species.[15]

Thus he was more of an empiricist than a theoretician. Applying these methods to the study of plague, he followed carefully every patient he could locate who was afflicted with some particular disease.

In his explorations in epidemiology, he gave detailed descriptions of a miscellany of familiar afflictions, including dysentery, scarlet fever, whooping cough, and others. He would also use opium as a painkiller, as well as quinine, quite correctly, to treat malaria, and would recognize the significance of iron in the diet for preventing anemia.

THEOLOGY AND SCIENCE

Scores of biologists and naturalists of this and former periods in the history of science saw Creation, as noted in the foregoing remarks, as proof of the wisdom and providence of a Divine Being. Indeed, this way of perceiving the relationship between science and theology persisted well into the nineteenth century and even into the twentieth. Ray, for example, later in life returned to a philosophy he had embraced in his youth. He wrote *The Wisdom of God Manifested in the Works of Creation,* where he asserted confidently that fossils are the preserved traces of animals of ages long past. Essentially, the book was an extended version of the "argument from design," or as some call it, the "teleological" argument for the reality of the Creator—the conjecture that the beauty and order in Creation constituted evidence for God's existence.

Ray, however, wanted to strengthen and popularize this alleged link more than any other biologist of the day. Perhaps his most significant effort here was the paper "Physical, Theological Discourses," published posthumously. In it he also supported the argument from design, claiming that the fossils he had found on a number of field trips were further evidence of the magnificent design of the universe by an omniscient Creator.

It may be of value to spend a moment on this famous argument, for it is unquestionably the argument religious apologists use most widely, and it is, in fact, one of the most attractive arguments for the existence of the Almighty. The overwhelming majority of scientists, from the Middle Ages throughout the Victorian era, enthusiastically accepted both this argument and the sweeping worldview that God had created the natural order. Isaac Newton was no exception. Nor was Einstein. This argument certainly was not created by Ray or anyone else in this period. It appears first in the Middle Ages, in St. Thomas Aquinas's classic *Summa Theologiae*. The Protestant theologian William Paley then revived it in the eighteenth century. In essence, St. Thomas argued, as did Ray, that the macrocosm was concrete evidence of the existence of an intelligent designer—God. St. Thomas argued that because of the high degree of order and beauty in the cosmos, the universe could not have come into being by chance. Rather, it had to be the result of the artistry of an infinitely brilliant mind. But, St. Thomas argued, this "cosmic mind" could not be any mere human mind, for it was evident that creating something like the cosmos as it exists is far beyond the capabilities of any human mind. Therefore, only a divine mind could have conceived of the universe.

ASSAULTS ON THE ARGUMENT FROM DESIGN

The argument has numerous weaknesses. Undoubtedly, one of the most sustained assaults on this teleological argument for the existence of a deity came from the pen of the Scottish Enlightenment philosopher David Hume. In his classic booklet, *Dialogues Concerning Natural Religion,* Hume assaulted the argument vigorously with a number of counterarguments, including pointing out that while there was much good in the universe, there was much evil as well. That, so argued Hume, could not be the product of a Christian God.

All this notwithstanding, William Derham, in 1713, tried to deal with this problem in his *Physico-Theology, or A Demonstration of the Being and Attributes of God from His Works of Creation.* Because of Derham's backing, this argument acquired still another name, the "physico-theological" argument. Derham tried to resolve this "problem of evil" by utilizing some recommendations of the German rationalist philosopher Leibnitz. Specifically, Derham tried to show that a measured amount of evil was actually necessary in the cosmos, so that one cannot fault God for allowing it. This solution to the problem of evil, nevertheless, while appreciated, may not be an inadequate answer to it, though a more detailed discussion would be beyond the scope of this book. In any case, the emergence of this controversy signaled what has come to be known as the "natural history" movement in the biological sciences, or the notion that the entire biological kingdom is one towering manifestation of the probity of God.

SUMMARY

Significant work in classification went on in this period in the laboratory of Ray—most notably his distinction between monocotyledonous and dicotyledonous plants. Camerarius also added to the science of plant taxonomy, while Hales added much to our understanding of plant physiology with his *Statical Essays* of 1733. Men like Charles Bonnet, Henry Duhamel, and the Dutchman Jan van Ingenhousz added still more to the science of plant physiology. Nevertheless, perhaps Robert Boyle and his myriad contributions to chemistry helped make that field a powerful rival to biology for stellar status in science in this epoch.

It was also about this time that philosophical theology began to suffuse biological speculation, principally with the work of William Derham and Leibnitz.

CHAPTER 14

The Age of Linnaeus and the Enlightenment

The Enlightenment denoted a significant break from the mythology and superstition of the past—a movement away from theologically based science and from the dominance of men like Galen and Aristotle. Philosophy, for instance, spawned considerable further progress during the Enlightenment by scrapping the tattered doctrine of rationalism—the theory that all genuine learning comes through reason, the five senses playing no role. Proponents of rationalism include men like Descartes, Leibnitz, and Spinoza. Instead empiricism, the idea that all knowledge comes via the five senses, took its place, with the advent of the skepticism and materialism of empiricists like Hume. Hume's renown in the annals of science and philosophy cannot be overemphasized. As noted earlier, he became known for his severe assaults on the "teleological" argument for the existence of the Creator. Hume's mighty arguments in favor of skepticism, the conviction that absolutely irrefutable knowledge does not exist, coaxed many scientists into following a similar track in their scientific thinking.

Developments in biology went hand in hand with developments in philosophy for a number of reasons. For one, the font of Western biology, Aristotle, was first and foremost a philosopher. Also, biology, since it included the science of man, inevitably collided with theological issues, such as whether man had an "immortal soul," whether he had free will, and so forth. Even so, Hume persuaded generations of scientists to trust no classical authors, no hallowed text, nor any rationalist philosophies. Since, as we have seen, so many biologists pictured the biosphere as one commanding manifestation of God's plan for the world, Hume's pronouncements assaulted the scientific estate like an icy blast. This, along with the persisting influence of Newton, led most scientists to try and perceive life as a series of mechanical systems governed by the laws of nature rather than the laws of the Lord. In a word, Hume had driven a wedge between the macrocosm and God—precepts that many previous scientists had thought were virtually synonymous.

Descartes would straddle the line between "vitalism," the idea that the phenomenon of life can be explained by spiritual forces other than ordinary physical and chemical processes, and "mechanism," the notion that physical and chemical forces alone can explain all the manifestations of life. Born in 1596, he was one of the most exalted philosophers as well as one of the most notable scientists and mathematicians of his time. He proposed that living organisms, or, more exactly, the behavior of living organisms was a matter of physics. That is, the principles of physics could explain, at least in part, all known animal behavior. He attended a Jesuit school as a boy, but while still an adolescent he began to depart from the Scholastic tradition. At seventeen he went to the Sorbonne, and he entered the military at twenty-one, stationed for two years at Breda. Frightened of possible trouble with Catholic leaders in his homeland, Descartes headed for the more congenial climate of Holland, educating himself in mechanics, mathematics, physiology, and philosophy.

FREE WILL AND DETERMINISM

In the seventh essay of one of his two exquisite tracts, the *Discourse on Method*, the other being the *Meditations*, Descartes outlined a materialist account of human and animal behavior. In these musings, he advocated the notion that it is unavailing to explore human behavior purely mechanically, though mechanics is an integral part of human behavior. Rather, the vital part of man is the "soul" which, in Descartes's philosophy, interacts with the body via the pineal gland. Doubtless his "rationalistic" approach to human knowledge hampered his explorations in biology and physiology. According to this tenet, information came not from the senses primarily, but from unaided reason. While this is indisputably true of, say, the knowledge derived from mathematics and logic, there is, at best, only an element of truth to this doctrine in science.

In Europe, Baron d'Holbach applied this "mechanistic philosophy" more rigorously to a collection of metaphysical problems, especially the ancient conundrum of free will. D'Holbach argued that determinism must be true, and, as a consequence, human beings do not control their own destinies as religion had always taught. To comment on just one thread in his analysis, d'Holbach argued that if human beings are really merely complicated mechanical systems obeying the laws of nature as much as a comet or a planet in orbit around the sun obeys such laws, then it must be possible—at least in principle—to predict the future behavior of any person. Accordingly, man cannot have free will.

The French biologist Buffon was thoroughly committed to a mechanistic picture of the universe as well, perceiving it, in the manner of Leibnitz, as one giant cosmic machine governed by the inexorable laws of physics. He did not tarry in extending the supposition to man: he conjectured that human beings were but another part of this grand machine. As such, he could not believe in free will. Since it was theoretically possible to divine all of the future movements of things like planets and comets, so too was it possible, in Buffon's worldview, to predict all of the future behavior of a person, if one had all of the particulars necessary—a conviction actually more in line with Spinoza than Leibnitz. In this way—and this perhaps is his most daring conjecture—he tried to view biology as simply a branch of physics, the laws of physics as applied to living organisms.

In 1748, Julien Offroy de la Mettrie upheld this tradition with the publication of his *L'Homme Machine* or "Man as Machine" (Leiden, 1748). In this essay he argued in much the same fashion as d'Holbach, that man is nothing more than a machine or a "mechanical organism," whose behavior is as predictable, in principle, as the tides.

Some, however, resisted mechanism. Occasionally, the English biologist John Hunter, in addition to sponging off his sister in Glasgow for some years, lapsed into already discredited doctrines, appealing occasionally to the idea of an *élan vital*, or "vital force." He believed that the blood was, in fact, just such a force. He supposed that even if one could give a complete description of the biological, physical, and chemical aspects of animal functioning, this would not yield an exhaustive explanation of what a "person" was. For that one needed the indefinable, nonlocatable but assuredly real "vital principle." He wandered further into bewilderment in claiming that all other parts of the body emerged from the blood. Unquestionably, this alone hindered scientific recognition during his lifetime.

The French biologist Charles Bonnet also conducted a stinging assault on the principle of vitalism. Indeed, his criticism reveals how committed Bonnet was to the methods of theoretical science. As a godly man, committed to the view that the Lord was the ultimate architect of the cosmos, one would expect Bonnet to support a doctrine like vitalism, since the latter proposal is quite close to the idea of a "soul." Yet he believed that the scientific facts made anything other than a purely mechanistic explanation of animals impossible. What he did believe was that the assumption of a soul could explain the fact that life goes on at all—what it could not explain were the detailed processes in the animal body. He was conceivably most influential on the Italian biologist Lazzaro Spallanzani, who is best known for his adherence to the preformation chimera.

At least three major philosophical creeds competed with one another during the eighteenth century—rationalism, empiricism, and the belief in the "great chain of being" derived from the German logician Leibnitz. Of these, the last two were the most authoritative. Rationalism was possibly the most ancient of the three conceptions; one sees it, after all, in philosophy even as early as the time of Plato, although it existed as a distinct and significant school only with the advent of Descartes. For Descartes and his legatees, Leibnitz, Malbranche, Spinoza and others, mathematics, the quintessentially rationalist science, was the paradigm of human knowledge.

Descartes's reasons for believing this were compelling; mathematical knowledge did not depend on sense experience, though some later philosophers, such as John S. Mill in the nineteenth century, denied this. Moreover, unlike sense knowledge, the verities that mathematics generated were infallible.

Nevertheless, rationalism as a comprehensive approach to truth is now a mere historical curiosity. With the advent of a dynamic group of scholars who opposed Cartesian thinking, John Locke, George Berkeley, and David Hume of the so-called British Empiricist school, rationalism gradually fell out of favor. Basically, their views about knowledge were the polar opposite of the Cartesian school. For these philosophers, most, though not all, genuine knowledge came from sense experience, and of this group, the most thoroughgoing empiricist and the most commanding was David Hume. Hume's attack on theology not only drew many scientists away from the archaic theologically based world view, as noted already, but caused numerous others to distrust many cherished suppositions in the sciences, such as the concept of causality. According to Hume, in his celebrated text, *A Treatise of Human Nature,* there are no "causal relationships" in the cosmos. All we ever observe is a "constant conjunction" of one event following another. In our commonsense way of looking at nature, Hume explained, we tend to think that there are "forces" acting between a cause and its effects. Thus we tend to believe that when we see a rock breaking a window, the rock is exerting a "force" that causes the window to shatter. According to Hume, this is an error. Since the force is not something we directly observe, it is thus a fiction. All we really see is that thrown rocks and broken windows tend to go together—are "constantly conjoined," in Hume's terminology.

OTHER CRITICS OF THEOLOGY

On the theological side, Hume questioned not only the existence of the Almighty, but the reality of the soul, eternal life, and many other abstractions. Consequently, Hume is the archetypal Enlightenment thinker. Soon, others would join Hume as the debunkers of theological cosmologies. In 1751, Robert Whytt of Scotland (known for describing the role of the spinal cord in reflex movements), in his *Essay on the Vital and Other Involuntary Motions of Animals,* offered strong arguments trying to refute the notion that the soul is the determinant causal factor in human action. Indeed, many of today's philosophers also still try to give purely mechanistic explanations of human action, though many others have marshalled powerful arguments against such mechanistic analyses. As the matter stands now, it looks as if the efforts of Whytt, Hume, and others to construct mechanistic explanations will not work. However, in the history of both philosophy and science, most failures are every bit as significant as the successes. Failures may give us a truer picture of how science and philosophy really make progress. They tell the historian of such subjects that progress is not a smooth, linear process but is instead marred by many ups and downs, deadends, and so forth. Beyond this, even a theory which is, in the final analysis, a failure, can still give us new information, new ideas, and suggest newer and more profitable directions. A good example is the field of behavioral psychology. While it is widely agreed to be a failure, it has nonetheless spawned better and possibly more profitable types of theories, such as the so-called computational picture of a person—the notion that we can profitably look at a human being as a kind of highly sophisticated computer.

Finally, there is the suggestion of a "great chain of being" from the philosophy of Gottfried Leibnitz. According to this concept, all of the terrestrial order is continuous from inert matter up to human beings. By "continuous" the defenders of this dogma meant that there is no separation or difference in kind between, say, a rock and any life-form. Therefore, a paramecium, a unicellular water-dwelling organism, is not a different kind of thing than a piece of stone—it is just slightly more complex. Although this proposal found its way into some of the earlier systems of classification in biology, scientists have long since discarded the ideology.

PHYSICS AND CHEMISTRY

In science *per se,* chemistry continued to lag behind other sciences because so many, even after the research of Lavoisier, resisted abandoning the "phlogiston" myth.

Physics, however, developed rapidly, and without progress in physics, progress in chemistry could not take place. The contention is that, ultimately, chemistry depends on physics for its most basic concepts. Scientists now know, for instance, that chemical reactions occur when the electrons in the outer "shells" of atoms are shared or exchanged with electrons in the outer shells of other atoms. The usual, if somewhat antiquated, analogy involves thinking of electrons circling the nucleus of an atom as resembling the behavior of planets circling the sun. Chemists such as Linus Pauling, Samuel Goudsmit, and Walter Heitler have shown that the precepts of quantum mechanics are vital in explaining how atoms bond together to form molecules.

But the more sophisticated concept of the atom that we take for granted nowadays had to await the genius of physicists like John Dalton and, in the twentieth century, Niels Bohr and Arthur Sommerfeld. Dalton, for instance, claimed publicly that atoms are the ultimate, individual building blocks of matter. He reasoned quite plausibly as follows: since chemicals combine only in multiples of whole numbers, elements must consist of discrete entities of some sort. Although twentieth-century physicists found particles much tinier than the atom, such as electrons, protons, and quarks (which carry the forces holding the atomic nucleus together), Dalton had come tantalizingly close to the truth. When the basic atomic structure emerged, and the relationship of electrons to the nucleus became apparent, chemistry began to advance briskly. The science of biochemistry would soon follow in its wake.

GAHN AND SCHEELE, CHEMISTRY AND ANATOMY

Appreciation of the role of chemistry in both anatomy and physiology increased still further in 1769, when Johann Gahn and Wilhelm Scheele began their investigations. Carl Wilhelm Scheele was one of the pioneers of chemistry, particularly the chemical behavior of gases and the science of chemistry as it applied to flora and fauna. He was born in Stralsund, Sweden, in 1742. In his teens he studied to be an apothecary, eventually coming to live in the village of Koping, where he died in 1786. Scheele, along with Gahn, discovered a number of chemical compounds that were essential ingredients of bone as well as many other tissues. Among the compounds he isolated were cyanuric acid, uric acid, lactic acid, hydrocyanic acid, citric acid, and others. From this point on, biology would never again neglect chemistry, especially when the illustrious Lavoisier himself, under the domination of Scheele, took up and extended these sorts of reflections.

TAXONOMY

In the science of taxonomy, the most illustrious name is still Carolus Linnaeus. He was born in 1707 near the city of Uppsala in Sweden, the child of a clergyman who was also an amateur biologist. According to most historians, Linnaeus developed a fascination with the wilderness through his many walks with his father in the resplendent Swedish countryside. At the age of twenty he enrolled in the University of Lund in Sweden, moving after a year to the University of Uppsala closer to home. At Uppsala, he became acquainted with the theologian and botanist Olaf Celsius. Celsius was sufficiently impressed with Linnaeus's abilities that he invited him to use his own private library. It is supposedly at this point in his career that Linnaeus began his first earnest probings into botany, even suggesting that botanists could efficiently identify plants by their reproductive structures.

At age twenty-eight he married Sara Lis Moraeus, the daughter of a physician in the town of Falun, Sweden. Oddly, Linnaeus had to promise to become a physician in Holland before Sara's father would acquiesce to the marriage. This he did, and he stayed in Holland for several years, more to pursue biology than to practice medicine.

His masterpiece is the 1735 *Systema Naturae,* or "System of Nature," in which he first introduced his classification system. In it he first distinguishes three separate realms: animal, mineral, and vegetable. The animal and vegetable realms he further broke into class, genus, and species. In his formative years in biology he had imitated the biologist Tournefort in that he labeled each genus with a single name and summarized the characteristics of the species in what amounted to brief "footnotes." Realizing the clumsiness of this sys-

Carolus Linnaeus (1707–1778). (Courtesy of the Library of Congress.)

tem, he devised the "binomial," or two-name, system still in use today, where each organism is assigned a generic and specific name. Accordingly, one has *Felis domesticus*, or the exceedingly unremarkable pussycat, *Homo sapiens*, or human beings, *Nerodia sipodons*, the common water snake, and *Felis tigris*, the tiger. In his *A Dissertation on the Sexes of Plants*, Linnaeus defines a genus succinctly; "A genus is nothing else than a number of plants [that] spring from the same mother by different fathers."[16]

Defensibly the most profound of his innovations consisted in his correctly categorizing, for the first time, whales as mammals—animals that nurse their young. He further subdivided whales by dental characteristics, a system that zoologists have long since abandoned in favor of external anatomical features, which are far easier to investigate. He also usefully categorized men with apes, calling both "primates," and this feature, too, scientists have retained. Linnaeus appeared to have held reptiles and other cold-blooded animals in unusual awe; in his *Systema Naturae* he refers to amphibians with the words "Terrible are Thy Works, O Lord,"[17] and he quickly adds that he did indeed intend this motto to apply to cold-blooded animals such as reptiles and amphibians.

He added to all of this when he penned the *Genera Plantarum*, or "Genera of Plants." In this tract he applies many of the same principles he used for classification of animals to botany. In it, he succeeds in classifying over 18,000 assorted types of plants. Although he never traveled to the United States, many of the plants he examined were New World plants that came to him via the generosity of his friend the American botanist John Clayton. Linnaeus recognized twenty-four classes of plants. The most primitive in his scheme were the Cryptogamia, a catchall term for plants that he felt had no really significant structural or reproductive features.

A few years later, Linnaeus switched roles from author to editor. With consummate virtuosity, he skillfully edited and published the writings of the eminent Swedish biologist Artedi, the *Petri Artedi Seuci, Medici Ichthyologia sive Opera omnia de Piscibus*, Artedi's complete works on fish. With this publishing event, the world had the first genuinely systematic ranking of fish. Petri Artedi straight away became known, not surprisingly, as the "father of ichthyology," the science of fish. Artedi was a close associate of Linnaeus throughout most of his years. Both cooperated in their mutual endeavors, Linnaeus in botany and Artedi in zoology. They would often swap ideas and writings to the mutual benefit of both. Now and then they even assisted one another financially, as neither had the resources of a university for most of their careers. Sadly, this esprit de corps ended in 1735, when Artedi drowned in a canal in Amsterdam.

Linnaeus completed his development of the archetypal classificatory system with his Species Plantarum, or "Species of Plants," published in 1753. This effort constitutes part of the theoretical foundation for grouping even today—for animals as well as plants—because the principles Linnaeus set down in "Species of Plants" are quite general and easily adapted to animals. Even so, his best-known work today is *Systema Naturae,* which zoologists primarily rely on, with modifications, as the foundation for modern nomenclature.

By this time, Linnaeus had long been among the most eminent biologists around. Among other universities, Oxford eventually invited him to its faculty; the man most directly responsible for the invitation was J. J. Dillenius, then professor of botany at Oxford, who wanted Linnaeus to teach his classificatory system there. He even offered to share his salary with Linnaeus.

But Linnaeus was not interested, preferring to remain in Sweden to continue his own private labors in biology. He shortly regretted turning down university bids, however, because his attempt to practice medicine had met with scant success. Most of the people of the epoch were so poverty-stricken that they were unable to pay for medical services. Finally, he seized the chance to enter the Swedish military, finally agreeing to be the exclusive physician to the queen. He passed away at home in 1778, near his beloved gardens.

The Swedish Linnaean Society has restored his gardens in what historians believe is pretty much the way he had them originally. Walking along one of the many garden paths, the visitor can see each plant explicitly identified by its genus and species name.

One thing many overlook about Linnaeus, given the fact that his system has survived so well, is that Linnaeus, in one sense, was inordinately unscientific. Like many others, he believed in Divine Creation. Such pious feelings may have played some role in the peculiar way he handled, in his old age, the classification of man. Throughout his career, he had always used external structure as the basis for his classifications. But with *Homo sapiens,* he used "emotional states" as the basis. Here his categories were bizarre in the extreme. He recognized four varieties of humans. The chief quality of "Homo sapiens americanus," or American Indians, according to Linnaeus, was that they were "contented." "Homo sapiens europaeus," or the people of Europe, were noted for being "lively." "Homo sapiens afers," or African Negroes, Linnaeus said were "slow." And finally, he noted that "Homo sapiens asiaticus," or orientals, were "haughty." It scarcely needs saying that practicing biologists mercifully pass over this facet of Linnaeus's work.

In his biological worldview, there was scant room for evolution, and, in fact, Linnaeus systematically ignored all evidence for it in his own deliberations.

THE TAXONOMY OF JUSSIEU

In 1789, Antoine-Laurent de Jussieu added his name to the roster of august classifiers. Born in Lyons in 1748, he came from a family of biologists of note, including Bernard de Jussieu, who had launched some investigations into botanical grouping. Antoine Jussieu began his education in medicine in Paris, later joining the faculty of the Jardin des Plantes. He lived a long and full life, passing away in 1836.

In his *Genera Plantarum* or "Plant Genera," Jussieu used a modified form of the binomial system that Linnaeus had proposed. He made the cotyledon the foundation of his system—a logical and tenable move inasmuch as the cotyledon, as a part of seed from which the mature plant arises, is an essential part of the plant. Here Jussieu was borrowing a bit from zoology. He believed that the cotyledon of the plant corresponded to the heart in the animal body. In his *Genera Plantarum,* he categorized plants into families so meticulously that most of his ideas are still in use today. Under the influence of both Linnaeus and Ray, he also adopted Ray's division of plants into monocotyledons, acotyledons—or plants without seed leaves—and dicotyledons as the primary divisions in the plant kingdom. He then further subdivided plants into orders. Some of his divisions are curious, including, for instance, his placing of ferns within the order Cycadales, a group of palmlike trees having large, leathery leaves and a thick trunk with no branches. Outside of his "order" subdivision, which botanists have since repaired, botanists quickly and universally accepted this classification and it remains intact today.

CUVIER

Following on the heels of Linnaeus, the legendary scientist Georges Cuvier of France turned his attention specifically to the knotty problem of categorizing a group of animals of bewildering complexity—the mammals. Georges Leopold Cuvier was born in 1769 in Montbeliard, a village near Basel in the Duchy of Wurttenberg. He was born into a family of French Huguenots who at an earlier period had been forced to flee from prosecution, as did all Huguenots during this era. His father had served in the French military, subsequently returning to Montbeliard where he lived out the balance of his days on his military pension.

Goerges Cuvier's abilities became evident when he was a child, and he graduated from the local schools with the highest honors. It was during his school years that he began studying the work of his French contemporary Buffon, whose research on classification would inspire and remain with him for the rest of his life. He soon entered the famous *Karlsschule* at Stuttgart. Once a military academy, it had evolved into a most august institution, which could boast of having schooled such luminaries as the German poet Schiller. His first biology teacher at this school was the German biologist Karl Friedreich Kielmayer, who later joined the faculty of the University of Tübingen. Among other things, Kielmayer persuaded Cuvier of the overpowering importance of comparative anatomy in biological inquiry.

Cuvier and Taxonomy

Cuvier contributed a substantial amount to the science of taxonomy. Early on he developed considerable adroitness as both a writer and an artist. Primarily because of this work, the biologists in Paris invited him to join the Natural History Museum there. At the height of his distinguished career, Cuvier published *The Animal Kingdom Distributed by Its Organization*. In this tract he cataloged the entire animal kingdom, subdividing it into four groups. These were the Vertebrata, Mollusca, Articulata, or animals with joints, and Radiata—starfish, for example. The Vertebrata included mammals, birds, reptiles and fish. The Mollusca included barnacles, clams, oysters, and the like. The third group, Articulata, encompassed crustaceans, spiders, and insects, while the fourth, Radiata, comprised all animals that show "radial" symmetry, that is, animals that are symmetrical at all points around a central axis. In this group he placed animals like starfish and polyps. Not surprisingly, given the increased speed and efficiency of scientific investigation and the resulting wealth of new information that has accumulated, scientists have altered these categories considerably in recent years.

Cuvier and Comparative Anatomy

After graduation at age eighteen, Cuvier turned to marine biology and authorship. An intense "researchaholic," Cuvier often had several projects going concurrently. He was one of the first to appreciate the pertinence and relationship of animal form to animal function, the theory that certain bones were formed the way they were in order to perform a characteristic function.

Certainly science knows Cuvier today primarily for his toils in comparative anatomy. In his *Lessons on Comparative Anatomy,* for instance, Cuvier all but solidified comparative analysis of animal anatomy as a crucial independent branch within biology.

ALBINUS AND ANATOMY

Anatomy continued to emerge from the background with the work of Bernhard Albinus, the son of a German doctor who himself was a scientist of considerable standing in Germany. His father had been on the faculty of such eminent institutions as the universities of Leiden and Berlin. Albinus was born in Frankfurt An der Oder in 1697, though his family finally took him to Leiden, where he spent the rest of his days. Something of a prodigy, he became professor of anatomy at age twenty-four at Leiden. He held this post until his demise, which occurred in 1770 in that city.

In contrast to scientists like Kolreuter, Mendel, and Sprengel, Albinus did not suffer from oblivion while he was alive. Indeed, most of the distinguished scientists of Germany revered him immensely. He was a man of broad interests. Among other things, he displayed some passion for philosophy as well as the history of science. He was also something of a biographer, having published books on the lives of Vesalius and Harvey, among others. However, Albinus's greatest work was *Tabulae Sceleti et Musculorem Corporis Humani*, or "Plates of the Skeleton and Muscles of the Human Body." This treatise describes, better than anything previously, the relationships between the skeletal system and the musculature. Albinus scrutinized both of these systems in the human embryo with meticulous care and accuracy. Yet he did not content himself with static descriptions of the skeletal system; he also tried to determine how it grew during embryonic development.

CUVIER, GOD, AND EVOLUTION

The idea of evolution was also making a leisurely entrance into scientific discussion, although the earth-shaking consequences of this proposal would not reach their apex until the reign of Charles Darwin in the nineteenth century. Also, for several years, outmoded dogmas held back the progress of evolutionary teachings. Many otherwise inspired biologists of the seventeenth and eighteenth centuries believed that God had created all species fixed and incorruptible. Even Linnaeus, to mention one prominent example, with the publication of his *Philosophia Botanica*, explicitly discarded the concept of evolution. (Later, even after Darwin, opponents of evolution persisted for decades, including such twentieth-century pillars of science as Robert Millikan of the California Institute of Technology. The issue is immensely complex, but philosophers of biology have charged, among other things, that since claims about the past cannot be directly verified, Darwinism is really not a scientific theory at all.)

A dazzling, though repeatedly overlooked, step in the inexorable march toward Darwin was an idea espoused by Pierre-Louis Moreau in his *System of Nature* that all life started by chance, rather than by the benevolent will of a grand designer. The physician Erasmus Darwin would add still more vitality to the evolutionary hypothesis when, in 1794, he published his *Zoonomia,* followed by his *Phytologia* in 1800. Although these books were more Lamarckian than Darwinian, in that they misunderstood and overemphasized the importance of environmental factors in generating evolution, they did bolster both popular and scientific interest in evolutionary questions and served as a foil for Charles Darwin.

Cuvier was also a figure, but a less important one, in the field of evolution. Despite the influence of David Hume, Cuvier, in fact, did not accept evolution at all, preferring instead the ancient doctrine of the "fixity of species"—the conviction that God created all species at once, fixed and unchangeable. Not surprisingly, he liked the postulate of "catastrophism," since it squared well with the biblical tenet that God could periodically punish humankind by wiping out all known species.

BUFFON

Not everyone agreed with Linnaeus or Cuvier. Notable among Linnaeus's antagonists was Comte de Buffon, perhaps the only other eighteenth-century biologist to rival Linnaeus in stature. He was born in the city of Montbard in 1707, in the province of Burgundy. He came from a wealthy and politically active family, his father having spent many years as councillor in the Parliament at Dijon. He was well educated in his hometown and was reconciled to following family tradition by entering politics when he abruptly had a change of heart. In large part, this abrupt change from politics to science came when he met and befriended Lord Kingston, an English *scholare vagrante.*

Buffon's ventures were in no way limited to biology. He translated Newton's *Fluxions,* and he also embarked on a thorough scrutiny of the school of philosophy known as "Continental rationalism." In particular, the philosophy of Leibnitz fascinated him. He accepted Leibnitz's notion that the earth had once been an awesomely hot, incandescent mass of matter which, after aeons, cooled. In his classic *Epochs of Nature,* published in 1779, Buffon took Leibnitz's musings about the earth's beginnings even further. He argued that the earth had once been part of the sun. But sometime in the distant past, a comet had collided with the sun, fracturing huge sections of it, which flew off into space, settled into permanent orbits and then cooled. He then hypothesized that after it cooled, the oceans thoroughly covered the earth. Not unexpectedly, he cited such evidence as the vast number of marine fossils in what is now dry land. He pointed out, correctly enough, that one can exhume marine fossils in mountains, for instance.

He then sorted the evolution of the earth into seven periods: (1) the period when the earth and planets formed, soon after spinning out from the primal collision of the comet and the sun; (2) the period during which the mountains gradually formed; (3) the period when the oceans completely covered the earth; (4) the receding of the oceans and the commencement of volcanic activity; (5), the emergence of the larger animals, such as elephants; (6) the period when the enormous land masses broke apart, forming the continents as they persist today; and (7) the emergence of the higher animals such as man.

The foremost argument of this book, still, was that the earth was about 75,000 years old. Although this was off the mark by several billion years, it was nonetheless significant insofar as it was the first time any trustworthy biologist had avowed that the earth was more aged than the biblical claim of 6,000 years. The motivating force behind Buffon's daring move was, of course, the fact that it was futile to try to reconcile geological evidence with biblical demands. Nevertheless, he was not the first to note such discrepancies. Steno had seen the problem but could hardly dare to throw off his own fervent Catholicism, preferring to believe instead that he may have misunderstood or botched his study of the geological evidence.

In Buffon's *General and Particular Natural History,* he was starting to accept the teaching that species can and do transmute, even though he had, earlier in life, accepted the antiquated teaching of the fixity of species along with Divine Creation. Even so, he never quite fully accepted evolution, although he admitted that it was conceivable that all creatures descended from one species.

His convictions on inheritance were far from orthodox. Among his more extravagant claims, he believed that overcrowding could lead to evolutionary change, and he also veered close to the Lamarckian notion that an organism could inherit acquired characteristics. More extravagant still were some of his succeeding canons, which arose as he tried to reconcile evolution and special creation. He ruled that the ape was an inferior version of man and the mule was an inferior version of a horse. That the special creation by a divine being of "inferior" organisms went against the omnipotence and benevolence of God did not appear to bother him very much.

Yet, whatever the merits of Buffon's thoughts, they had comparatively few adherents during his lifetime. He was the most theoretical of the scientists of this period—not surprising considering his allegiance to the anti-empirical philosophy of Leibnitz. Inevitably, however, such a revered biologist as Cuvier endorsed Buffon's teachings and became the first to truly appreciate their significance.

ERASMUS DARWIN

Erasmus Darwin is often and unjustifiably overlooked in discussions of evolution. Born in 1731 in Nottingham, Darwin studied medicine at Cambridge and Edinburgh, gradually starting a flourishing practice as a physician at Lichfield. He was also a most prolific author, writing a great number of papers for the prestigious Royal Society. He also had a literary flair. In the following lines from Erasmus Darwin's *The Loves of the Plants* we see covert references to the biological laws and also the pistils and stamens of plants symbolized by human females and males respectively:

> A hundred virgins join a hundred swains
> And fond Adonis leads the sprightly trains;
> Pair after pair, along his sacred groves
> To Hymen's fane the bright procession moves
> . . . Wide o'er the isle her silken net she draws,
> And the Loves laugh at all, but Nature's laws.[18]

Of his many offspring, one was the father of Charles Darwin, and a daughter was the mother of the distinguished geneticist Francis Galton. Erasmus Darwin's most important book is assuredly his *Zoonomia,* which represented his attempt to formulate the laws of biology in much the same manner that Newton had formulated the laws of physics. He asserted in this renowned publication of 1794, his well-known declaration that

> The Great Creator of all things has infinitely diversified the works of his hands but has at the same time stamped a certain similitude of the features of nature, that demonstrate to us that the whole is one family of one parent. On this similitude is founded all rational analogy.[19]

Still, this was far from Erasmus Darwin's most fundamental contribution to biology; he was one of those who were beginning to believe that species could and did change. In fact, many of his grandson's famous insights had already surfaced in the writings of Erasmus Darwin, such as descendence from common ancestors and survival of the fittest. Otherwise, his evolutionary views were more like Lamarck's; that is, Erasmus Darwin believed that animals and plants could inherit acquired characteristics.

Fortunately, Erasmus Darwin enjoyed the recognition of his toils during his own lifetime. Many German scientists and experimental philosophers roundly proclaimed their significance. Such accolades, notwithstanding, his contributions seem to have gone unnoticed in the modern era, and the cause of this is surely the sad fact that Charles Darwin far overshadowed him in renown. In much the same way, the more celebrated investigations of the English empiricist and moralist John Stuart Mill overshadowed the equally vital, but less well-known, philosophizing of his father, James Mill.

BUFFON'S TAXONOMY

Because he was superbly gifted and published on a massive scale, Buffon made a mark in the field of taxonomy, though historians of science pretty much agree that his contributions to the science of classification hardly matched those of Linnaeus. In 1739, he consummated his career by winning election to the French Academy of Sciences, a post that conferred immediate prestige. Consequently, he was something of a Renaissance man in that he was a writer as well as a statesman, in addition to being a biologist.

Buffon sanctioned the belief that internal rather than external characteristics of organisms should be the basis of any systematic selection. As a consequence, he introduced the now-familiar idea that scientists should base the definition of species on the possibility of interbreeding. That is, biologists should consider two animals to be two separate species if the two either cannot mate at all or produce only sterile offspring when mated. Hence a donkey and a horse are different species since although they can mate, the resultant mule is sterile.

Buffon was a fine historian of science as well. In 1749, he published the first volume of his *Histoire Naturelle Générale et Particulière,* or "General and Particular Natural History," the writing of which would occupy many years and produce a highly readable fourty-four-volume chronicle appearing from 1749 to 1778. An encyclopedic work, Buffon tried to include everything he or anyone else knew about animate or inanimate matter. He anticipated present biology to the extent that he was probably the first to appreciate how close the relationship was between humans and the lower primates.

THE SOCIAL ORGANIZATION OF SCIENCE

It is worth noting that the systematic organization of science itself was proceeding as fast as the development of the various branches of science. In 1739, Scotland founded her own Royal Society, as did Sweden. America bolstered the trend when Benjamin Franklin founded the American Philosophical Society in Philadelphia, although the society has never had much to with philosophy, and is really only a scientific organization. America also added to its own burgeoning scientific industry when Thomas Jefferson produced some reasonably careful analyses of plants and fossils.

Later on, Erasmus Darwin would found the Lunar Society, so-called because it held meetings during a full moon, only because moonlight helped some, who lived some considerable distance, to return home easily. Soon, the Lunar Society could boast as members James Watt, inventor of the steam engine, Ben Franklin, and Joseph Priestley, lauded for his discovery of oxygen.

The year 1742 was one of enduring professional growth in biology. In that year, Danish scientists founded their own analogue to the Royal Society and called it the Royal Danish Academy of Sciences and Letters. Science was spreading in other ways as well.

HALLER AND PHYSIOLOGY IN THE ENLIGHTENMENT

The science of physiology made innovative and rapid progress in the Enlightenment as well. The Swiss biologist Albrecht von Haller wrote his *Prima Lineae Physiologiae,* or "First Introduction to Physiology," apparently the first book that really anticipated twentieth-century physiology. Haller was born in Bern, Switzerland, in 1707, the progeny of an influential attorney who was something of a fanatic on the subject of education. Understandably, he gave his son the most exceptional education his money could purchase. Albrecht studied at the University of Tübingen and then at the University of Leiden. Haller appears to be among the most gifted scientists of this period, displaying amazing gifts almost since toddlerhood. Like the British scientist and philosopher John Stuart Mill in the nineteenth century, he knew a considerable amount of Greek by the age of ten, and within two more years began composing poetry and tragedies. He earned his doctoral degree in medicine before he was out of his teen years.

After receiving his degree, Haller headed to Paris and then to Bern in pursuit of his growing fascination with pure science, as opposed to medicine. He eventually opened a medical practice in Bern, Switzerland, but again, his real interest remained with pure biology rather than the "applied " science of medicine. One of his most able publications was the *Fuselage Corporis Human,* or "The Human Body as a Subject of Physiological Experimentation." Although this offered little that was new, it did serve to organize and collect in one volume a substantial amount of the then-current physiological knowledge.

Haller was one of the first to describe the nervous system as "irritable," a notion still in use today. Generally, he theorized that there were two fundamental "sensations" in the human body—those which were "irritable," and those which were "nonirritable." Oddly, he held that science could not explain this, but must content itself with merely proving that these phenomena existed. He defined a part of the body as irritable if it contracted when in contact with some foreign object; a part of the body was "nonirritable," or "sensible," if, upon being touched, it produced an "impression," on the mind, an idea much in line with the empiricism of the British philosophers such as John Locke and George Berkeley. An example would be the archaic Greek idea that "fine atoms" emerge from objects in one's environment to produce an "impression" on the mind. So if one is looking at a couch, the fine atoms emanating from the couch produce the mental image of a couch on the brain that the observer sees.

Haller postulated that the phenomenon of muscle contraction was related to this in the sense that irritations to the organism would provoke such contractions. He had at least a rudimentary understanding of the function of the nervous system as the carrier of irritations to the brain, which, in turn, "told" the body to respond in various ways to what was going on in its environment. He followed this study with further research in which he showed, or tried to show, that muscle was irritable not because of anything having to do with muscle tissue as such, but just because of its connection with the nervous system. He further tried to prove that muscle movement began with signals emanating from the brain, which then traveled through the nervous system, ultimately exciting muscle tissue—a concept that came moderately close to the truth.

Descartes too made a passable guess at the function of the nervous system. In his philosophy, "sensations" emanating from the senses impinged on the sense organs. The nerves were hollow tubules through which sensations flowed to the master gland, the pineal. Still, Descartes went awry in arguing that sensations activated and controlled various "valves" in the nerves that monitored the flow of sensations.

GALVANI

By 1772, the eminent Italian biologist Luigi Galvani would have speculated that electrical shocks and certain metals could stimulate muscle movement in frogs. That same year, he proved this through experimentation.

Galvani dug ever more deeply into the mysteries of physiology when, in 1790, he identified the physiological "reflex," an involuntary movement due to some outside stimulus. (However, that term appears for the first time in 1833 in the writings of the British physiolo-

Luigi Galvani (1737–1798). (Courtesy of the Library of Congress.)

gist Marshall Hall.) In 1791, he finally publicized his momentous study of electrical stimulation of nineteen years before. Galvani also made some technological contributions. He invented, for instance, what physiologists today call the "galvanometer" in his honor—a device to measure electrical current in the nervous system.

Galvani's studies in the domain of the physiology of tissues inevitably led to further involvement on the part of other biologists. Up until this point, although biologists had begun to appreciate how tissue functioned, as well as its relationship to the nervous system, they knew exceedingly little about the *varieties* of tissue found in either animal or plant bodies.

BICHAT AND THE PHYSIOLOGY OF THE NERVOUS SYSTEM

At this point the French biologist Marie Bichat came to the rescue. Marie Francois Bichat was born in 1771 in the canton of Jura, Switzerland. His father was a well-known physician and pressured his heir to follow the same profession. Marie began studying surgery at a Lyon hospital, but his studies were interrupted when the hospital was destroyed during the French Revolution. He later studied in Paris, there befriending the well-known surgeon Desault, with whom he developed a lifelong bond. Under Desault, Bichat studied both anatomy and surgery. After Desault's passing, Bichat edited his compositions and remained a close confidant of Desault's widow.

Despite the fact that he lived through the most turbulent years of the French Revolution, he was able to continue his own research virtually uninterrupted. He departed this world in the spring of 1802, after a short but productive career. He was apparently an exceptionally kind and genteel man, possessed of not the least trace of arrogance.

In 1801, Bichat wrote the *Anatomyie Generale Applique à la Physiologie et à la Medicine* ("The Study of Applied Anatomy, Physiology and Medicine"). Much of this volume consisted of an analysis of the tissues. In fact, Bichat created a system that classified tissues as either skeletal, cartilaginous, or muscle. The sum total of the information about tissues he called "general anatomy." He appears to have adopted something like a mystical attitude toward them, believing that the life-force that keeps the body alive and healthy resides in tissue. Indeed, although he did other reliable work, including accurate explorations of the brain, his exploration of tissues is his enduring contribution, and historians of science therefore consider him one of the founders of the modern science of histology. He described and categorized over twenty different tissues, including blood, striated muscle, cardiac muscle, bone, cartilage, and the mucous membrane tissue of, for example, the mouth.

RÉAMUR AND DIGESTION

The biologists of the Enlightenment did not restrict their physiological investigations just to the nervous system. In 1752, the French scientist René Ferchault de Réaumur again ennobled himself, adding to his previous investigations in entomology. He began to fathom the process of digestion in higher animals when he found that digestion is not merely mechanical, but also chemical. He discovered this when experiments on hawks showed that the secretion of gastric juices in the stomach caused meat to dissolve. Later, the Italian biologist Spallanzani would also study digestion.

The British biologist John Hunter added to our knowledge of digestion in what is perhaps his most discerning contribution, *Philosophical Transactions,* which appeared in 1767. This essay issued from a analysis of the recently dead, and suggested, correctly enough, that digestion was a function of the stomach and gastric juices.

HALLER'S EMBRYOLOGY

Haller too, like Leeuwenhoek, examined the development of the chick embryo. It was not clear which side Haller took in the preformationist controversy between ovists, those who thought that the intact organism lived in the egg, and spermatists, those who believed that the entire formed organism abided in the sperm. On the one hand, he said that the egg was essential in the production of the embryo, in this way appearing to side with the ovists. On the other hand, he also said that the "spermium," or the "vermiculus seminalis," in his words, created the man, which allows the interpretation that he took the spermatist side. Sometime later, others, such as the French biologist Charles Bonnet, announced their allegiance to the ovists much more decisively, believing that an entire miniature organism existed preformed inside the egg. Throughout history, Haller's conclusions have been controversial, though less so in more recent years. His detractors, in particular, assaulted him for a lack of care and detail in his hands-on work. This, nevertheless, misses the point, since Haller was above all a theoretician, believing that lesser minds could take care of the detailed lab maneuvers. No one today would deny that he contributed an enormous amount to physiology.

BONNET AND EMBRYOLOGY

Charles Bonnet was born in 1729 to a wealthy family that had left Paris at around the time of the persecution of the French Huguenots. In his teen years, he trained for the bar, and after completing his schooling, his hometown appointed him to the town council. Nonetheless, the law ultimately lost its fascination, and the sciences took over his mind unconditionally. Initially, he was a pupil of Réaumur, focusing, naturally enough, on insect studies. He soon began losing his vision and could no longer practice practical science with any efficiency. So he decided to turn his attention to pure science, albeit without venturing beyond the confines of orthodox Christianity—a feature that is evident in his repeated references to a deity. He died in 1793 in Geneva.

In 1764, Bonnet, in his *Contemplation of Nature,* submitted his own version of the homunculus, or "preformation" supposition, as later biologists liked to call it. He followed this work with his *Philosophical Palingenesis.* This treatise showed the same obsession with preformationism, by claiming, without any solid evidence, that it was only the females of the species that contained the seed of generations to come. As usual, what it offered was insignificant compared to the legacy of the Greeks.

Bonnet held more tightly than his fellow German, biologist Haller, to the concept of preformation. Still, in attributing so much of creation to the pre-activity of the Almighty, without providing evidence for this conviction, he found himself in heated disputes with many other thinkers, including Needham and Buffon. In his own even more bizarre version of preformationism, Bonnet believed that the female of every species contained all the eggs of every future member of that species that would ever be born, until the very end of time itself. Thus Bonnet felt that the eggs of all future human beings existed preformed inside of Eve from the dawn of creation when God created Adam and Eve. By 1758, Bonnet had consummated this investigation and concluded that he had all but proven that the embryo survived preformed inside the egg.

Haller had an enchanting argument in favor of ovism over spermatism. He observed, as had Bonnet in 1745, that an egg can develop even though it has not been fertilized. However, what he had noticed was not ovism, but a demonstration of parthenogenesis. His belief stemmed from his earliest research with aphids. He saw, for instance, that with the right prerequisites, female aphids would yield live offspring without these being fertilized. The problem was that while the phenomenon of parthenogenesis was genuine, as in aphids, eggs in the overwhelming majority of higher animals cannot start developing without fertilization. What he observed really could not be construed as evidence for a universal phenomenon of ovism in the biological world.

INSECT REPRODUCTION

Bonnet followed up on the work of a number of biologists who had discovered that certain animals could regenerate missing parts of their bodies. He studied some primitive animals, including coelenterates, a group of invertebrates that includes corals, jellyfish, and others, and bryozoans, a phylum of water animals with gelatinous bodies, as well as a selection of worms. He noticed that even after cutting such organisms in half, each part could regenerate an entire organism. He steered his investigations into the metamorphosis of insects, trying to learn exactly how the various parts of the body changed when the organism was undergoing its metamorphosis.

DISEASE

No one was neglecting the study of disease during the Enlightenment either. While biology progressed rapidly, medicine moved even faster, because of its life-and-death practical character. Physicians of the eighteenth century discovered, among other things, the technique of autopsy as a way of acquiring medical knowledge, advanced the science of histology via the microscope, and made progress in grasping the importance of blood pressure. News of inoculation found its way from the Near East to England, thanks to Muslim physicians. The American Puritan leader Cotton Mather then became the first to systematically employ inoculation in Boston, hoping to curb a flare-up of smallpox.

Once again, Marie Bichat is a pivotal figure. Because of his medical training, Bichat was as interested in the biology of disease as he was in the biology of healthy organisms. He consequently devoted a great deal of time to postmortem examinations, particularly on those who had succumbed to various odd infirmities, such as smallpox and typhus.

Perhaps the most fascinating medical tale of this epoch centered around the life of the British biologist John Hunter, the brother of another distinguished biologist, William Hunter. He was born in Scotland in 1728 to an impoverished farming family and endured the further misfortune of being orphaned almost in his infancy. In 1771, John Hunter penned his important treatise *Natural History of the Human Teeth,* which established the foundations for the evolution of contemporary dentistry, dental anatomy, and dental physiology. In addition to the above-mentioned work, John Hunter also contributed to dentistry with his book *A Practical Treatise on the Diseases of the Teeth,* which not only advanced the practice of medicine, but added to the science of taxonomy by categorizing teeth as incisors, molars, bicuspids, or cuspids. He also looked at the origins, anatomy, and the maturation of teeth.

The most remarkable thing about Hunter's exertions is what he succeeded in accomplishing despite having had a meager formal education. He never learned to read or even write his native language very credibly. For this reason he did not enter the professions that scientists of the day usually did. Even so, his brother William, who was able to become a physician, did assist his brother considerably, repeatedly providing the support needed for John to pursue his work. Others also helped him, due in large part to the fact that he did have rather obvious talent and because he was apparently an exceptionally likable and affable man, though occasionally given to fits of temper. As a consequence, despite wretched conditions, he at last became a surgeon, assigned to treat the British navy during the Seven Years' War.

Ultimately, through shrewd saving, he was able to construct a home museum and devote himself almost entirely to his own research. He eventually became a surgeon at St. George's Hospital, and almost immediately became fascinated with museums. He traveled widely always in search of specimens for his museum. Perhaps the most renowned of his acquisitions was the skeleton of the legendary giant O'Byrne, who measured nearly eight feet in height.

His habits were engaging. Always a believer in direct experience, his faith in previous book "knowledge" was infinitesimal. He was an indefatigable scientist, rising at daybreak to begin his comparative anatomical observations. He then treated patients for most of the afternoon and resumed his own inquiries in the waning hours of the day. Conceivably what was most arresting about John Hunter's life was his death. In effect, he committed suicide, not because of depression, but because of self-investigation. Starting in 1767, he began experimenting on himself, a practice which was a lot more common than many sometimes realize. At the time, there was a substantial amount of controversy within the medical and biological community over venereal disease, principally gonorrhea and syphilis. Many suggested, erroneously, that these were only two labels for the same infirmity. Somewhat capriciously, he decided to follow the progress of the illness by injecting himself with the pus of a patient failing from syphilis. He theorized that one virus caused the symptoms of these two sicknesses, but in unlike ways. For example, if the virus chanced to invade the mucous membrane of a host organism, then the symptoms of gonorrhea would appear. But, if it landed on the skin, the symptoms of syphilis would appear.

Unfortunately, he derived this baseless notion from a patient who by a fluke had the misfortune of suffering from both ailments at once. But Hunter turned out to be a remarkably resilient and hardy individual. Even without proper medical treatment, he did live on for over twenty-five more years. Nevertheless, the untreated syphilis finally took its toll, and he finally died on October 22, 1793, although his mental faculties had begun to desert him a great deal earlier. From the viewpoint of epidemiology, the real tragedy was that his erroneous findings thoroughly derailed researchers for numerous years. Part of the problem was that he published his faulty conclusions in the book *Treatise on the Venereal Diseases*. It took over half a century for biologists to untangle the confusions and realize that syphilis and gonorrhea were distinct ailments.

Even so, other of Hunter's works were more reliable. In 1762 John Hunter served with the British army in Portugal. With the field training that provided, he was able to compose his revered *Treatise on Blood Inflammation and Gunshot Wounds*.

Hunter's brother, the British biologist William Hunter, also made his presence felt. He was interested in both comparative biology and reproduction and in fact, was a practicing obstetrician. He trained at Glasgow and Edinburgh and quickly established a formidable presence in London's medical circles. His magnum opus was *Anatomy of the Human Gravid Uterus*, a masterpiece of biological illustration, although the data it contained was old hat. Nevertheless, directly because of his efforts to make the public more aware of proper nutrition and childcare generally, the rate of infant mortality as well as the death of the mother at childbirth greatly decreased.

SUMMARY

Philosophical work in science reached a fever pitch in this period, principally with the work of Descartes, Leibnitz, Spinoza, Malbranche, Offroy de la Mettrie, Buffon, and others. But the spotlight was on David Hume, who would mount an immensely powerful assault on the argument from design, offering arguments still widely discussed today. Of course, this was the age of Linnaeus, though others also contributed to classification, such as the Frenchman Antoine-Laurent de Jussieu. Comparative anatomy began to take shape with the work of Cuvier, and Erasmus Darwin began to anticipate his more famous relative in conceiving of the then-heretical notion that species could change. Erasmus Darwin also added to the social organization of science at this time. In physiology, Haller is preeminent, while modern dental science began to form with the work of John Hunter. At the same time, William Hunter made his contributions to obstetrics.

CHAPTER 15

Lamarck and His System

Lamarck deserves a better place in biology. It is true that he surmised, mistakenly, that an animal could pass on to coming generations characteristics it acquired during its existence. Yet despite Lamarck's well-known fantasy, he did spawn further interest in evolution when, at the turn of the century in 1801, he wrote his classic *Systemé des Animaux sans Vertebres*, or "System for Animals without Vertebras." This volume presented the scientific world with some of his preliminary thoughts on evolution and, equally significantly, implied a taxonomic system for invertebrates. Also, one must not forget that he coined the term "biology" as well.

Jean-Baptiste Lamarck was born in Picardy in northern France in 1744, the youngest of eleven children. At age sixteen, because of the loss of his father and the family's impoverished circumstances, he had to leave school to enter the military, and landed in the midst of the infamous Seven Years' War. He performed heroically, despite the fact that the military was scarcely his calling. During one battle he actually took over his regiment and held out after the enemy had wiped out all of the senior officers. For this he received an officer's commission.

During this period, though not clearly because of the war, he injured his lymphatic system, leading to his departure from the military. When he had recovered sufficiently, he began to study biology and medicine in Paris while employed at a bank. It was at about this point that he met the famous Jussieu, certainly one of the most renowned botanists of yesterday. Lamarck endured four unhappy marriages, doubtless in part because he never seemed to be able to escape the web of poverty that had haunted him since boyhood. His old age was particularly sad; although two of his daughters did assist and stay with him throughout, his eyesight rapidly failed. Still, he did manage to carry on some biological inquiries even during his last years.

He spent many years traveling through Europe, gradually accumulating the material that would comprise his *Dictionary of Botany,* as well as the *Flora of France.* It was really at around fifty years of age that Lamarck began the work that would bring him immortality. He became prominent enough that the French Academy of Sciences admitted him, and following his tenure there, during the French Revolution, he obtained a chair in botany at the University of Paris.

LAMARCK AND TAXONOMY

He became a pupil of the French biologist and geologist Compte de Buffon early on, even following him when the latter took a position at the Jardins des Plantes in Paris. Beyond the well-known oddities of his conceptions, Lamarck's contributions to the science of classification, at least, were far advanced over those of his predecessors. He differed from Linnaeus, for instance, in basing categorization on function rather than structure and saw more distinctly than Linnaeus how closely related structure and function were to one another. Consequently, those organisms which were capable only of reflex motions, such as amoebae, were at the bottom of his ladder type of arrangement. Animals with a primitive nervous system, such as sponges, occupied the middle, and those with intelligence of any kind were on top. By 1822, Lamarck had completed his classic *Natural*

Jean-Baptiste Lamarck (1744–1829). (Courtesy of the Library of Congress.)

History of Invertebrates, which brought together much of this accumulated knowledge between two covers, making the material easily accessible to others.

LAMARCK AND THE INHERITANCE OF ACQUIRED CHARACTERS

By 1809, Lamarck was enjoying the kudos of his newly published *Zoological Philosophy.* It was here that he stated the now-notorious delusion that made him famous. Parts of his system, specifically the notion that higher forms evolved from more primitive life-forms, were true enough; also, the fact that he did not accept the idea of "catastrophes" was further evidence that his merits have not been sufficiently appreciated. Beyond this, he was possibly the first to realize that mere geographic isolation can radically modify an organism, while two species living close to one another, because they had to struggle with similar problems, would begin to develop similarities—these are all modern and accurate propositions.

Admittedly, what was off the mark was his opinion that animals could inherit qualities acquired by their parents. He held that an alteration in the environment could cause, for instance, an animal to develop new habits, which the organism inevitably reflected in anatomical changes. The animal would then transmit these changes to offspring, so that, after several generations of accumulated changes, an entirely unique species of animal would result. He also surmised that factors other than environmental constraints could induce change. Disuse, he felt, could cause the decay of organs. This latter tenet is, of course, closer to the truth.

During his own life, most of the distinguished scientists of the day recognized his eminence, primarily because of his contributions to classification. The fact that he received no acknowledgment for any contributions to "natural philosophy" was perhaps foreseeable, since he adopted a version of materialism that was no longer in fashion by the opening decades of the nineteenth century. This was the age of John Stuart Mill and Romanticism, not the age of David Hume or Baron d'Holbach. Fortunately, such eminent scientists as the nineteenth-century German biologist Haeckel would, years afterwards, revive Lamarck's precepts. In fact, Haeckel was certainly his most outspoken champion during the Victorian era, which lasted roughly from 1840 to 1900.

ROBINET AND EVOLUTION

In 1761, French philosopher and grammarian Jean Baptiste René Robinet added to evolutionary theory when he published in his *De la Nature,* or "On Nature," his speculation that it was possible to place all species—plant and animal—on a single scale of linear development that would, over aeons, depend on the same random external factors. He was a materialist early in life, but veered towards a crude idealism later. Although his importance as a philosopher is nil, his biological views are worth a brief mention.

OKEN AND THE FORMATION OF THE SKULL

Another arresting, albeit ultimately erroneous, conviction appeared in 1807, when the German biologist Lorenz Oken entered the biological cosmos. Born in 1779, he was the son of an indigent German family. Even so, he did succeed in obtaining at least a tolerable education, eventually becoming a physician. By 1807, the University of Jena had appointed him to its faculty, primarily because of his involvement in the skeletal system.

It is likely that Oken contributed more to the history of science than to the furtherance of science *per se*. For many years, to give one example, he was the editor of *Isis*—a journal that still exists today in the United States as a publication of the History of Science Society.

As for science directly, his only studies worthy of mention were his summations of the development of the intestine in vertebrate embryos, which do contain some correct observations. Also, early in his career he formulated his notorious and wrongheaded thesis about the skull. Oken took a close look at Goethe's claim that the vertebrate skull evolved from lower forms via a progressive fusion of the vertebrae.

Like Kant, Fichte, and Schelling, he was more of a philosopher than a scientist. But regardless of his broad scholarship, he did not even begin to compare in importance as a philosopher to these men—certainly not to Kant.

WOLFF AND MYSTICISM

Scientists returned to the study of embryology during the Enlightenment. Nonetheless, the historical period notwithstanding, some of the "intellectuals" of this era were less than enlightened. Indeed, one still beholds a strange fusion of up-to-date scientific methods and occult mysticism. The eclectic German biologist Caspar Wolff of the Academy of St. Petersburg, for instance, although he conducted some careful and useful studies of the developing embryo and rejected the obsolete "homunculus" tenet, lapsed into mysticism when he suggested the infamous dogma of the "vis vital," or "vital force." He argued that it is this vital force, or "vis essentialis," that is responsible for both the nutrition and growth of an organism. Wolff came closer to real science with his *De Formatione Intestinarum,* or "About the Formation of the Intestines." This composition, more than any other, put embryology on a solid scientific foundation. The most consequential of his positive contributions was the doctrine of "epigenesis," which asserted that organs and tissues of an animal or plant emerge from a more basic, undifferentiated tissue during embryonic development.

Nevertheless, despite superb accomplishments, the mystical dimension to his meditations probably contributed more than anything else to the comparative neglect of his work during his own time. Even Haller, to whom he dedicated his *Theory of Generation,* paid him scant attention, although he publicly expressed his appreciation of the dedication.

SPALLANZANI AND OVISM

The Italian physiologist Lazzaro Spallanzani, an otherwise superb scientist, nonetheless joined Wolff in adopting absurdly mystical theories. Spallanzani was born in 1729 in Reggio, Italy, the son of an attorney. As was customary in Europe during this period, he initially accepted the idea that he would follow his father's profession. So he began his preparation for the law at the University of Bologna. But once more, another would-be attorney abandoned the laws of man for the laws of the cosmos. He began mastering biology on his own and soon engineered a post for himself as professor of philosophy at the University at Modena. He drifted into mysticism with the preformation hypothesis. Although he made thorough studies of the function and activity of sperm and appreciated their significance in reproduction, he could not bring himself to spurn the "ovist" notion that the egg contained the complete organism. Instead, he held that the sperm merely facilitated the development of the organism already inside the egg.

Spallanzani came closer to modern science when he demonstrated that fertilization occurs only when sperm cells come into actual physical contact with an egg. He pushed back the frontier of reproductive biology still further when, in 1785, he performed the first artificial insemination on a dog. He could have hardly surmised the immense controversy that would surround this technique in the inventive reproductive biologies of the twentieth century.

SPALLANZANI AND SPONTANEOUS GENERATION

Spallanzani came still closer to modern science in his assault on spontaneous generation. He added another level of complication to the seemingly insoluble debate over this doctrine—the idea that life can arise out of nonliving matter—when he followed his studies of preformation with a brilliant series of experiments. It was here that he began to hack away at preformation. Like Redi, he realized that verifying theories required experimentation. In 1768, he finished a book that had occupied many years and had wrought devastating changes in his private life. After successfully overcoming a bout of sustained depression, he published his *Prodromo d'un Ouvrage sur les Reproductions Animales,* or "Preface to Studies on Animal Reproduction." In this treatise, he judged that he had finished the demolition of spontaneous generation; in one of his most significant trials, he boiled a beef broth concoction for one hour and then placed the broth in containers which he hermetically sealed. Since no life arose, he concluded that spontaneous generation was a myth. Previous to this work, however, he had not been so certain it really was a myth; he describes his initial uncertainty about spontaneous generation in *Saggio di Osservazioni Microscopiche Relative al Sistema della Generazione dei Signori Needham e Buffon,* or "Discussion of Microscopic Observations Relative to Generation in Needham and Buffon's Work":

> I sought to discover whether long boiling would injure or prevent the production of animalcules in infusions. I prepared infusions with eleven varieties of seeds, boiled for half an hour. The vessels were loosely stopped with corks. After eight days I examined the infusions microscopically. In all there were animalcules, but of differing species. Therefore, long boiling does not of itself prevent their production.[20]

Other talented biologists disagreed with Spallanzani's findings on grounds that they smacked of the old mythologies. John Needham, for instance, a Catholic priest sharing Spallanzani's interest in biology, claimed that Spallanzani had used such extraordinarily high temperatures that he had destroyed the "vegetative force," preventing spontaneously originating life from arising. Ensuing findings, nevertheless, proved that Spallanzani was entirely on target.

PANDER AND CHICK EMBRYOLOGY

By the opening decades of the nineteenth century, Russian biologist Christian Pander would discover the three basic forms of tissues constituting the chick embryo, another step forward in the blossoming science of embryology. Pander was born Hans Christian Pander in Riga, in what is now the independent nation of Latvia. His father, being a most influential and wealthy banker, was able to send his son to the most renowned schools, including the universities at Dorpat, Berlin, Göttingen, and Würzburg. It was at Würzburg that Pander began the observations of the chick embryo, which he subsequently published. By 1826, he had joined the faculty in St. Petersburg, though he almost immediately became disenchanted with the politics there and returned to his native city of Riga. Although he studied geology, he contributed comparatively little to this field. First and foremost, he influenced upcoming generations of biologists via his studies of the chick embryo, as one of the founders of the present-day science of embryology.

DUJARDIN AND INVERTEBRATE ZOOLOGY

The French zoologist Félix Dujardin followed Lamarck's investigations on invertebrates with some of his own. Born in 1801, he became professor of biology first at the University of Toulouse, and then at the University of Rennes. He was primarily an invertebrate zoologist, and made the Infusoria—a heterogeneous group of minute single-celled organisms—the principal object of his study. He became the first scientist to realize that many such primitive single-celled invertebrates, including amoebae and paramecia, do not have anything comparable to the organ systems found in the more evolved vertebrates. Dujardin discovered that these organisms were made up of a homogeneous mass, which he called the "sarcode," though that term is no longer in use today, having yielded its place to the term "protoplasm." He noticed that though there was an assortment of vacuoles and granules in the sarcode, there appeared to be no fixed organs. He also realized that despite a prima facie similarity, the cilia he found in many of the Infusoria had nothing to do with the hair that covered the higher vertebrates. Terminology notwithstanding, many of the theories and pronouncements he nurtured survive today. Later, Darwin would all but enshrine the rule of the chance introduction of life. (Not to imply that Dujardin first introduced this concept: It goes back to the Greeks.)

Lazzaro Spallanzani also contributed to invertebrate zoology. His earliest inquiries were directed at amphibians, primarily salamanders. He paid close attention to the most minute details of the tail, muscles, nerves, bones, and practically every other part of the anatomy. He also examined the phenomenon of regeneration, or the ability of some lower animals to reproduce lost body parts, paying particular attention to how variations in temperature and food consumed affected the rate of regeneration.

GOETHE AND THE EVOLUTION OF PLANT STRUCTURE

Any historical account of biology should not overlook the epochal work of Johann Goethe. Johann Wolfgang Goethe was born in 1749 to a wealthy family in Frankfurt am Main. He began his career by studying law at the University of Leipzig and then at the University of

Strasbourg. Although he practiced law, his real passions lay in both literature and the sciences. In 1786, he traveled to Italy, remaining there for two years. It was there that he began his scientific toils in earnest. He lived a long and full life, passing on in 1832.

Initially, Goethe explored not only biology but concepts in physics such as magnetism and electricity. While he was a sort of poet-in-residence at the court at Weimar, he conferred often with scientists at the University of Jena and with the eighteenth-century scientist Johann Herder. Although it held his curiosity, evolution was not his dominant concern in much of his biological career; indeed, some of his published writings indicate he was interested chiefly in advocating a theory of plant structure. In his *Versuch, Die Metamorphose der Pflanzen zu Erklaren,* or "Attempt to Explain the Metamorphosis of Plants," he put forth the clever but utterly wrongheaded view that every part of a plant is just a modified leaf. Despite this erroneous suggestion, the book clearly shows that he favored the hypothesis of the evolution of species rather than special creation by a deity. Biology was getting ready for Darwin.

As has happened so tiresomely often, most of Goethe's contemporaries failed to certify the excellence of his scientific work, at least while he was alive. On the other hand, his reputation as a phenomenal poet and writer prevented any great amount of opprobrium from falling upon him.

Ultimately, such authorities as Haeckel began to appreciate Goethe's scientific contributions. In fact, Haeckel believed that, despite Goethe's lack of interest in the issue, he had come as close to the principle of evolution as anyone had before Darwin. His deliberations on the nature of the mind also influenced such thinkers as the German physiologist Johannes Muller and the Czech biologist Jan Purkinje.

KOLREUTER AND PLANT REPRODUCTION

Yet another pivotal Enlightenment figure at this time was Joseph Kolreuter. He was born in 1733 in Sulz, in the Württemberg region in Germany. He began his schooling at the universities in Berlin and Leipzig as well as in St. Petersburg. In 1764, he was awarded the post of historian and curator of the botanical gardens in Karlsruhe.

His first interest was reproduction, and this continued to monopolize his research for most of his professional career. Based on microscopic observations, he argued that the fertilizing qualities of pollen stem from an oily fluid that it emits. Kolreuter also studied the myriad ways in which nature fertilizes flowers. He was almost certainly the first to show how insects fertilize some varieties, and he also analyzed the wind's role in carrying pollen from one flower to the next. He discovered in his research on tobacco that hybrid plants could resemble either of the parents or neither in particular. As obvious as it may seem today, Kolreuter disclosed that both parents contribute equally to the final form and shape of the offspring in mixed crosses. For example, if one crossed a male from variety X with a female from type Y, the ensuing plant was the same as if one had crossed a female X with a male Y. He soon extended these observations to plants such as mullein, pinks, and many others, always noting the same phenomenon. In his *Preliminary Report on Experiments and Observations of Certain Species of Plants,* he discussed his probings into plant inheritance. He also anticipated some of Mendel's proposals on the genetics of plants and may even have had some rudimentary grasp of the phenomenon of "mutations."

Unfortunately, oddly anthropomorphic views marred his studies. He claimed, for instance, that the female sexual element in plants was "mercurial," without providing any explanation of what this "property" was supposed to connote. Because of this, he received no real recognition during his lifetime, a fate he shared with Mendel. Fortunately, future scientists revived his findings and began to appreciate that beneath the bizarre manner of expression, there were at least a few ideas of real gravity in his writings.

POISEULLE AND PLANT HYDRAULICS

The French physician Jean Poiseuille, using a mercury manometer, distinguished himself in 1842 by formulating a set of laws governing the way the blood travels through the capillaries. He applied these thoughts to any sort of liquid coursing through any type of tube, thereby freeing the way for the application of these laws to fluids moving through plants.

PROUST AND PLANT PHYSIOLOGY

By digging ever deeper into the internal secrets of the plant world, the French chemist Joseph Proust found that sweet vegetable juices contained the three sugar variations fructose, lactose, and glucose. He also showed that grapes and honey contain identical sugars. Years later, the French biologist Henri Braconnot triumphantly gleaned something of the biochemical character of glucose when he effectively extracted it from ordinary materials such as sawdust and bark. In this way, subsequent generations of scientists were able to prove beyond any reasonable doubt that the biochemical precursor of glucose was the common plant compound cellulose. Similarly, the French amateur naturalist Bernard Courtois would discover that seaweed contains iodine, though it required the efforts of Humphrey Davy to appreciate, in 1814, that iodine was a hitherto-unknown and fundamental chemical element.

TECHNOLOGY

In keeping with the spirit of the Enlightenment, technology was making rapid strides forward as well. Among other things, technicians and enthusiastic amateurs were improving the science of lens-grinding. In 1768, the French physicist Antoine Baumé added to the technological armamentarium of biology as well as other sciences when he invented the hydrometer. He first published a description of it in the journal *L'Avant*. Eventually, this device would prove to be of inestimable value in measuring the specific gravity of liquids or solids.

Also, Samule Klingenstierna of Sweden captured the prestigious first prize of the Russian Academy of Science for designing better lenses and optical instruments. Klingenstierna was born in 1698 in Sweden and ultimately achieved the rank of professor of physics at Uppsala. His "formula" for improved lenses consisted of further minimizing chromatic aberration. This improvement resulted in better microscopes and telescopes, with consequent rapid advances in both sciences.

Still other technological breakthroughs arrived which greatly aided the maturation of physiology. In 1797, for instance, Jean

Poiseuille again distinguished himself by anticipating the sphygmomanometer—the unpretentious cuff device physicians use today for measuring blood pressure. Beyond its obvious medical applications, the device proved to be indispensable whenever physiologists were trying to measure the effects of stress, chemicals, the environment, and the like on the functioning of the body. It merits noting, however, that the actual sphygmomanometer in use today came from the inventive intellect of biologist Étienne-Jules Marey in the 1860s.

SPALLANZANI AND DISEASE

During the Enlightenment, scientists first began to suspect that disease resulted from spoilage in food. Although they had no clear grasp of what, exactly, was wrong, they tried as best they could to prepare foods in myriad ways to prevent spoilage. The Italian physiologist Lazzaro Spallanzani once again occupied center stage when he took the first step toward any real understanding of food spoilage. In 1765, he began to suspect that food might cause sickness if air could enter it.

In the opening decades of the nineteenth century, a French chef, Nicolas Appert, would apply some of Spallanzani's ideas in a different way, when he offered the world a medley of methods for sterilizing food by using heat. This effort brought him practically instant fame. Others swiftly translated his writings into French and other languages. Then, in 1780, Spallanzani's *Dissertazioni de Fisica Animale e Vegetale*, or "A Treatise on the Physiology of Animals and Plants," uncovered untouched ground in understanding the process of digestion.

PHILOSOPHY

Given the influence of the Scottish Enlightenment figure David Hume, it is scarcely jarring to find out that there were a number of peculiar philosophical speculations espoused during the Enlightenment. Hume's philosophical and scientific ideas, while few accept them today, do at least remain part of the philosophical canon studied by every graduate student in philosophy. The tenets of some others, on the other hand, are comparatively minor. Notorious here is Denis

Diderot, one of the French encyclopedists, called so because they were immersed in assembling comprehensive tracts on various areas of human knowledge. Early on, French authorities had tossed Diderot into the dungeon for his assault on organized religion. Still, a man of unrivaled courage, he never hesitated to advocate his philosophical view of the cosmos, which was entirely shorn of spiritualism. Like Hume, Diderot was a "mechanist" through and through. He believed that it was not God but only the precepts of physics that governed the universe. In his *Le Rêve d'Alembert*, or "The Dream of d'Alembert," Diderot engaged in metaphysical musings on the nature of knowledge as well as on the constitution of the external world. As in all his writings, he combated what he conceived of as superstition, advocating instead the virtues of science and technology as a panacea for all human miseries. Philosophers today virtually universally reject his metaphysical ideas as well as his ideas on the nature of knowledge, although many still cling to the notion that science can solve all human problems.

SUMMARY

The work of the thinkers discussed in this chapter show that the late eighteenth and early nineteenth centuries were very much transitional periods. Lamarck demonstrated a grasp of true science in his great work on taxonomy, *Natural History of Invertebrates,* while achieving eternal notoriety with his doctrine of the inheritance of acquired characteristics. Also, Pander and Caspar Wolff did some quite respectable work in embryology, while at the same time, the latter, at least, clung to the mythical "vis vital" tenet. Spallazani too had a foot in both the old mystical views and in modern methodology. Although he embraced "ovism," he approached genuine science with his views on fertilization and with his work on artificial insemination and food spoilage. In technology too there were significant advances, with the work of Samule Klingenstierna and Antoine Baumé. In botany the research of Poiseulle and Proust added still more to our knowledge of plant physiology, while Dujardin broke new trails in invertebrate zoology.

CHAPTER 16

The Rise of Paleontology

Up to and through the seventeenth century, paleontology, or the science of fossils, had expanded, but only slowly. Buffon had identified organisms embedded in fossils, as had Thomas Jefferson. By the nineteenth century, however, the progress of the field was more rapid. Biologists were starting to realize, even before Darwin, that the crust of the earth was seething with the remains of plants and animals that had evolved, lived, and perished eons ago.

George Cuvier was among the important scientists working in this field. At the onset of his research, studies of prehistoric reptiles cultivated his appetite for paleontology. During the French Revolution, someone had sent a skull of what seemed to be a crocodile to the Jardin des Plantes. Explorers had found it many years earlier in the Netherlands on some land belonging to a "Dr. Goddin." Cuvier, in fact, proved that it was not a crocodile but an ancient lizard related to the modern monitor lizards. Cuvier incorporated many of his findings into his book *Research on Fossil Bones*. It was this book that, in effect, established paleontology as a separate science. Beyond this, Cuvier identified fossilized remnants from both oceanic reptiles and a potpourri of winged reptiles such as the pterodactyl, an extinct flying reptile. The keenness of Cuvier's "eye" was manifest when he identified a pterodactyl from only a drawing of the fossil remains.

In 1780, scholars identified the skull of a previously unknown dinosaur, which some amateur explorers had found in a quarry in the Netherlands. Eventually, science called this creature a "mosasaur." At this point Cuvier really began to exert his authority in paleontology. According to Cuvier's measurements, the reptile stretched the measuring tape to nearly fifty feet. As is often the case in paleontology, the discovery was serendipitous. In fact, most scientists were unable to identify it, beyond realizing that it was an animal, until Cuvier realized, in 1795, that it was in fact a dinosaur. In the mid-nineteenth century, the paleontologist Richard Owen would propose the neologism "dinosaur" to describe the increasing number and diversity of such large reptile fossil discoveries paleontologists were unrelentingly uncovering. In 1799, a team of geologists on a routine expedition discovered an entire mammoth completely and faithfully preserved in the Siberian ice.

This was not to be the last of the colossal animals of prehistoric times that would come to light in this era. In at least one case, Cuvier, again, demonstrated that some earlier discovered fossilized bones, long thought by religious apologists to be the residue of an animal that had allegedly drowned in the great biblical Flood, were actually bones from an enormous salamander that had probably died from natural causes. Paleontology was ripening into a mature science. Well into the nineteenth century, the German naturalist Herman von Meyer found the first feather fossil. Scientists soon followed this bombshell with another, a nearly complete fossil of the same winged animal—archaeopteryx. In his own speculations about evolution, von Meyer surmised that there were no literally "extinct" animals—simply animals that paleontologists imagined to be extinct but which were lurking elsewhere waiting to be discovered.

THE TYPE THEORY

Still another of Cuvier's enduring legacies was the "type" axiom. The type approach disallows comparisons between advanced lifeforms of independent species. For, according to Cuvier, all animals fall into one of four main "types" or groups: *Vertabrata, Mollusca, Articulata* and *Radiata*. Since the four groups are radically different in their overall design, one cannot understand an animal in one group by comparing it to an animal in another. Furthermore, Cuvier's various opinions on embryological development also contributed substantially to the contemporary dogma of evolution. Inevitably, Cuvier himself finally accepted the fact of species extinction and moved even closer to orthodox evolutionary teaching, although he never entirely relinquished his belief that the cosmos and species were evidence for the benevolent hand of the Almighty.

THE RISE OF CATASTROPHISM

Despite all of this, Cuvier certainly held a spiritual worldview that he sought mightily to reconcile with the scientific data. Today, most scholars agree that this reconciliation had taken place—at least in Cuvier's own mind—by 1812, when Cuvier published his multivolume series, *The Fossil Bones of Quadrupeds*. In it he puts forth the "catastrophe" theory of evolution and presents guidelines for the study of comparative vertebrate paleontology. Cuvier became more and more consumed by fossil work as he developed his views on "catastrophism," the suggestion that periodically, the biosphere had eliminated entire species of plants and animals which the heavenly Father in his wisdom replaced in the ensuing years. This elegant philosophy, still prized even today in some quarters, urges that, periodically, some unprecedented catastrophe such as a major flood, an ice age, or the like annihilates every living species on earth. Professor Peter Ward of the University of Washington, for example, believed as late as the 1980s that ammonites, an ancient mollusc, had become extinct very gradually. Recently, however, even he has changed his mind, arguing that their extinction was in fact very sudden and probably due to a catastrophe of some kind. (Intriguingly, the story of Noah's ark was the prototype for all "catastrophe" ideas.)

UNIFORMITARIANISM

The scientists Hutton and Lyell, and Darwin himself, favored a competing hypothesis called "uniformitarianism." According to this idea, all of the observable features of the earth appeared bit by bit over vast stretches of time. Darwin incorporated a similar proposal into evolutionary teaching, arguing that the recasting of species also occurred gradually over inestimable numbers of years. Possibly because of Darwin's prestige, uniformitarianism would dominate geological and evolutionary thinking for years to come. Born in 1726, James Hutton, the child of a wealthy landowner, added still more fuel for Darwin, and science gained a priceless biological legacy in his geological commentary *A Theory of the Earth.* He argued that, because the evidence pointed so clearly to gradual revisions in living species, geology could offer no support whatsoever for the biblical doctrine of creation. In his view, the horizontal layer of fossils he found were the result of gradual deposits. The biblical Flood could not have produced them.

EXPLORATION OF THE SEAS

Soon scientists were starting to consider the possibility that biology had neither discovered all forms of life on the earth's surface, nor even begun to fathom the enigmas that lay beneath the seas. So it was that in 1815, the biologist Edward Forbes of Durham, England, demonstrated in a series of ocean voyages that living creatures thrived deep within the ocean. Scientists would do no further significant investigations in this arena until past the midpoint of the nineteenth century, when the zoologist Sir Charles Wyville of Scotland would start scavenger operations on the ocean floor that unearthed still more forms of life. And in 1872, the Scottish zoologist Charles Thomson was ready to begin a term as scientist-in-residence on board the *Challenger,* which was really the first significant oceanographic voyage in recent biology. Thomson spent four years on this trip gaining unique information not only on new life but about the depth and composition of the ocean itself.

THE RISE OF GEOLOGY

The year 1799 was, defensibly, *the* most significant year for archaeology. In that year, soldiers in Napoleon's army uncovered a tablet that carried three varying forms of script. The Rosetta stone had returned to civilization. Soon, the verities scientists obtained from it would unlock the door to the translation of both Egyptian hieroglyphics and Sumerian cuneiform writing. One could scarcely consider it a surprise, accordingly, when, in 1802, the Austrian archaeologist Henry Rawlinson first translated a piece of cuneiform writing dating back to the ancient Persian empire, although no one got around to publishing his epochal translation until 1846.

WILLIAM SMITH

A commendable figure in connection with the field of geology, though he often receives scant credit, is civil engineer William Smith. Author of *The Geological Map of England*, Smith was born in 1769 to an impoverished family. Without obtaining any formal education, he became a surveyor, which spawned his lifelong search for biological knowledge in the strata of the earth. He was certainly the first scientist to use fossils for purposes other than for identifying organisms. He realized that fossils yielded another pattern of features altogether. If properly examined, it was possible to correlate individual fossils with certain "strata" or layers of the earth. By scrutinizing them and relying on the assumption that discrete strata correlated with specific periods in the earth's history, geologists were able to date both the strata and the fossils themselves. Smith eventually published his views and theories in a huge atlas, and he is plainly the first scientist to appreciate the importance of fossils in dating distinct layers of the earth's crust.

THE CONTINUING RISE OF CHEMISTRY

Developments in other branches of science were occurring during this period as well. The physicists Henry Cavendish and Daniel Rutherford, for instance, both discovered nitrogen in 1772, while Scheele discovered the gas oxygen, which he dubbed "fire air," since fires thrive in oxygen. Although scientists had already begun to appreciate the connection to biology of these elements, both would prove to be of even broader significance for biology later on, when biochemists found that both nitrogen and oxygen were integrally related to protein structure. It should be noted that, although convention generally credits Joseph Priestley with the discovery of oxygen, recent historical reassessment has shown that Scheele probably did so concurrently, though he did not publish his inferences until after Priestley did.

ASTRONOMY

In astronomy, Johan Bode of the Berlin Observatory devised a relation which stated that the distances of planets to the sun follow a predictable mathematical ratio, later called Bode's law. Bode also developed a theory about the make-up of the sun similar to William Herschel's. Additionally, Pierre Simon Laplace of the French Academy of Sciences, famous for his work in probabilities, proved that even the entry of another planet into the solar system would produce only an infinitesimal deviation in the orbit of any other planet. In the ensuing years, a decade after the French Revolution, Laplace would collect, in several volumes, all of the advances in astronomy to date. They would appear in his magnificent *Méchanique Céleste* ("Celestial Mechanics").

PHYSICS

Henry Cavendish also proposed that electricity was in reality a motion of a "fluid," a dogma that the American physicist Henry Rowland discredited in the nineteenth century. Later, Cavendish would measure the gravitational constant "g." The idea involves suspending two smaller lead weights next to two larger lead weights. The gravitational force of the larger attracts the smaller weight, thus twisting the suspending wire. The force required to twist the wire yields the gravitational constant.The ratio of the large to small objects yielded g. Consequently, Cavendish concluded the research in physics that Newton had begun. Using the gravitational constant, he revealed that the earth is about six times denser than water. As in prior epochs, the imperial progress of physics only inspired biologists to continue to try and apply the precepts of physics to biological phenomena.

MEDICINE

In medicine it was the age of Mesmer, the German physician who first used hypnotism—then called "animal magnetism" or "mesmerism"—to treat illness, an approach which Sigmund Freud took up years after, in the nineteenth century. By the mid-nineteenth century, the Scottish physician Sir James Braid had renamed the effect "hypnotism," and even tendered a theoretical mechanism explaining how it functioned. Hypnotism would rapidly find a secure niche in the physician's repertoire. Roughly coincidentally, Sir Percival Pott of St. Bartholmew's Hospital anticipated the modern science of environmental medicine when he found an unusually high incidence of scrotal cancer among British chimney sweeps, guessing correctly that soot was the culprit.

THE AGE OF KANT

For philosophy, this period was emphatically the most critical in history. It was in 1781 that the German philosopher Immanuel Kant published his august *Critique of Pure Reason,* which triumphantly fused the empirical tradition of Locke, Berkeley, and Hume with the rationalist tradition of Descartes, Leibnitz, and Spinoza.

BOTANY AND THE RISE OF POPULATION THEORY

In the ensuing years, the Reverend Gilbert White added to the development of botany with his *Natural History and Antiquities of Selborne*—a superbly organized group of descriptions of flora and fauna of the English town of Selborne. Published in 1789, this volume helped establish the importance of environmental and ecological issues in biology. Soon would follow the magnificent reflections of Thomas Malthus in 1798. Malthus was born in 1766, the offspring of a well-to-do landowner in England. He prepared himself for the priesthood at Cambridge, although his professional passion was economics. His father had worked with the French Enlightenment philosopher Jean-Jacques Rousseau and had taught his son the value and justice of redistributing wealth for the net benefit of the masses. Thomas Malthus's *Essay on the Principle of Population as It Affects the Future Improvement of Society* appeared under a pseudonym in England, as Malthus was troubled by the possible repercussions of such a drastic social tenet. His thesis in this book was that the principal cause of misery in the world was the careless and wasteful use of the planet's resources. He proposed that only famine or war could control world population—a population that was rapidly spiraling out of control. By the 1970s this book would evolve into a "bible" for the so-called zero population growth movement of that decade.

From the vantage point of biology, the salient issue in this book was Malthus's notion that competition in society would inevitably and methodically weed out the substandard and less "fit" classes through crime and evildoings—a recommendation that Darwin would exhaustively analyze.

ENTOMOLOGICAL TAXONOMY

No one had done anything of striking merit in entomology since the epochal findings of René de Reaumur in 1734. The zoologist Johann Fabricius (a different scientist from his namesake of nearly two centuries past) ended the drought in 1775, when he published his *Systema Entomologiae*, or "Systematic Entomology." Born in Denmark in 1745, Fabricius was a physician's son. He enrolled at the University of Copenhagen in 1763. Later he went to the university at Uppsala, where he befriended Linnaeus, creating a relationship which would last for many years. Sometime afterwards, the University of Kiel appointed him to a professorship. Even with the low salary they offered, Fabricius somehow managed to spend most of his semesters reflecting and visiting his friends in Paris

With his *Systema Entomologiae*, Fabricius had taken an impressive step forward in two branches of biology—insect investigations and taxonomy. In large part he applied Linnaeus's methods to entomology. Even the titles of his publications were similar to those of Linnaeus. In this dissertation he was able to productively and usefully classify species and varieties of insects using variations of mouth structure rather than wings as his basis.

SPRENGEL AND INSECT REPRODUCTION

Without letting the momentum ebb, the German biologist Christian Sprengel, continuing the work of Kolreuter, explained still more of the ways in which both insects and wind could, by carrying pollen, fertilize plants. Christian Conrad Sprengel was born in 1750 at Brandenburg to a pious and devout family, his father having served as a lay vicar for many years. His early education was in language and theology as well as philosophy. For many years thereafter he served as a headmaster of a Berlin school. However, his disposition was mercurial and not easily tolerated. His personality conflicted with both parents and other faculty at the school. He was also, so far as anyone can tell, a perfectly awful teacher. Inevitably, the school had to ease him out, albeit with a full pension, in 1794. He lived most of his years alone, passing away in Berlin in 1816. His studies, conducted mostly at home, consisted of assembling volumes of information on florescence in plants as well as a number of anatomical descriptions of a multiplicity of flora. Most of his ideas on plant fertilization, particularly the above-cited conjecture that insects and wind could carry pollen, were more or less in line with Kolreuter's and still hold true today. As Sprengel himself describes things in his book *The Secret of Nature Discovered in the Structure and Fertilization of Flowers*:

> In the summer of 1789 I examined several species of iris, and found that these flowers could be fertilized in no other way than by insects. Then I searched for other flowers whose structure should show the same thing. My studies convinced me that many, perhaps all of the flowers which secrete nectar are fertilized by insects.[21]

As was true of so many scientists of this period, Sprengel saw the created order as one grand proof of the reality and munificence of the Lord. Like Kolreuter and Mendel, Sprengel, too, had to stomach neglect at the hands of his fellow scientists. In part this may be due to the rather obvious theological messages in his writings, which served only to alienate thinkers who, in this period, were desperately trying to comprehend the heavens "mechanistically." The supposition that some deity had designed all of nature to reach some final state was too reminiscent of Aristotle and medieval theology. Fortunately, Darwin himself rescued Sprengel from oblivion. He based many of his own conclusions about evolution and natural selection on Sprengel's investigations of the anatomy of flowers and plants as well Sprengel's observations on the relationships between insects and flora.

SUMMARY

Above all else this was the era of Cuvier, which saw the rise of paleontology and the discovery of innumerable important animal remains such as the mosasaur. This era also witnessed the gradually escalating competition between the "catastrophist" and "uniformitarian" ideologies in evolutionary theory—a controversy that has not resolved itself even today. Not surprisingly, the "cousin" science to paleontology, geology, began to evolve with the work of Jean Guettard, who first demonstrated the existence of an ancient ice age.

Other sciences too were advancing rapidly. The discovery of Bode's law in astronomy, the work of Cavendish in physics, and the work of Scheele in chemistry all bear witness to that. In medicine it was the era of Mesmer, and the German virtuoso Immanuel Kant dominated philosophy. Botany too made progress in this period with the work of Gilbert White, while Thomas Malthus would begin his epochal reflections on population growth. Invertebrate zoology, particularly entomology, pushed ahead with the efforts of Sprengel and Fabricius.

CHAPTER 17

Biology in the Victorian Era

Throughout the eighteenth and nineteenth centuries, scientists made astounding progress in most branches of science and, to some extent, even in psychology. Many, such as Hans Oersted, André Ampère, Faraday, and Allesandro Volta grasped and harnessed the force of electricity. Oersted would discover that electricity can generate a magnetic field. André Ampère demonstrated that if several wires each carry an electric current, they will attract one another if the currents are flowing in the same direction and repel one another if the reverse is true. He further theorized that magnetism results when "small electrical charges" travel through objects. In fact, this is exceedingly close to the truth.

In the Victorian era specifically, doubtless the primal and truly revolutionary scientific achievement was the brilliance of the British physicist James Clerk Maxwell in unifying electricity and magnetism. From this time onward, scientists would speak of one force—the "electromagnetic" force—where they had previously spoken of two. Although the term figures loosely and easily in contemporary discussion, Maxwell's deed really does deserve to be called a "paradigm shift." For his toils started other scientists on the trail of unifying all of the rest of the forces of the universe: the "weak" force, responsible for radiation, the "strong" force, responsible for holding the nucleus together, electromagnetism, and gravity. Though science is only part of the way toward that unification goal, it was Maxwell who started it all.

Much else occurred during this epoch that was truly ennobling. Just after the Victorian period ended, the British physicist Ernest Rutherford hypothesized that the nucleus of the atom was concentrated in the center of the atom—the so-called compact nucleus. In short, to an exceptional extent, physics took the spotlight during the Victorian era. Nevertheless, there was one respect in which all the sciences stayed on an equal footing for quite a while. Scientists would function, with a few exceptions, essentially as amateurs. The day was not yet at hand when eager young scientists would receive travel fellowships, lucrative university posts, government and foundation funding, and so forth.

Among nations, Germany was the first to develop anything resembling the modern "perks" considered above. With such credentials, it is no wonder that well into the twentieth century, the scientific community still considered Europe the place to learn, principally right after graduate school. The Yale physicist J. Willard Gibbs went to Europe for further schooling, as did Henry Rowland and the acclaimed A. A. Michelson. In chemistry, men like Linus Pauling and Irving Langmuir would someday make this trek, as would John Slater and J. Van Vleck in physics. The orthodox explanation for this migration was that the United States offered little in the way of "real" science during the nineteenth century, and that it was actually only a nation of "tinkerers," albeit brilliant ones such as Edison, Westinghouse, the Wright brothers, George Eastman, and so forth. This viewpoint is often true, but there were exceptions. To name a few of the United States' most remarkable scientists, A. A. Michelson of the Case Institute and the University of Chicago measured the speed of light precisely for the first time. The American physicist Henry Rowland was conducting his epochal experimentation on the makeup of electrical currents, and the theoretician J. Willard Gibbs was busily pioneering the laws of thermodynamics at Yale University. The list goes on and on.

PHILOSOPHICAL FOUNDATIONS OF SCIENCE IN THE VICTORIAN AGE

As with any period in science, the nineteenth-century scientists did their work under the auspices of certain philosophical preconceptions. Much of the philosophy emanated again from Germany and the so-called German naturalist school of Hegel, Fichte, Schlegel, and others. Under the influence of Immanuel Kant and his unrivaled *Critique of Pure Reason,* the emphasis was on the rationalism of Descartes and Leibnitz rather than the empiricism of Locke and Hume. While it was hardly feasible to turn science into a rationalist discipline in this sense of the term, it may well be that the German openness to abstract speculation, even in the absence of hard empirical and observational data, may have caused them to develop into the leaders in abstract science for so many decades.

Be that as it may, physics, if it overshadowed other sciences at all, did so only a bit. Above all, there was Darwin. Chemistry too had its own title to fame. Organic chemistry emerged from a vacuum during this period, and the record is replete with the brilliant innovations of people like Marie Curie, Hugo de Vries, whose mutation hypothesis we will soon look at, and many others. Curiously, up to this point scientists had devoted negligible attention to the analysis of differing races, at least not as part of any genteel biological system of classification.

THE EMERGENCE OF RACE

Shortly that began to change. In 1819, the British scientist William Lawrence, who first used the term "biology" in an English-language work, *Lectures on Physiology,* delivered a series of lectures arguing, innocuously enough, that humanity consisted of several races. Though there was nothing particularly macabre or offensive in Lawrence's proposals, others later managed to misinterpret these explorations in various deranged and racially biased "scientific" tracts aimed at showing that some groups were more intelligent than others. The infamous French anthropologist Paul Broca, for instance, weighed the brains of men and women in France. From this he made the deduction that since women's brains were lighter than men's, women must be less intelligent than men, forgetting that women also generally weigh less than men. His pupil Le Bon carried on the "tradition," arguing that the brain of a woman was more like the brain of a gorilla than the brain of a man. As Le Bon further concluded,

> All psychologists who have studied the intelligence of women recognize that they represent an inferior form of evolution and that they are closer to the child and the savage.[22]

EUGENICS

Assuredly the most unfortunate aftermath of this inquest was the "eugenics" movement, which Francis Galton discussed in his *Inquiry into Human Faculty* of 1883, his *Hereditary Genius* of 1869, and several other books. In a letter written to his biographer Karl Pearson, he discusses an interesting application of I.Q. studies to the Indian Civil Service:

> I have in view now [a study], that I began some years ago, but found that not enough years had elapsed since the experiment began to draw useful conclusions . . . but . . . there ought to be enough now. It is the correlation in the Indian Civil Service between the examination place of the candidate and the value of the appointment held by him.[23]

Galton declared that science could systematically improve humans through selective breeding. Knowing what we do about his home environment, it is not hard to grasp Galton's penchant for the "eugenics" creed. He had all of the finest advantages at home, intellectually and culturally. He himself was far above average in intelligence. Some biologists, such as Lewis Terman, claim that his I.Q. was approximately 200, but this is most likely fanciful. His parents catered to his every whim, and, as so repeatedly happens, he all but appointed himself a paradigm of the "perfect" human being. Part of his proposal involved the notion that "undesirable" qualities could be somehow empirically discovered, a notion that would seem to indicate that despite his alleged high I.Q., Galton had few gifts for abstract thought, or else he would have realized that "undesirable" is not the label of a scientifically measurable property.

Nevertheless, many supported the proposal at the time, so much so that an actual journal, *Biometrika*, was founded exclusively to support eugenics. Galton believed further that the "desirable" traits would always dominate over the "undesirable" traits. Of course the doctrine has few adherents today.

To be fair, however, Galton intended no massive genetic "rebuilding" of the human race; he intended the term eugenics to refer solely to the science of genetics as applied to man. Perhaps the problem was that his thoughts too easily lent themselves to, indeed invited, misuse. Some applications of his concepts were better than others of course. Social reformers have suggested that society should sterilize rapists and other sex offenders, for example, which is at least an arguable position. The coup de grace for the eugenics movement, however, was the Nazi movement of the 1930s and 40s, whose horrors scarcely need amplification here.

RETZIUS AND THE MORE SCIENTIFIC STUDY OF RACE

Finally, at least some of this research took a slightly more credible turn with the research of the Swedish biologist Anders Retzius. Anders Adolf Retzius was born in 1796 at Lund, the offspring of a minor, but competent, scientist. Indeed, his early instruction was under his father, although he ultimately studied at Copenhagen under Ludvig Jacobson. The launching of his career as a scholar began when the Veterinary Institute at Stockholm appointed him professor of biology. His style of investigation was a bit peculiar—most of it consisting of rather preliminary reflections, notes, and so forth. Still, much of this was important. In his mature years, after some minor contributions to dental anatomy, as well as some probing into the circulation of the blood and the nervous system, particularly in the Myxinoidei, he turned to what the academic world today calls comparative anthropology. He began by looking at vestiges in antediluvian graveyards in Scandinavia. He quickly found that skull modifications were easily classified as long or short, that is, as dolichocephalic or brachycephalic. Slavs, for illustration, were "short-skulled," while the Germanic races were "long-skulled." Furthermore, he devised a rather comprehensive classification of the bones of the face.

Soon however, in 1842, Retzius encouraged another more radical approach to intelligence. According to this concept, there were intellectual variations among the races. The anatomical structure of the cranium was the key to such differences. He still did not actually try to *rank* the respective intelligence of any races. Broca himself produced results that were considerably more scientific when he theorized that individual human abilities are located in the brain. Broca realized that the ability for speech, for instance, was so located when, on doing an autopsy on a man with a major speech affliction, he found a lesion on the brain. Sadly, this approach led many scientists into deadends when they speculated that many mental faculties, such as memory, were also located in the brain. (Norman Malcolm of Cornell University, for instance, cast considerable doubt on this view in his 1974 book *Memory and Mind.*) In still another slightly more promising—if also unclear—line of exploration, many biologists and psychologists of the Victorian era were haunted by the curious notion that one could "locate" various mental faculties in different parts of the brain. Some supposed, for example, that "intelligence" was in the prefrontal lobe. Nevertheless, even this experimentation promptly decayed into efforts intended solely to confirm preconceived biases. To systematically twist what is clearly evidence *against* one's view so that it appears to be evidence *for* the view is a paradigmatic example of the logical fallacy of a "self-sealing" argument.

Eventually the tumult ebbed, and it began to appear as if this sort of pseudoscience would never reappear. However, in the twentieth century, the Nobel–prize-winning physicist and inventor of the transistor, William Shockley, returned to Broca-type arguments intended to reveal that blacks were inferior to whites. In the 1990s, the City College of New York philosopher Michael Levin supported Shockley's positions. Virtually no other academicians of any significance, however, have accepted their views.

THE WORK OF MENDEL

In 1857, far more worthwhile probings into genetics began. In that year, the legendary Gregor Mendel of Austrian Silesia began tinkering with peas in his backyard. Born in 1822, his enormous contributions stem from his early experiments on *Pisum sativum,* the common garden pea. Mendel had exhibited a fascination with biology since childhood, primarily because he grew up on a farm with a father who was also an avid amateur disciple of the outdoors. Realizing

Mendel's interest as well as his talent, the local priest Pater Johann Schreiber encouraged him to continue his education. In fact, he took it upon himself to teach Mendel farming and botany.

Mendel entered the Altbrunn Monastery in 1843, a monastic community in Czechoslovakia that embraced the ideology of predestination and versions of Christianity based on the philosophy of Plato. As is well known today, the monks of such orders conducted a huge amount of scholarly activity. Much of it was in translating, but much was also in the sciences. Therefore, it is not surprising that Mendel used his spare moments to master both zoology and botany. It was during his years at Altbrunn that he began his legendary experiments on garden peas, which virtually created the modern science of genetics.

In his middle years, Mendel became increasingly fascinated by genetics, though, of course, the science did not actually exist yet. During his life he scrutinized heredity phenomena in pumpkins, beans, peas, in an ample variety of fruits, and in the farm animals around him. Mendel wisely chose to study garden peas on grounds similar to the reasons why geneticists today utilize the fruit fly, *Drosophila*. Both reproduce quickly, in substantial numbers, and display their traits very clearly, in part because both organisms are relatively simple. Additionally, since the crossing of garden peas with other plants is uncommon, the characteristics of the garden pea are markedly different from those of most other plants. The qualities of the plants Mendel paid the most attention to were height, seed color, whether the outer seed coat was smooth or wrinkled, and whether the flowers of the plant displayed axial (found all along the length of the main stem) or terminal (collected together at the top of the plant) distribution.

One of his shrewdest experiments had to do with garden pea plants of dissimilar heights, which Mendel called "tall" and "dwarf." He first allowed the tall version of the plant to self-fertilize (which was possible because male and female reproductive parts occupy one and the same plant in garden peas). He found that all of the offspring were tall. Similarly, the dwarf plants produced all dwarf offspring. Although under poor conditions the tall plants would be less than the standard height of about seven feet, they would always be taller than the short plants raised under the same circumstances. Mendel immediately realized, that despite constant and irregular changes in temperature, weather, and other environmental factors, this basic pattern did not vary, indicating conclusively that whether a garden pea plant was tall or short was entirely a matter of heredity.

THE LAWS OF HEREDITY

Soon, Mendel would formulate the most basic laws of heredity. In his next step he crossed the tall plants with dwarfs, producing all tall plants—no dwarfs whatsoever. He next allowed these new tall plants to self-fertilize. The results of this second generation of plants did again show some dwarfs. By repeating this experiment over and over again and by keeping an accurate tally of the outcome, he proved that, on average, in the second generation there were about three tall plants to one dwarf. He then began theorizing, trying to explain these facts. The result was a now famous tenet of genetics: some traits are dominant while others are recessive. Mendel himself described these ideas in a paper read to the Brunn Natural History Society in 1865:

> Henceforth in this paper those characters which are transmitted entire, or almost unchanged in the hybridization, and therefore in themselves constitute the characters of the hybrid, are termed the *dominant*, and those which become latent in the process *recessive*.[24]

He also advanced the idea that the parent plants transmit genetic characteristics to their offspring in the form of "particles" in pairs. Although Mendel knew nothing of genes or chromosomes, this was at least a step in the right direction. To that end there were four possibilities: a given offspring could have both recessive characteristics, a dominant and a recessive characteristic, or both dominant characteristics.

So as long as a dominant trait joins with a recessive trait, the offspring will exhibit the dominant trait. Therefore, since tall was dominant over dwarf, any plant with one dwarf gene and one tall gene would be tall. The plant would be a dwarf only when both members of the gene pair were recessive. Since the odds against that were 3:1, on average, tall plants appeared three times as frequently as dwarf ones. Barely past the midpoint of the nineteenth century Mendel had made substantial progress toward formulating these laws, laying out the hypothesis of "segregation"—the separating out of genes in egg or sperm reproduction and the subsequent independent recombination of such genes to form a unique organism.

MENDEL'S WORK FALLS ON DEAF EARS

At this point Mendel in some sense went awry—not in his research but in the way he promoted it, or worse, failed to promote it. He was inherently shy and reluctant to tout his studies. This problem compounded itself when he failed to confirm his own conclusions in later experiments. Why the latter occurred is easily explicable today in the following way: We now know that the plants he used in his ensuing trials were utterly self-fertilizing, so the production of hybrids was impossible. But it was only through hybridization—the crossing of one species or variety with another—that Mendel's ratios would consistently appear. It is also true that, during this time, botanical and genetic learning was so primitive that other scientists did not clearly understand how to interpret experimental data.

About the only thing he did do was communicate his early outcomes to the German biologist Karl von Nageli, who was not interested, probably for the reasons stated above. Mendel did ultimately publish his conclusions in the *Transactions of the Brunn Natural History Society.* The early twentieth-century British biologist William Bateson immediately realized the power of Mendel's work and, in fact, discussed them in front of the Royal Horticultural Society under the rubric of a more comprehensive talk entitled "The Problems of Heredity as a Subject for Horticultural Investigation." After a considerable span of time, Mendel's paper appeared in the 1901 *Proceedings of the Royal Horticultural Society*, but that is about the end of any promotional efforts.

Eventually, however, scientists like the French biologist Charles Naudin of the Jardin des Plantes in Paris became enthralled with Mendel's work. He began experiments involving hybridization, and he was able to build on Mendel's research to ferret out the regularities in the way plants inherit characteristics from a previous generation.

PROGRESS IN GRASPING CELL DIVISION

Still, real understanding of genetic transmission of traits lay in the future, with twentieth-century work on genes and chromosomes. Yet before that could occur, scientists would have to understand more about the cell itself, the mysterious entity in which the genes were "housed." A step in that direction occurred when the Polish biologist Eduard Strasburger of the University of Bonn, along with his colleague Walther Flemming of the University of Kiel, began delving into mitosis, or the process by which cells divide. Flemming was born in 1843, and after receiving his graduate training joined the biology faculty at Prague and Kiel. Especially noteworthy was his use of exceedingly inventive techniques for fixing and staining tissue. Many biologists, especially Flemming, pioneered the use of killed rather than living material for microscopic analysis. They found that this allowed for a much more reliable examination of metabolic and physical transformations in cells. They also added to the body of information on cell division when they began to realize that chromosomes line up right before a cell divides in mitotic division. In fact, Flemming coined the new terms "prophase," "metaphase," "anaphase," and "telophase," to denote the four successive stages of mitotic division as well as the terms "chromatin," "aster," and others. Scientific knowledge of chromosomes increased when the Belgian biologist Edouard von Beneden found that the number of chromosomes is distinct for every animal or plant species.

By the last decade of the Victorian era Walther Flemming had penned his monumental treatise on cytology, *Cell Substance, Nucleus, and Cell Division*. It was here that he revealed his ideas on chromosomes as well as on mitosis. Still undiscovered however, was meiosis, the process by which reproductive cells divide. Flemming also discovered the centrosome (independently of von Beneden).

BOVERI AND MITOSIS

In 1888, while on the faculty at Würzburg, the German zoologist Theodor Boveri gave a highly exact description of spindle formation in mitotic division. Spindles are entities that appear in the cell only when it divides, and which radiate out from the chromosomes. Most strikingly, he found that these did not exist in plant cells, something scarcely anyone expected. He explained that the cytoplasm itself does the same thing in plant cells that the spindles do in animal cells. Boveri also found an object now called the "centrosome," which shows up only when a cell divides and which is responsible for the formation of spindle fibers. He theorized correctly that the centrosome controls the entire process of cell division, by "telling" the chromosomes to line up alongside one another just before the cell divides. In fact, it was Boveri, along with the Utica, New York biologist Walter Sutton, who postulated the existence of chromosomes in the first place, as the carriers of genetic information.

THE DISCOVERY OF MEIOSIS

By 1887, the Belgian E. J. Beneden had discovered the alternate process of cell division known as meiosis. He was the first to realize that during meiotic, as opposed to mitotic, division, the number of chromosomes in a cell reduces by half. This halving is essential in sexual reproduction; since each parent contributes half of the number of chromosomes, the resultant new being will again have the characteristic number of chromosomes for the species. Also, in his 1903 paper, "The Chromosomes in Heredity," Sutton says:

> It has long been admitted that we must look to the organization of the germ-cells for the ultimate determination of hereditary phenomena. Mendel fully appreciated this fact . . . to those who in recent years have revived and extended his [Mendel's] results the probability of a relationship between cell-organization and cell division has repeatedly occurred. . . . Nearly a year ago it became apparent to the author that the high degree of organization in the chromosome-group of the germ-cells as shown in *Brachystola* could scarcely be without definite significance in inheritance.[25]

Sutton also recognized early on the enormity of Mendel's efforts. He was himself one of the seminal theorists on the role of chromosomes in heredity. Also, he contributed to our understanding of how genes and chromosomes behave during cell division.

Closely linked to this work was the research of the German Albrech Kossel, who, in 1879, started investigating the structure and properties of "nuclein," found in the protoplasm of cells. (Nuclein was later renamed "cytoplasm" by the German biologist Eduard Strasburger, who also coined the term "nucleoplasm" for the cytoplasm inside the nucleus of the cell. However, the liquid inside the cell is now called "cytosol.") Within a few years Kossel's research would lead him to a discovery of colossal significance in the new science of genetics—the nucleic acids. Their importance would shortly become manifest to the scientific world when scientists realized that the nucleic acids DNA and RNA were "the molecules of life," the constituents of genes and chromosomes, and the particles charged with transmitting genetic traits from one generation to the next. For his epic studies in the area of nucleic acids, Kossel captured the Nobel prize in 1910.

WEISMANN AND THE GERM PLASM

One cannot forget August Weismann of Germany. He adopted and, in fact, improved upon Darwin's ideas, which means that he rejected the Lamarckian concept of the inheritance of acquired characteristics. He conducted explorations designed to show that Lamarckian inheritance was biologically impossible, if only because body cells and reproductive cells were quite unlike one other, so it would be unthinkable for changes in one to affect the other. In 1892, he became the first to unveil the significance of the "germ plasm," an element that remains unaltered from one generation to the next and that carries genetic information to the next generation. He further proposed that the germ plasm was located in the chromosomes. In 1909, he published his masterpiece, *Die Selektionstheorie,* or "On the Theory of Selection," where he vigorously defended the concept of natural selection with arguments similar to Darwin's. He also here utterly rejects the Lamarckian notion of the inheritance of acquired characteristics.

THE BEARERS OF DISEASE

The year 1892 was a watershed in biology. Previously, scientists had known that bacteria, protozoa, and a few other organisms could cause disease. No one truly knew, though some had suspected, that there were such things as viruses. Microscopic and filterable, viruses were discovered when Russian botanist Ivanovsky was looking at mosaic disease (what is now called TMV, tobacco mosaic virus), a common blight afflicting tobacco plants. Then, in 1898, the botanist

Martinus Beijerinck made the important connection that a virus caused this affliction. In fact, this was the first time anyone had ever identified a specific virus. However, the mechanism by which viruses acted stayed hidden for several decades until, in 1936, the biologist Wendell Stanley separated the nucleic acids from the tobacco mosaic virus. And in 1901, Reed and Carroll showed that yellow fever was caused by an ultramicroscopic agent that could pass through any filter they had.

SUMMARY

The Victorian period was surely one of the preeminent periods in the history of science generally as well as of biology specifically. In physics one sees such titans as Oersted, André Ampère, Marie Curie, Faraday, Volta, the legendary James Clerk Maxwell and, in America, A. A. Michelson. In the same era, Rutherford would lay the foundations for nuclear physics with his discovery of the "compact nucleus." In technology, rather than "pure" science, such stellar names as Edison, Westinghouse, the Wright brothers, and George Eastman appear. Immanuel Kant and his unrivaled *Critique of Pure Reason* came to dominate philosophical thinking.

In biology, everything was set for the emergence of Darwin and his epochal views on evolution. On a darker note, the eugenics movement of Francis Galton cast a cloud over the mantle of scientific objectivity. Fortunately, this cloud would soon lift with far more scientific work of Retzius, who encouraged another and more radical approach to intelligence.

Above all, there was Mendel and his fabulous inquiries into the laws of heredity. Others, like Strasburger and Flemming would probe deeper into the mysteries of the cell with their research on mitosis while, by 1887, the Belgian E. J. Beneden unlocked the secrets of that uniquely reproductive process of cell division—meiosis. Weismann would complement these efforts with his studies of the germ plasm. It can be no exaggeration to suggest that the Victorian era was one of the most dramatically productive of all previous periods in science, eclipsed perhaps only by the appearance of quantum theory and the theory of relativity in the opening decades of the twentieth century.

CHAPTER 18

Darwin and His Age

THE PROFESSIONALISM OF SCIENCE GROWS

By 1866, Othneil Marsh of Lockport, New York, a graduate of Andover Academy, became professor of vertebrate paleontology at Yale, as the number of professional, academic positions for scientists gradually, if sluggishly, continued to expand. He was one of the very first to be appointed to such a post, as well as to the U.S. Geological Survey. He dedicated the balance of his days to searching for dinosaur fossils in the American West and noting their implications for the ancestry of all species, including man. He described many of his ideas in his book, *Dinosaurs of North America*.

During the initial decades of the Victorian era, scientific institutions were proliferating and thriving. The Royal Society's membership list registered strong gains throughout the nineteenth century, as did that quintessentially American scientific institution, the Smithsonian. In 1823, for instance, the American zoologist Spencer Baird, founder of the Woods Hole Laboratories, while head of the Smithsonian Institution, systematically compiled and preserved huge numbers of animals in the Smithsonian collection, a collection that is recognized today worldwide. Baird is also the author of the *Catalogue of North American Birds*.

Yet while the social structure of science was growing more organized and professional, other forces acted to retard the growth of science itself. Undeniably, theology had not yet lost its grip on science. In his inaugural lecture at Oxford University in 1820, for instance, the eccentric naturalist William Buckland strongly insisted that the goal of geology is and ought to be nothing else but to confirm biblical claims—an event suggestive of the controversies that arose during the twentieth century over evolution versus "creation science." In 1823, he published his *Observations on the Organic Remains Contained in Caves, Fissures, and Diluvial Gravel and on Other Geological Phenomena*. Predictably, this tract "confirms" Buckland's preconceptions. By systematically twisting evidence to his favor when it could be twisted and ignoring evidence that could not be twisted to bolster his argument, he "proved" that the biblical Flood had happened, using water-level marks in caves to prove it. He also "proved" that it had occurred about 6,000 years ago. When Darwinism first appeared, and for many decades after, religious fundamentalists would try to force it out of the curriculum of the public schools because it appeared to contradict Genesis.

MECHANISM

This era witnessed an ever-escalating controversy over mechanism, the idea that all biological phenomena can be explained purely through the laws of physics. Scientists like Claude Bernard of the Sorbonne, in his *Introduction to the Study of Experimental Medicine*, would argue that the philosophy of mechanism is true and that the concept of a "vital force" is clearly so much useless, abstruse baggage. Science plainly did not need it.

Darwin himself made his own contribution to this controversy. In 1872, his book *The Expressions of the Emotions in Man and Animals* appeared. In it Darwin added his name to the list of defenders of the "mechanistic" analyses of the evolution of the mind. Here he defends the notion that human emotions, for instance, descended directly from analogous behaviors in more primitive species. And, in his studies of the circulation of the blood, Karl Witzenhausen of Germany argued that bodily activity was explicable on a purely mechanical basis.

Philosophically, the German theorist Ernst Haeckel, too, was a materialist in the spirit of the pre-Socratic philosophers. There was no deity, and the universe was fully explicable with the laws of physics. Life too began through blind chance, probably from some random combination of chemical elements. Haeckel further angered traditional fundamentalists when, like Darwin, he proclaimed that there was no vital difference between the mind of man and the minds of lower animals, simply a distinction in degree. In fact, psychology was not and could not be anything more than a branch of physiology.

As things stand today, the controversy over mechanism has hardly abated. Many important philosophers, such as the British analytic philosopher Elizabeth Anscombe and D. M. Armstrong of the University of Adelaide, still argue that the old Cartesian idea of the "soul" has been completely refuted. Yet many philosophers believe today that this is not necessarily true. This group of thinkers, many heavily influenced by the legendary philosopher Ludwig Wittgenstein, have offered enormously powerful arguments against mechanism in any form.

PALEONTOLOGY AND RICHARD OWEN

In 1822 in Tilgate Forest, the amateur Zoologist Mary Mantell, a confidante of Charles Lyell, found the fossil of an ancient lizard, the later-named *Iguanadon.* This was a pivotal point in the development of paleontology. For although dinosaur remains had appeared earlier, no one had recognized them as such. In fact, the term "dinosaur" did not even exist at this point. It was Richard Owen who coined that term to apply to the giant reptiles that had ruled the earth many millions of years ago.

Richard Owen was born in 1804 in Lancaster, England, the offspring of a local retailer. Distinguished as he became in succeeding years, he gave no signs of special talent in early life. Expecting to go no further in his vocation than the local pharmacy, he began an apprenticeship in that field, yet a passion for science and medicine drove him constantly. Finally, he headed for the University at Edinburgh to learn medicine, spending his spare hours on anatomy.

As an anatomist he was without peer in his native land. His anatomical descriptions were scrupulously accurate, as were the drawings that accompanied and recorded them. He studied biology in Paris under Cuvier and became director of the natural history section of the British Museum in London, remaining in that position until past eighty years of age.

Unlike so many of his predecessors, he was not only an illustrious experimenter but a theoretician as well, and he indulged in a wide array of speculative explanations of what he had seen in his investigations. Following Cuvier, he scanned the same organ throughout the animal kingdom to try and understand how the organ had recast itself in the course of evolution. In this manner, he established that despite remarkable changes that evolutionary processes had caused, the same organ could do the same thing or perform the same function in animals separated by immense evolutionary gaps. Both the flying fish and birds used their limbs as wings, for instance, and the lungs of mammals and swim bladder of some fish also performed similar functions.

OWEN ON HOMOLOGY AND ANALOGY

In 1858, Owen published *On the Classification and Geographical Distribution of the Mammalia* and, two years after this, *Systematic Summary of Extinct Animals and Their Geological Relations*. He also made a valuable contribution when he published *On the Archetype and Homologies of the Vertebrate Skeleton*. Here he envisioned the skull as emanating from a series of adaptive overhaulings of the vertebral column. This is where he made his momentous distinction between homologies and analogies. In his view, and in the view of all subsequent biologists, two body parts are "homologous" when they have the same evolutionary heritage—even if their functions have little to do with one another. On the other hand, they are simply "analogous" if they have similar functions but divided evolutionary histories. The penis and the clitoris are homologous, for instance. This view was strikingly similar to that of Goethe's at the dawn of the nineteenth century.

These notions still play an integral role in modern comparative anatomy, and any sophomore year biology major will very quickly gain fluency with them. Beyond this, Owen's anatomical concepts of homology and analogy appear, quite soundly, in most of the initial arguments in favor of evolution. It is also arresting that Owen had originally been in favor of Darwin's views. What he could not allow, in spite of this, was the further contention that man was part of the evolutionary pathways. Thus, he differed more with Darwin's book *The Descent of Man* than with *The Origin of Species.* Inevitably, this preoccupation with the distinctive status of man, traceable, no doubt, to Owen's religious beliefs, led him to break completely with Darwin.

AGASSIZ'S CONTRIBUTIONS TO PALEONTOLOGY

Paleontology pushed further ahead when the Harvard zoologist and geologist Jean-Louis Rodolphe Agassiz, a student of Cuvier's, published the book *Recherches sur les Poissons Fossiles*, or "Studies of Fossil Fishes," again unveiling the strength of geology to illuminate biology. In fact, Agassiz compiled a massive work on glacial geology, titled "Studies on Glaciers," in which he waxes eloquently on the formation and movement of glaciers, revealing as well his bias in favor of a biblical view of creation:

> . . . gigantic carnivora, was suddenly buried under a vast mantle of ice, covering alike plains, lakes, seas, and plateaus. Upon the life and movement of a powerful creation fell the silence of death.[26]

THE GEOLOGICAL WORK OF LYELL

Closely related to paleontology is geology and the work of Charles Lyell. The son of a Scottish farmer, Charles Lyell entered the world in 1797. Because Lyell's father was also a student of biology, he provided an atmosphere conducive to developing his son's boundless native talent. Lyell graduated from Oxford University and practiced law for a while, until geology posed too strong an enticement to him.

Louis Agassiz (1807–1873). (Courtesy of the Library of Congress.)

Of particular note is the fact that he went on many voyages, surveying the fossilized forms of life that dotted the earth's crust. That, in turn, provided the theoretical basis for Darwin's revolutionary conviction that higher forms of life descended from lower ones.

Lyell studied the potent forces that had served to shape the nature of the earth over eons. He also concluded that the same forces were operating at the same rate during his time as they had millions of years ago. That fact, he argued, precluded any reason to accept the biblical tenet of "catastrophism," or the conviction that the earth and the many forms of life on it had altered drastically and abruptly at various stages in the earth's history. For he insisted, in fact, that the fossil record showed that the reverse was the case—the history of species was a study in gradual change. Many of these suggestions appear in his *Principles of Geology* (1830–1833). In these volumes, he also disavows the Lamarckian chimera of the inheritance of acquired characteristics.

Several years later, Lyell added substantially to the biological and geological sciences when he found indications of the existence of an extraordinarily early form of human being in the unearthing of some stone tools.

Not surprisingly, such an astute scientist as Lyell had powerful supporters. The second part of Darwin's geological studies, for example, was even more significant than the first in that Darwin offered evidence confirming the views of Lyell about the existence of the Pleistocene, a subdivision within the Cenozoic period, the most recent era of geologic time. The Pleistocene is also the era where most modern flora and fauna appeared. Darwin also confirmed Lyell's views about other periods in the earth's history. This reassessment also tended to confirm the existence of an ice age in the earth's distant past.

CHAMBERS AS AN INFLUENCE ON DARWIN

Another classical predecessor of Darwin was amateur geologist Robert Chambers (author and publisher of *Tracings of the North of Europe*) with his book *Vestiges of the Natural History of Creation*. Although there were several inaccuracies in this book, the deliberations

Asa Gray (1810–1888). (Courtesy of the Library of Congress.)

in it were sound enough. His support gave encouragement to both Darwin and Wallace to push on with their development of the theory of evolution.

PROGRESS IN BOTANY

Botany was adding to its store of data as well. In 1823, the German botanist Nathanael Pringsheim became, arguably, the first biologist to make the mastery of algae his exclusive preserve. With his research, science for the first time came to know something about the biochemistry as well as the taxonomy of algae. In his classic study of the algae *Vaucheria,* he clearly set out how fertilization occurs. In addition, a quiet New Englander named Asa Gray was completing his own analysis of plant structure and function. Soon he would publish the *Elements of Botany,* which ultimately established him as one of the preeminent botanists of the past. He followed this with his *Manual of Botany,* which included a catalogue of all of the known plants in the northern United States. Beyond this, Gray was a well-known supporter of Darwin, opposing even Gray's own colleague at Harvard, Louis Agassiz. After many years of tireless exertion, he collected all of the evidence that he believed supported Darwin in his book *Darwiniana.* Notably, Gray avowed that there was no real conflict between evolution and theology.

A BOTANIST DISCOVERS BROWNIAN MOTION

The early part of this era was one of the exceedingly few times in history when a scientist made a major discovery in a field other than his own. In 1827, the Scottish botanist Robert Brown, while doing some routine microscopic analysis, noticed that particular liquids contained vanishingly small particles that remained in constant motion without any outside force, such as stirring, being applied to them. Science in succeeding decades dubbed this movement Brownian motion, which is how science still refers to it today in Brown's honor. During the early decades of the twentieth century, scientists and physicists would offer this phenomenon as proof of the reality of molecules.

Brown was born in 1773, the son of a Scottish clergyman. He began his training in medicine at the University of Edinburgh and served as a military physician. This phase of his career over with, he devoted himself to botany and did a considerable amount of investigation in Australia.

Brownian motion was not Brown's only contribution; among other accomplishments he described the cytoplasm of cells quite precisely in his monograph *Microscopic Observations on the Pollen of Plants*. He is also well known for having discovered the cell nucleus. Another of his celebrated contributions consisted of his meticulous accuracy in cataloguing the various families of plants. An excellent example of this is his exploits with the Asclepiadaceae—the milkweed family. There are a bewildering number of plants in this family, found mainly in Africa, and some of them are cactuslike. Brown was also what one might call a "geographer" of plants, having studied the way plants are scattered under assorted climactic conditions.

But Brownian motion endures as his most heralded contribution. Oddly, the actual size of his published conclusions is modest, though some of his papers were significant enough to be published in Germany—then the world leader in science. Nonetheless, the scientific community fittingly regards him as one of the notable botanists of history.

DARWIN AND PLANT EVOLUTION

Many tend to overlook the fact that Darwin concerned himself not only with animal but also with plant evolution. Indeed, soon after he published *The Origin of Species* he insisted that specialized parts of flowers must have evolved to allow the most efficient pollination by insects. Eventually, he followed this argument with *The Different Forms of Flowers on Plants of the Same Species*, in which he elaborated on the idea that one could resolve the radically distinctive variations in flowers by suggesting that plants had divergent evolutionary "purposes."

DARWIN AND THE RISE OF EVOLUTION

One of the first major steps toward the legendary theory of evolution took place when the entomologist Henry Bates engineered his epochal studies of insects in South Africa. His finest work here is his study of the *Heliconidae*, a type of butterfly. Though today's generation of biologists perhaps does not give him the appreciation he merits, they concede that he contributed to the scientific world's eventual endorsement of Darwin's tenets.

Darwin was born in 1809 at Shrewsbury in western England, the heir of the physician Robert Waring Darwin. Charles Darwin was one of eight children, and his education was traditional. He studied the classics and later went to the University of Edinburgh to begin his schooling in medicine. But he shortly left to train for the ministry at Cambridge. Natural history soon absorbed most of his attention. At Cambridge University he studied theology, but during his spare moments he collected insects and read virtually everything he could find on geology. He had the happy fortune, too, to apprentice at Cambridge with Adam Sedgwick, a distinguished teacher of geology during his time.

Doubtless Darwin's first important move toward formulating the theory of evolution, influenced somewhat by the geological findings of the Victorian geologist Charles Lyell, was Darwin's decision to travel on the *H.M.S. Beagle*. Starting in 1831, the journey was destined to last five years. England herself had commissioned the *Beagle* to sail around the tip of South America and the globe. The British Crown had charged its staff with drawing maps, discovering untouched locales, measuring distances, and so forth. Recognizing Darwin's eminence in science, the ship's officers immediately appointed him "scientific officer," though the position carried no salary. By 1835, they had arrived at the Galápagos Islands, where Darwin noted that certain finches appeared to have evolved from some common ancestor other than what then existed in South America.

In 1839, Darwin married his cousin Hannah Wedgwood. Her vast fortune, spawned by her family's thriving ceramics industry, allowed Darwin to spend the remainder of his life in security, able to pursue his interests full time, despite the fact that he suffered from ill health off and on for much of his life.

Charles Darwin (1809–1882). (Courtesy of the Library of Congress.)

In the ensuing years he published his epochal *Journal of Researches into the Natural History and Geology of the Countries Visited during the Voyage of H.M.S. Beagle.* This book describes in painstaking detail Darwin's voyages over the five-year period, including his surveys of an enormous variety of animals and plants as well as numerous fossils. Recognizing the significance of geology for biology, Darwin had made it a point to thoroughly explore the geologic features of all of the areas he had visited. He reported these features in his book as well. When the voyage ended, Darwin read Malthus's august writings on population and pondered his thesis that innumerably more individuals are born than can survive. Accordingly, the weaker ones had to perish in the competition for food. In 1840, Darwin published *Zoology of the Voyage of the Beagle,* in which he outlined in vivid detail all of the fauna and flora he had assembled on his travels.

By 1842, Darwin was confident and ready to publish his book *The Structure and Distribution of Coral Reefs, Being the First Part of the Geology of the Voyage of the Beagle.* Here he added, quite inadvertently, to the science of taxonomy, when he segregated coral reefs into three categories. There was, however, beyond the above inquiry, comparatively little advance in the science of classification that came from Darwin's pen. In this treatise, he also stated his own beliefs about the formation of atolls and coral reefs, via subsidence of islands in those locations. He theorized that atolls or coral reefs had formed when the land on which they were built gradually sank deep into the sea.

THE ORIGIN OF SPECIES APPEARS

In 1859, the consummate scientific event of the nineteenth century occurred—the publication of *On the Origin of Species by Means of Natural Selection, or the Preservation of Favoured Races in the Struggle for Life.* Here Darwin explains in painstaking detail the premise of natural selection, mentioned earlier. He argued that nature, rather than a deity, selects among the offspring those best fit to survive. He noted, among other things, that the parents of most species produce more offspring than can possibly survive. That, in turn, leads to competition for shelter, food, water, and so forth. Then a random factor surfaces. Given that offspring vary in their properties quite haphazardly, those with the qualities exceptionally suited to the environment will survive. Some might be stronger, or have better hearing, or a body color which better blends in with the environment to hide them from predators. As a consequence, plants and animals with the more suitable qualities will transmit them to their offspring. In his second major tract, *The Descent of Man*, published three years after *The Origin of Species*, Darwin argued that man first descended from an apelike ancestor. He also shows how the operation of natural selection affected the actual development of various known species.

A surprisingly cosmopolitan aspect of this book is that Darwin announced that the "mental life" of *Homo sapiens* is not as unique as many theologians had always presumed. He argues legitimately that such psychological properties as memory, imagination, curiosity, and so forth exist in varying degrees in lower animals as well. Recent experiments with chimpanzees and language-learning inform us that Darwin's thinking here was both lucid and profound.

SEXUAL SELECTION

The Origin of Species did not end Darwin's labors. A few years later he published *The Variation of Animals and Plants under Domestication*, which was a further elaboration and discussion of chapter one of *The Origin of Species*. *The Descent of Man and Selection in Relation to Sex* shortly complemented *The Origin of Species*. In this treatise, he focused on the evolution of human beings from lower life-forms. He also demonstrated the truth of sexual selection. He devised this notion because he believed that ordinary selection processes could not elucidate secondary sexual qualities, for example, brilliant colors in an assortment of animals including butterflies and birds, antlers in the stag, and the like. Instead, a unique kind of competition between males emerged to win favor with females. In this way, the strongest and most handsome males were able to propagate and transmit their characteristics to the next generation. As Darwin described the phenomenon:

> This form of selection depends, not on a struggle for existence in relation to other organic beings or to external conditions, but on a struggle between the individuals of one sex, generally the males, for the possession

> of the other sex. The result is not death to the unsuccessful competitor, but few or no offspring. Sexual selection is, therefore, less rigorous than natural selection.[27]

Later, in his declining years, Darwin also began his *Animals and Plants under Domestication,* a two-part text that included a detailed description of his biological investigations on both plants and domestic animals. Beyond any doubt, the most important aspect of this effort was that it appended some valuable ideas to the suggestion that higher forms descended from lower ones, even hinting at a rudimentary notion of heredity. In it, he implied that parents could transmit qualities directly to their progeny under the control of a conglomeration of environmental factors, including infrequent use of a limb, climate, the food supply and so forth—a remarkably Lamarckian strand in Darwin's thinking.

A STORM OF CONTROVERSY

The immediate aftermath of Darwin's ideas was not entirely favorable. It scarcely needs saying that the religious ramifications were and still are disturbing to many. How could one now reconcile Darwin with Genesis? Yet Darwin was not the only source of trouble for believers. The German theologian David Strauss and the French thinker Ernest Renan had both disclaimed the historical accuracy of the Bible, exposing many of its inconsistencies. Without reservation, nevertheless, the most troubling aspect of Darwinism for the orthodox was the direct insinuation that no deity governed the cosmos. Instead, the universe resulted from blind, random chance. There were no assurances, no fixed celestial truths—not even fixed moral truths.

The social implications of Darwinism went even deeper. Ultimately, a doctrine would appear called "social Darwinism" which applied Darwinian principles to the evolution of societies and moral codes. Those moral codes were "true" which assisted a human being or society to survive; they were not true because a deity had decreed them to be true from on high. Among other things, the Social Darwinists swore that the white race had proved its superiority to the other races by the "success" of Western Civilization. Among the exponents of this teaching, though he never espoused the most racial applications of the view, was Herbert Spencer. Like Huxley and Haeckel, Darwin transformed Spencer's thinking, and he universalized the concept of evolution so that, in his view, everything in the universe was the product of evolution.

HUXLEY SUPPORTS DARWIN

The tempest swirling around Darwinism soon profoundly touched men like T. H. Huxley, who, along with Asa Gray, became one of Darwin's most conspicuous defenders. Huxley was of inestimable help to Darwin and added some much-needed rationality to the whole question of evolution versus special creation when he began his crusade in favor of Darwinism. Darwin himself had laid down the gauntlet in *The Origin of Species.*

> I can entertain no doubt, after the most deliberate study and dispassionate judgment of which I am capable, that the view which most naturalists until recently entertained, and which I formerly entertained—namely, that each species has been independently created—is erroneous.[28]

Huxley entered the world in 1825 and, in childhood, displayed stupendous scientific as well as philosophical talent, devouring every book he could locate. He began systematic biological study at Sydenham College, shortly receiving his bachelor's degree in 1845 from the University of London. Two years after that he completed his medical degree.

He immediately joined the British Navy as a medical officer stationed on the *Victory,* transferring after less than a year to the *Rattlesnake* because of hostilities on the former ship as well as because the latter ship was heading to Australia. That nation would provide him with a chance to inspect a range of flora and fauna in its rich biological environment. He accumulated enormous amounts of biological data on this trip, and in 1854 he published *Oceanic Hydrozoa* under the auspices of the famed Royal Society of London.

On the one hand, Huxley at first did not accept evolution. On the other hand, he did not think that Lamarckianism was a practical alternative either. Nor could he accept the biblical view. This philosophical stalemate ended, nonetheless, when he met the great philosopher Herbert Spencer in 1852. Since Spencer was a staunch advocate of evolution, Huxley's many discussions with Spencer shortly turned him toward Darwinism. In the end, he became such an advocate of evolution that he essentially became Darwin's publicist. When *The Origin of Species* appeared in 1859, Huxley's name would be everlastingly linked with Darwin's.

By 1860, the controversy over Darwinism had reached its zenith. So high did passions run that in 1860 Huxley had it out with Bishop Wilberforce, then Bishop of Oxford and a professor of mathematics, at an impassioned meeting of the British Association for the Advancement of Science. The Bishop scored first when he asked Huxley whether such evolution was through their grandfathers or grandmothers. Although he forswore an actual proclamation of atheism, Huxley replied that he would rather have an ape as an ancestor than the bishop, for such a beast "would not misapply his intelligence to joke about such a serious matter."[29] That reply drew loud and long applause from the spectators, hinting, perhaps, that the grip of biblical Fundamentalists was beginning to weaken.

Beyond this, Huxley did say that "there is no evidence of the existence of such a being as the God of the theologians." He then declared that Christianity was

> a compound of some of the choicest and some of the worst elements of Paganism and Judaism, molded in practice by the innate character of specific people of the Western World.[30]

In fact, it was Huxley who coined the term "agnosticism." With it he declared his hatred for the blind, dogmatic certitude he saw around him. Because of his uncompromising defense of Darwin, he earned the nickname "Darwin's Bulldog." Still, Huxley did much elegant research of his own in biology over a span of many years, chiefly in comparative anatomy. But his vigorous and long defense of Darwin is one of his chief legacies to biology.

The bishop was not the only one to vent his feelings about the villainous teachings. On the other side, some scientists sided with Wilberforce. Among others, the American biologist Louis Agassiz attacked the theory, asserting that it was absurd and impossible. Instead, he proposed reinstating the venerable theological notion that God created all of the species separately (though his own findings actually substantiate evolution).

WALLACE AND EVOLUTION

Alfred Russell Wallace undeniably deserves mention as the co-creator of the theory of evolution. Born in 1823, he lived a long and full life, dying in 1913 after spending his years in the service of science. Besides evolution, he devoted a considerable amount of attention to the distribution of animal species all over the globe. At the age of fourteen, he and his brother William began to master surveying, a prevalent occupation at the time. However, his surveying projects soon led him far afield, into both astronomy and botany. By the mid-1800s he had become a disciple of the brilliant Thomas Malthus, coming, like Darwin, under the influence of Malthus's essay *On Population,* thereby further cementing his obsession with biological questions. In April of 1848, his thirst for biological knowledge was nearly insatiable; he left England with his younger brother Herbert to explore the flora and fauna of South America. Sadly, disaster struck when his ship caught fire on the way back to England. Although he survived, he lost virtually all of the specimens he had collected as well as his own diaries. Undaunted, he relied as best he could on memory and authored both *Travels on the Amazon* and *Palm Trees of the Amazon.*

At forty-six, he married and began further biological quests in London, where he gave final form to his views on evolution. He wrote these up in his book *Contributions to the Theory of Natural Selection,* published in 1870. Possibly his most renowned volume, the above book notwithstanding, was his *Geographical Distribution of Animals,* which he finished in 1876. In this book he divided the planet into six land masses, all having their own characteristic life-forms. The "Paleartic," which encompassed a substantial part of Asia and almost all of Europe, was home to reindeer, pigs, hawks, pigeons, dogs, rats, cows, cats, other varieties of deer, oxen, bears, and other animals. The "Neartic" included North America and housed large animals like foxes, bears, and elks. He differentiated "Ethiopia," the rustic habitat of gorillas, lions, tigers, giraffes, and rhinoceri. The

Oriental region included China and Southeast Asia and its natural inhabitants—elephants, flying foxes, and orangutans. In Australia and New Zealand he surmised there dwelled abundant marsupials. Lastly, in the "Neotropical" zone, consisting primarily of South America, dwelled sloths, tapirs, monkeys, bats, and other animals. Of course, he thoroughly scanned the accomplishments of his compatriot Darwin, even writing a book about him called *Darwinism*.

Wallace would achieve a level of recognition equal to Darwin's when, in 1858, the secretary of the Linnean Society read from several letters traded between Alfred Wallace and Darwin, as well as from an unpublished paper of Darwin's. In all of these manuscripts the authors described the thesis of evolution in considerable detail, even devoting some considerable space to the concept of natural selection. These letters and other evidence made it clear that Wallace had come up with the main ideas of evolution independently of Darwin.

TAXONOMY: INVERTEBRATE

Debates over evolution did not comprise the whole of nineteenth-century biology—even if it sometimes seems that way. Since Linnaeus's day, scientists had been gradually building on the foundations of taxonomy set down by Linnaeus. One significant extension of that science occurred in 1872, when the Breslau botanist Ferdinand Cohn, a pupil of Johannes Müller, published the first of a three-volume series devoted to arranging bacteria into genera and species titled *Bacteria, The Smallest of Living Organisms*. This volume summarized all of Cohn's epochal research in the embryonic field of bacteriology to date. The classification of bacteria continued with the deliberations of the Danish bacteriologist Hans Christian Gram, who, in 1884, invented a stain that he used to split bacteria into a gram-positive group and a gram-negative group as he called them. The former group absorbed the stain, and the latter group did not. The immediate practical significance of this admittedly odd principle of grouping was not clear. By the 1940s, however, bacteriologists found that the classification was indeed critical: the two groups reacted drastically dissimilarly to antibiotics, making it imperative to discover to which group a bacterium belonged before trying to develop a vaccine.

LAMARCK ON INVERTEBRATES

Lamarck kept working during the Victorian era as well. In 1822, he published his *Natural History of Invertebrates.* In this volume, he made a major contribution to biology when he pointed out the distinction between vertebrates and invertebrates, or animals with backbones and those without.

TAXONOMY: VERTEBRATE

Very much in the spirit of Lamarck's work was that of Balfour. In 1880, Francis Balfour of Scotland, a zoologist and apostle of Haeckel, would open more fresh avenues in the field of biological classification when he argued that the notochord was not a true backbone, as previous biologists had thought. Instead, he argued that biologists should place animals with a notochord in the phylum Chordata, comprising animals with either a backbone or a notochord, while animals with a true backbone should be placed exclusively in the subphylum Vertebrata—taxonomic groupings still extant today. Balfour incorporated many of his observations and ideas in his two-part volume *Comparative Embryology,* which is an extended comparative analysis of both invertebrates and vertebrates.

THE CIRCULATORY SYSTEM

Maybe the most solid piece of evidence suggesting that biology was truly transmuting into a science was the experimentation of the German biologist mentioned earlier, Karl Witzenhausen. In 1847, he contrived a prototype of the sphygmomanometer, a device to measure blood pressure. With this he was able to study the circulation of the blood more thoroughly that anyone had previously done. Soon, the Danish physiologist S. A. S. Krogh, a student of Christian Bohr, demonstrated that ultratiny blood vessels called capillaries monitored the flow of blood through the body. For this find, he would garner the Nobel prize in 1920.

EMBRYOLOGY AND REPRODUCTIVE BIOLOGY

Another field that began to move more rapidly in this era was reproductive biology. In scrutinizing the female reproductive system, the Berne biologist and author of *De Phaenomeno* Gabriel Valentin and Czech scientist Jan Purkinje found that cilia lining the oviduct cause ova to move through the reproductive system in vertebrates, a finding which the biologist von Siebold later confirmed.

In the closely related field of embryology, the zoologist Karl Gegenbaur of Germany proved beyond all doubt that every cell in the body of any vertebrate emanates from the successive divisions of a fertilized ovum, though no one in Gegenbaur's day fully understood the stages of these divisions—blastula, gastrula, and so forth. Within twenty years, the Swiss biologist Rudolf von Kolliker would prove with absolute certainty that the egg is a cell and that all subsequent cells in any biological entity materialize from egg cells.

Rudolf Albert von Kölliker first mastered elementary zoology in Zürich, the town where he was born, in 1817. The son of an eminent businessman, his first teacher was Lorenz Oken. He later supplemented his studies in Berlin under the tutelage of Johannes Müller. He was on the faculty of Würzburg from 1847 until his retirement in 1902. He passed away in 1905, the year that Einstein's imperial special theory of relativity appeared. Von Kölliker was responsible as well for adding to both the substance and prestige of still another field of biology which had been stalled for some years—histology. He significantly widened the application of the ideas of cell theory from individual cells to embryonic development and entire tissues.

Proceeding relentlessly, Kölliker even faintly anticipated some of the classic conceptions of future genetics as well as cell theory. He prophesied that the cell was the carrier of genetic information, although, of course, he had scant understanding of genes, chromosomes, and so forth. After still more research, he would confirm the notion that axons and dendrites, fibers leading to and from nerve cells, or "neurons," were merely extensions of the basic material of the nerve cell itself. Kölliker then applied this to reproductive cytology (the study of cells) when he showed that both spermatozoa and ova were also cells. In his inspection of the ovum, he subscribed to the notion that since the ovum is also a cell, the behavior of the nucleus, dividing before the entire cell divides, was the critical phase in cell reproduction. The German physician Remak would subsequently fully confirm this view. Former thinkers believed ova were the consequence of fermentation of organic substances. Kolliker was also certainly the first to notice and describe cellular tissue in the so-called smooth muscles, such as those lining the intestinal tract. By just past the half-way point of the nineteenth century, Kölliker managed to explain a substantial amount of embryonic development in terms of the successive and progressive division of cells.

A LINK BETWEEN EVOLUTION AND EMBRYOLOGY

Just past the midpoint of the nineteenth century, a remarkable link developed between evolution and embryology. The ingenious German biologist Ernst Haeckel appeared with thoughts that would further shake the religious world. He had started his career as a physician but later became a university professor of biology. Though Darwin had his British supporters, Haeckel was among the first reputable scientists in Europe to unflinchingly adopt Darwinism. He introduced his own thoughts on Darwinism in his book *The Riddle of the Universe.*

A principle that caused even greater commotion than Haeckel's adherence to the philosophy of mechanism was embodied in his slogan "ontogeny recapitulates phylogeny"; that is, the developing embryo goes through all of the evolutionary stages that led to the organism in its present form. As he put it in his epic *Generelle Morphologie* of 1866:

> Ontogeny is the short and rapid recapitulation of phylogeny, conditioned by the physiological functions of heredity (reproduction) and adaptation (nutrition). The individual repeats during the rapid and short course of its development the most important of the form changes which its ancestors traversed during the long and slow course of their paleontological evolution.[31]

He noted, for instance, that embryos of all vertebrates will have something looking much like gill slits at a given stage in embryonic development. Though hardly anyone takes this to have any profound overtones today, Haeckel redeemed himself a bit when he coined the word "ecology." Part of his belief, too, was the idea that not only did heredity modify upcoming generations, but the environment did so as well. In this he anticipated what psychologists soon called the "nature–nurture" controversy.

SUMMARY

All in all, the nineteenth century was a period of profound change and new insights. The expansion of the social organization of science continued in the nineteenth century with the rapid growth of the Royal Society and the Smithsonian. Arguably, for the first time, science, with the work of Bernard, Witzenhausen, and Haeckel, was beginning its final abandonment of vitalism and beginning to accept mechanism. Paleontology also continued its ascendance with the work on ancient fossil reptiles of Richard Owen, who also formulated the doctrines of homology and analogy. With that came the destruction of still another venerable notion, the principle of catastrophism, when Charles Lyell forcibly argued that the fossil record showed that species had changed gradually over millennia. Related to this was the work of G. Valentin of the University of Berne, Purkinje, and Kölliker in reproductive biology and, especially, Haeckel's infamous doctrine "ontogeny recapitulates phylogeny."

CHAPTER 19

Embryology and Biochemistry in the Darwinian Era

Just before the time of Darwin's legendary work, a major breakthrough occurred in chemistry which would eventually have profound implications for biology. In 1834, the French chemist Jean Dumas stated for the first time his "law of substitution." This edict stated that under the right conditions of temperature and pressure, a variety of chemicals, including fluorine, chlorine, bromine, and others, could take the place of hydrogen in any organic compound. This made it possible to synthesize, in the ensuing years, an enormous array of other types of materials, including a number that were indispensable to living organisms. An illustration of the phenomenon of hydrogen substitution is the well-known refrigerant freon, found in many freezing units. Chemists construct the freon molecule by substituting fluorine and chlorine atoms for the hydrogen atoms in a molecule of methane gas.

Shortly afterwards, another French scientist, Auguste Laurent, proved that it was thinkable to substitute chlorine for hydrogen in individual substances without altering a substance's properties very much. Since this ran counter to both plain intuition and the attitude of the era, few accepted Laurent's proposals. Later, virtually everyone accepted them.

Two other pioneers in chemistry were the noted Italian Stanislao Cannizzaro and his colleague Avogadro. Avogadro endorsed the proposition that any gas at the same temperature would have the same number of particles in it, whether it was fluorine, oxygen, carbon dioxide, or whatever. Although Avogadro did not know about molecules at this time, Cannizarro submitted some of the earliest evidence that molecules really did exist and were, in fact, the "particles" Avogadro was talking about. Some years after this, Cannizzaro would continue supporting Avogadro's hypotheses, most notably at the International Chemical Congress of 1860.

Another chemical concept vital for biology was the concept of chemical bonding. One of the first to formulate this idea was the Scottish scientist Archibald Couper of the University of Edinburgh, who proposed that carbon chains of indefinite length form the central axis of all organic compounds. He was also the first to publish a structure for aromatic compounds. Although science had to await the likes of G. N. Lewis, Linus Pauling, and Irving Langmuir for any really significant penetration into the mysteries of chemical bonding, the supposition of bonding itself was a major step forward. As it turned out, the bonds in biological molecules such as proteins, amino acids, DNA, and others were put together, to a large extent, with the bonds Couper had discussed.

THE EMERGENCE OF NUTRITION AS A SCIENCE

Equipped with many of these new details, Thomas Osborne of New Haven, Connecticut, showed that there were a huge number of proteins in the body. Beyond this he was the first to discover and recognize the importance of vitamin A in human metabolism, though he did not fully appreciate the far-reaching import of vitamins as an entirely novel class of crucial food elements.

The science of nutrition was making its debut. Further progress would be minuscule for the next forty years. It would not be until 1875 that the Virginia biologist Henry Sherman of Columbia would take another significant step. Sherman managed to show that it is not only the amount of nutrients in the body that is crucial, but the proportions as well. As an example, he proved that the proportion of calcium to phosphorus was as pivotal as the total quantity of each for optimal metabolic function. While at the Carnegie Institution, Sherman also began some inquiries into the amounts of vitamins needed in the diet. Generally, the figures he came up with were too low, and more precise recommendations had to wait, again, until the twentieth century. Still, in honor of his work, he was named president of the American Institute of Nutrition

PROGRESS IN PHYSIOLOGY

Physiology began to grow and divide more sharply than ever during the Victorian era, from about 1840 until the turn of the century. Eventually, there were two discrete fields—animal and plant physiology. Earlier physiologists scarcely bothered to try and identify themselves as experts in one field or the other. Now they did so regularly. One of the first topics physiologists surveyed in this epoch was the phenomenon of sensation. The German biologist Ernst Weber, for instance, noted that if he pricked two points on someone's skin with a pin, and if the pricks were close enough together, the person would be unable to discriminate between them. The two pricks would be perceived as a single sensation.

By 1833, the field had advanced so fast that Johannes Müller felt ready to put together a gigantic tome. In that year he published his *Handbook of Physiology,* which tallied all of the physiological commentaries of that period. He also championed some ideas of his own, intimating that every nerve has its own "energy," though he was scarcely able to make much sense out of that suggestion. With further exertion, the German physiologist would come a bit closer to modern precepts regarding nerve cells when he demonstrated that nerve fibers emanate directly from nerve cells or "neurons." Interestingly, progress in this field was slow enough that no one felt ready to write a book on the topic of sensation until 1927, when the British neurophysiologist Edgar Adrian's book *The Basis of Sensation* appeared.

Not too many years later, the German physiologist Emil Heinrich Bois-Reymond of the University of Berlin opened further windows in physiology. Of French extraction, he began his tutelage in medicine at the University of Berlin, the town in which he was born in 1818. Soon he became a pupil of Johannes Müller, subsequently following Müller as professor of physiology at Berlin in 1855 until his passing in 1896. In this period, he proved using a galvanometer that the nervous system uses electricity to send messages (in "waves," as he called them) to different parts of the body, a precept now a part of the hypothesis behind the transmission of nerve impulses. Most of his work focused on electrical activity in muscles. This fact, understandably enough, tended to further erode the moribund "vitalistic" hallucination that the "essence" of animal existence was inherently undetectable and absolutely immeasurable. In Bois-Reymond's view, it was the electric current he measured that was the real life-force, or *élan vital,* so beloved by ancient biologists. He was right in recognizing the importance of electricity in the physiology of living organisms, but he went awry in speculating about the existence of "electric molecules." While the thought isn't altogether preposterous, scientists have, of course, long since forgotten it.

At about the same time, the Scottish physician Sir David Ferrie conducted some monumental studies on the brains and nervous systems of a range of animals. His goal was to work out the locations of motor nerve impulses—impulses leading to muscle motion—and sensory nerve impulses, which carried outside stimuli to the brain and nervous system. Still more gains came when, in 1850, the German physicist Gustav Fechner submitted that there is no simple linear relationship between the strength of a sensation and the intensity of the stimulus. Rather, as the strength of the stimulus increased, the sensation would increase far faster.

By 1852, the German physiologist Hermann von Helmholtz added to the conclusions of the fabled eighteenth-century Italian physiologist Galvani. Hermann Ludwig von Helmholtz is unique in the archives of science in that he was one of exceptionally few who made important contributions to both physics and biology. He was born in 1821 near Potsdam, the child of a teacher in the local gymnasium school. He apprenticed with Johannes Müller in Berlin, and soon obtained his medical degree. In the years to come he served as a military physician and eventually became a member of the faculty at Köningsberg, the home of the extraordinary philosopher Immanuel Kant. Ultimately, he became professor of physics at Berlin and at the new physico-technical institute in Charlottenburg. He held these positions until his demise in 1894. Although he took notice of such fields as biology, philosophy, physics, and mathematics, he accomplished nothing of consequence in philosophy and mathematics.

One of his choicest contributions complemented Galvani's findings. He was the first to measure the velocity of a "message" along a nerve fiber. He accomplished this by paying close attention to the neurons of frogs. He discovered that such impulses travel at about thirty meters per second, and he showed further that chemical changes are critical to the health of the nervous system. He would later go on and apply this knowledge to the study of the sense organs.

HELMHOLTZ AND THE EAR

In 1856, Helmholtz proposed the "resonance" principle of hearing, when he suggested that membranes in the cochlea of the inner ear act as resonators to amplify sound. Understanding of the hearing mechanism took another surge forward in 1863, when Helmholtz supported the assumption that via a set of resonators on the cochlea, the ear determines the pitch of the sound an animal hears. Yet it would not be until the 1950s that physiologists, taking the right clues from Helmholtz's findings, could claim that they had indeed grasped the mechanism of hearing. In 1961, the Hungarian physicist Georg von Békésy captured the Nobel prize for his work in this area. In his theory, "waves" that appear on the fluid in the cochlea carry the external vibrations which are then processed by the brain to produce hearing.

BELL AND NEUROPHYSIOLOGY

Evidently, the understanding of physiology was advancing rapidly, and Charles Bell explored this field still further. Born in 1774 in

Edinburgh, the son of a preacher, he spent his formative years in the worst sort of poverty imaginable. His father was barely able to feed the family from one day to the next. Even so, he was able to enroll at the University of Edinburgh, eventually getting his medical degree.

He later became a curator of the hallowed Hunter Museum and eventually a professor of anatomy in Scotland. In 1830, he penned *The Nervous System of the Human Body,* in which he distinguished a miscellany of separate types of nerves including sensory and motor nerves, a find later confirmed by Magendie. In "Idea of a New Anatomy of the Brain," a short paper which he wrote in 1810, he described the effects of severing the medullary nerve. He had discovered that the medullary nerve functions in two ways, depending on whether the experimenter tampers with the anterior or posterior root. If he touched the posterior root, the muscle connected to it could not contract. If, on the other hand, he touched the anterior root, the muscle would go into spasm. This explained a problem no previous biologist had been able to explain—how it was that nerves could both send *and* receive nerve impulses. Although Bell's explanation is far from complete, he did apparently understand that the double roots made the double function possible, adding still more data to the growing fund of information about muscle anatomy and physiology. He, like so many others before and after, also tried to localize "mental states" in various parts of the brain in the spirit of modern inquiry into the philosophy of mind. In this manner, he exerted considerable leverage over recent academic surveys of the nervous system as well as on speculations about the constitution of the human mind that are in vogue today.

HELMHOLTZ AND BOTANY

Like Purkinje, Helmholtz became a most acclaimed and popular teacher as well as researcher of his generation. Among his research contributions are experiments, performed in 1862, in which he coated plant leaves with wax and exposed them to sunlight. He promptly realized that only the plants with uncoated leaves were manufacturing starch, a clue that sunlight was necessary for photosynthesis. Both German botanists Julius von Sachs and Nathanael Pringsheim would confirm that chlorophyll exists in the plastids of plants and is not diffused throughout the tissues.

By 1840, the French botanist Jean Boussingault swelled the already large corpus of facts on plant physiology when he proved that the amount of nitrogen and carbon in plants could not be explained merely by referring to the sum total of these elements in manure, which is a source of carbon, hydrogen, oxygen, and nitrogen. He showed that plants obtained the nitrogen they needed for maturation from nitrates found in the soil. He also demonstrated that carbon came directly from the carbon dioxide in the air.

THE RISE OF BIOCHEMISTRY AND SOME PRACTICAL APPLICATIONS

This progress in botany led others to search for practical applications which, in turn, yielded still more knowledge of biochemical processes. The German biologist Justus von Liebig, in his *Applications of Organic Chemistry to Agriculture and Physiology,* believably started the modern branch of biology known as biochemistry. Born in 1803, he was the son of a Darmstadt merchant. His initial fascination with chemistry stemmed from helping in his father's shop. He considered becoming a pharmacist for a while, but, ultimately, in the laboratory of the eminent chemist Gay-Lussac in Paris, he became completely absorbed in chemistry. Eventually, the University of Giessen appointed him to the faculty as professor of chemistry. Arguably, the creation of biochemistry in Germany and even much of Europe was due to his efforts. His principal interests were in organic chemistry, and he contributed much to the techniques for analyzing organic compounds.

He did the spadework for biochemistry when he explained how flora and fauna swap carbon for nitrogen, thereby taking a major step toward comprehending the nitrogen cycle—the process whereby green plants convert nitrogen in the air and in soil into material they can use. He accomplished this when he began his appraisal of manure, showing that plants derived their carbon from carbonic acid and that ammonia served as their principal source of nitrogen. This research constituted a major contribution to plant physiology as well.

His inquiries were not without either inaccuracies or criticism. Among others, the nineteenth-century German biologist Matthias Schleiden criticized him, particularly for denying that manure was a source of nitrogen for plants. Criticism also descended upon him for ignoring the role of plant respiration and for his odd fantasy that crops would thrive better if farmers supplied them with insoluble phosphoric acid. Current agricultural science now knows this is quite wrong. It is vital to grasp, nonetheless, that Liebig evidently brought much of this opprobrium upon himself because of his overbearing and critical demeanor—he was very likely one of the most unpleasant biologists in history.

The American biologist Samuel Dan would later extend biological comprehension of the role of manure by revealing that phosphates are the primary compounds in manure that act as a fertilizer of flora.

THE EMBRYOLOGY OF VON BAER

Then came the important work of the Russian biologist von Baer in the field of embryology. Karl Ernst von Baer first saw the light of day in 1792, in the modern state of Estonia, though he was of German lineage and German nobility. This status aided him immensely in his later life. He first attended school at Revel, where he began his education in natural science. In 1810, he began studying medicine at the University of Dorpat. He says in his *Autobiography* that he would have liked to have studied biology as such, rather than medicine, but that, in those days, the only route to science was via medicine. This remained true until late in the nineteenth century. Many of his early quests were in epidemiology, or the science of the spread of disease, where he focused on the people of his own country.

After leaving medical school he headed for Würzburg, where he began investigating comparative anatomy. He purchased a leech for his first dissection. The preeminent faculty member at Würzburg during this period was Ignaz Döllinger, a former apprentice of the German philosopher Schelling. Döllinger, with his combined ardor for philosophy and anatomy, managed to instill a lasting interest in the origins of human life and the evolution of animal species in von Baer.

At the university, von Baer also began to read widely in physiology and anatomy. However, money rapidly became a problem and he traveled to Berlin in hopes of creating some kind of medical practice to bring in badly needed funds. Inevitably, his talent became surpassingly clear, and the University of Königsberg appointed him professor of biology, charged with taking care of the university museum—the perfect environment to support his own interests. After a series of brilliant studies, he became dean of the medical school and then rector of the entire university. In 1827, he found that all mammals originate from eggs stored in the ovaries of the females of the species. Much of this research appeared in his 1828 book the *Epistele,* in which he described his discovery of the mammalian egg.

He followed this volume with *The Developmental History of Animals,* which subsequently became a standard embryological text. In this, a work of vast importance, he recounts his finding of the ovum as well as the notochord—a primitive structure seen only in the embryos of vertebrates which gives rise to the vertebral column. Also, he cautiously set out his "germ-layer" theory of embryonic development. According to this conjecture, the fertilized egg in mammals develops into four distinct layers of tissue (later corrected to three by the German physician Remak). Out of each, distinct organ systems would develop.

VON BAER AND THE BIOGENETIC LAW

This was the first appearance also of what biology now calls the "biogenetic law," the belief that a higher animal passes through states that resemble stages in the development of lower animals. More accurately, this law says that the general characters of an organism appear before the more specialized ones. Also, the specific and detailed features develop out of the general ones. Although von Baer did more or less cogently describe these modifications, he had no notion of evolution and made no attempt to relate these findings to evolution.

Von Baer also aided the evolution of embryology as a science in more sweeping ways. For one thing, he transformed embryology, following the lead of anatomy, into a "comparative" science by emphasizing the advantages of comparing the embryological development of various species with one another. He also offered some of the earliest explanations of the development of the amniotic sac and the urogenital system, as well as the digestive and nervous systems.

Von Baer was also mesmerized by nature and the wilderness. At various times he headed expeditions to Lapland and Nova Zemlya and sailed the Caspian Sea to research marine flora and fauna. In his last years he founded the Society of Geography and Ethnology at St. Petersburg and supported the launching of the German Anthropological Society. He received many honors throughout his career, including the prestigious Copley medal as well as an appointment to the Paris Academy of Sciences. His influence on future biology is unquestionable. In modified form, his germ-layer concept still exists today. Beyond this he was something of a minor prophet. He predicted, among other things, that a comparative anatomical analysis of the great variety of distinct organs in the animal kingdom would be a critically important investigative implement. This prophecy has long since been fulfilled.

REMAK'S WORK IN EMBRYOLOGY

In 1845, the German physician Robert Remak would bring all of this to the highest level of perfection. Born in Posen in 1815, Remak was the offspring of a Jewish family. He studied biology under Müller, who, in recognition of his talent, aided him in achieving faculty rank. Nonetheless, Remak earned his keep for most of his life as a medical doctor.

Remak found some comparatively minor blemishes in von Baer's work when he noted that there were actually only three, rather than four, primitive layers of tissue—the ectoderm, mesoderm, and endoderm. The general idea and other details of von Baer's conclusions turned out to be approximately valid as stated before. Remak also recast some of Schleiden's research, showing, in 1841, and contrary to Schwann's conviction, that all cells came from previously existing cells. In the process of attaining this he scrutinized the development of frogs' eggs, proving quickly that the egg is itself a cell which continues to divide, forming new "daughter" cells. He testified too that cell division begins in the nucleus. He bolstered this work with further studies of the comparative embryological development of both frogs' and birds' eggs. He then coined the terms "holoblastic," to refer to eggs that divide into two equal parts—typical of frogs' eggs—and "meroblastic," whereby only partial division of the egg occurs, resulting in two unequal parts—characterizing birds' eggs. It was at this point that he corrected von Baer's work and thereby distinguished the now-familiar three germ layers in the embryo. The outer, or ectoderm, he hypothesized gave rise to the skin and the nervous system, the middle, or "mesoderm," to the musculature, and the inner, or "endoderm," to the digestive tract. We know today that all of this is roughly correct.

KOVALEVSKI AND THE EMBRYOLOGY OF INVERTEBRATES

After this work, the Russian biologist Alexander Kovalevski of the University of St. Petersburg observed other vertebrates and invertebrates, quickly finding that von Baer's findings applied to invertebrates as well. Thus, even invertebrates began the life cycle from the three basic layers in the embryo. Kovalevski would also do a considerable amount of work on the notochord, or backbone, as well as promote Darwinism in Russia.

MÜLLER AND FURTHER REFINEMENTS OF VON BAER

Yet, still more probings into von Baer's results were imminent. Toward the middle of the nineteenth century, the German physiologist Johannes Müller made his claim to everlasting renown. The son of a comparatively affluent shoemaker, he started his schooling at the University of Bonn, a moderate distance from the town of Coblenz on the Rhine where he was born in 1801. After obtaining his medical degree, he headed for Berlin, and in 1830 became a member of the faculty at the University of Berlin.

Historians universally embrace him as one of the great physiologists of his day. Among his many publications is the *Handbook of Physiology,* mentioned earlier, published in 1840. Many scientists and scholars used this tract very widely during the nineteenth century. His concerns were sweeping, extending from physiology and chemistry to comparative anatomy and physics. Müller refined von Baer's ideas, further describing as well as emphasizing the existence of the three basic layers of embryonic tissue mentioned above. He was responsible too for his doctrine of "specific nerve energies," a somewhat contentious belief at the time which stated that nerve impulses make their debut in the sense organs and then travel through the central nervous system. In general, he added much to biological comprehension of how the immediate environment affects the nervous system and, of course, the whole organism.

More enigmatic was his *Handbook of Human Physiology,* in which he tried to explain human thought "mechanistically." Again, this approach ran into extreme opposition, primarily because such a view was inconsistent with the belief in the human soul as something separate and distinct from the body and which is the locus of thinking. Another line that branched off of this was the odd notion that the mind is a "computer" and that all mental states are really "computational" states. (In the twentieth century, all of these versions of "mind–brain materialism" appeared with reckless abandon. From the 1960s on, nonetheless, many philosophers, inspired by the writings of the Austrian theoretician Ludwig Wittgenstein, cast considerable doubt on all these propositions). As Müller's career neared its finale, he trained a number of followers who carried on his work, among them Theodor Schwann, who formulated the cell theory, and the German biologist Rudolf Virchow, who conducted important investigations in pathology.

Müller spent his last years plagued by, ironically, the very phenomenon that had catapulted him to stardom—nerves. He had suffered from depression since childhood. The debilitating effects of age, plus the trauma of suffering through a shipwreck, inevitably overcame him. After a certain point he was never able to rebound from his melancholia. Although no one has ever proven it, his own recognition of his waning powers plus his inherent melancholia make it plausible that he committed suicide. He was simply found dead in his bed one morning, although he had not been suffering from any apparent illness. How exactly he killed himself—if indeed he did do so—is not known.

ADVANCES IN TECHNOLOGY

Others were improving the microscope still further. In 1830, Joseph Jackson developed an "achromatic" lens, a lens made of two or more materials so that different colors focused at the same location, thereby avoiding distortion of the image. Although such lenses already existed, no one had fabricated any expressly for the microscope. Toward the close of the century, microscopes with such achromatic lenses were growing common, avoiding the devilish problem of chromatic aberration that had haunted the older microscopists. Before long, many universities and museums as well as wealthy amateurs were able to purchase these spectacularly altered microscopes. By 1840, the popularity of the microscope had swelled so much that the Italian physicist Giovanni Amici was also busily improving the relevant techniques. He was able to devise a so-called oil-immersion microscope which, because of the optical properties of oil droplets, had an enlarging power of 6,000×—far greater than anything previously. With this microscope he became the first scientist to actually follow the process of plant fertilization.

Soon, the amateur microscopist John Dolland appeared. A novice of Klingenstierna's, he was a brilliant mathematician and physicist. In 1844, he fulfilled an ambition by constructing the so-called water-immersion lens; that is, he used water as a medium to separate the lens and the glass slide. After his work came the contemporary microscope, with the ocular eyepieces and the objective separated by a movable tube, the variety familiar to college undergraduates everywhere.

GOLGI AND THE RISE OF CYTOLOGY

With the aid of these magnificent technological refinements, the inspired Italian biologist Camillo Golgi rapidly entered the front ranks of the world's great biologists. Born in 1844, he got his medical degree at the University of Padua in Italy. Then, in 1875, he became professor of histology at Padua. Among other things, he streamlined scientific investigations of the nervous system and cytology by devising a technique for staining cells using silver salts, which allowed the investigator to see the cells much more easily. With these tools, he became the first to discover the synapse—the gap between nerve fibers. In a word, Golgi showed that nerve fibers do not constitute a continuous thread but are, instead, separated by these synapses. He ventured into the field of immunology as well, examining a large number of malarial parasites.

Much later, in 1883, he found a type of cell organelle essential to the functioning of the nervous system. He confirmed and elaborated on his discovery in 1898 when he described in detail this structure found in the cytoplasm of cells. Subsequent generations of biologists named these organelles "Golgi bodies," in his honor. In his essay "What Is Its [cytoplasm] Structure, and How Does It Work?" Professor John E. Harris describes the Golgi apparatus:

> A more elusive cytoplasmic inclusion is the so-called Golgi apparatus, a complex structure of water vacuoles partially sheathed with fatty material and revealed in a protean diversity of form by various complicated histological methods.[32]

Golgi would enrich the world still more with his research on malaria and pellagra. His toils revealed that of the two brands of malaria, so-called intermittent and pernicious, the former develops when parasites invade only the blood; but when they invade the brain and other organs, the pernicious form of malaria appears. He showed also that the severity of the disease varied directly with the number of parasites in the blood. Golgi was predestined to win the 1906 Nobel prize for his research on the nervous system and olfactory system, as well as for distinguishing between a variety of different nerve cells.

PFEFFER AND SEMIPERMEABILITY

Related to this was the work of the German botanist Wilhelm Pfeffer. Pfeffer had begun, in 1878, his epochal probing into "semipermeable" membranes, which are tissues that allow only molecules below a certain size to pass through them. He used this concept to ascertain the weight of a single protein molecule. However, further insights into the detailed architectural structure of proteins would have to wait until the twentieth century with the offerings of thinkers like Dorothy Wrinch, Irving Langmuir, Erwin Chargaff, and several others.

EHRLICH AND THE STUDY OF DISEASE AND BLOOD

The German physician Paul Ehrlich would push ever deeper into the nature of disease. Ehrlich is, defensibly, one of the initial heroes in the fields of hematology—the study of blood formation and function—as well as chemotherapy. He was born in Strehlen in upper Silesia to a Jewish family. He became enthralled with chemistry in childhood, but his attention soon turned to biology. He received his medical degree from Leipzig in 1878. Even as a undergraduate he began investigating new frontiers. Using a type of dye called an aniline dye, he found and categorized nearly every discrete difference among white blood cells. By 1881, he was using the more efficient dye methylene blue. Its effect on some bacteria provided him with a clue that this dye might be medically useful in combating certain bacterial infections. In fact, he found the dye methylene blue was effective against malaria. He then generalized his reasoning as follows: if certain tissues were selectively affected by certain dyes, perhaps other tissues would also be selectively affected by different drugs. He had hit on the concept of the "magic bullet"—a drug that would selectively destroy certain invading organisms while leaving the host tissue unharmed.

His work on dyes eventually led him to experiment with other chemical compounds. By tinkering with various drugs previously found ineffective against the disease trypanosomiasis, he found a new and colossal tool in the war on syphilis. After years of struggle, pharmaceutical companies marketed his innovation as Salvarsan, although since the advent of penicillin physicians rarely use Salvarsan. For these efforts, Ehrlich would receive the 1908 Nobel prize.

MORE WORK ON DISEASE

In 1894, the University of Tokyo pathologist and student of Koch, Shibasaburo Kitasato and the French bacteriologist Alexandre Yersin of the Pasteur Institute would each discover the bacterium answerable for bubonic plague, that infamous scourge of civilization that wreaked havoc in ancient Rome, the late Middle Ages, and other periods of the past. Kitasato also probed such scourges as anthrax, tetanus and diphtheria. (Yersin also studied diphtheria.)

Just before the turn of the century, the French physician Paul-Louis Simond discovered that rat fleas are the vehicle of transmission of many illnesses to human beings, a discovery that emerged while he was fighting the plague in India. In the twentieth century, of course, other scientists would follow similar paths and develop "vaccines" for an assortment of animal and human disorders, including anthrax, polio, smallpox, yellow fever, diphtheria, and many others. Paul Ehrlich, as suggested above, all but created the field of immunology when he perfected a vaccine against diphtheria.

Subsequently, in 1913, Bela Schick devised the now well-known Schick test for diphtheria, allowing physicians to quickly and easily diagnose the dreaded ailment. The beginning of the end for yellow fever as a slayer of millions came in 1900, when Walter Reed began his probings into the epidemic of that malady in Cuba, where he and his team found that the *Aëdes* mosquito (more accurately, the *Stegomyian* mosquito, a subgenus of genus *Aëdes*) transmitted yellow fever.

SUMMARY

In sum, there was tremendous progress in chemistry with the work of Dumas, Couper, and Cannizzaro, as well as Avogadro's famous research. Physiology also made gains with Weber's and Müller's work on sensation. In the adolescent field of cytology, the legendary Camillo Golgi made his momentous discovery of Golgi bodies. Medicine, too, was rapidly moving toward the twentieth century, as Ehrlich found his "magic bullet," while Kitasato and Yersin solved the mystery of bubonic plague and Bela Schick developed the Schick test for diphtheria. No one was overlooking technological advances either, as Amici developed the "oil-immersion" microscope while Klingenstierna nurtured the water-immersion lens.

CHAPTER 20

The Age of Pasteur and the Development of the Microscope

Although Pasteur's name all but outshone other scientists of this period, it is evident that Pasteur built his research upon earlier titans of science. Robert Koch added to the biological understanding of disease, for instance, when he found *Vibrio cholerae,* the bacterium that causes cholera. Like Pasteur, he averred that contaminated food and water were necessary elements in spreading the malady.

Koch was born in 1843 and took a degree in biology and chemistry at the famed University of Göttingen under the pathologist Jacob Henle. The latter had, among other things, also anticipated the germ theory of disease, despite years of persecution and torture in Prussia because of his Jewish background and "anti-establishment" views. The university awarded Koch his doctor of medicine degree in 1866, and for a number of years thereafter he did little work, preferring the quiet practice of medicine. He even offered his services to the government during the Franco-Prussian war. Though it is not absolutely certain, it seems that it was during the war that he became interested in sickness and microbiology—which is not surprising given the amount of severe illness he would certainly have seen around him. By 1876, he had successfully followed and described the entire life cycle of the anthrax bacillus, ultimately leading to Pasteur's development of a treatment for that dreaded ailment. In following the anthrax microorganism in this way, he saw that infection could spread from one animal to another.

Still another of Koch's contributions was the isolation of the tubercle bacillus. He accomplished this in 1882, and it is certainly one of the celebrated moments in the annals of bacteriology. For the very first time, science had a clear case of a well-known disorder proven to be transmitted from one individual to another by a microorganism. For this work he garnered the 1905 Nobel prize.

Then came Louis Pasteur's most awesome legacy. First and foremost he substantiated beyond doubt that life comes only from life, which is to say that the idea of spontaneous generation was mere mythology. Pasteur was born in 1822 to a middle-class family in Dole, a village in the French province of Franche-Comté. His father, who had served as an officer under Napoleon, entered the tanning trade after leaving military service. Pasteur began his university instruction at the University of Paris and fully intended to be nothing more than a teacher of science. He accomplished this when he joined the gymnasium school at Strasbourg, and he soon married a daughter of the school's rector. It was here that he began his exploration of chemistry. This led him to become professor of chemistry at the University of Lille and then to hold the same post at the École Normale in Paris.

His life was a harsh one, partly because throughout his years he clung tenaciously to his Roman Catholic faith—beliefs which were not popular among many in the scientific world. He was also politically conservative at a stage in history when political radicalism was deeply rooted in the public consciousness. No one had forgotten the French Revolution.

Although Pasteur originally studied chemistry, he soon became equally enthralled with biological issues. He became absorbed in the problem of spontaneous generation through scanning the scientific memoranda of others, including his countryman Félix Pouchet, a professor at Rouen who was already a leading biologist because of his contributions to both zoology and botany. So Pasteur designed some experiments to test the hypothesis. In broad outline his technique was similar to Pouchet's, except that he made some substitutions in materials, using a sugar and yeast medium in the flasks instead of rotting meat. Some of these flasks were sterilized and sealed, while others were allowed to remain open in a variety of settings. Perhaps the most telling result occurred when Pasteur found that every single flask left opened in a dirty building as well as on the streets of Paris contained living organisms. But the sealed flasks, when opened, showed no life whatsoever. If life could arise "spontaneously," there should have been living organisms in the flask which had arisen from the nonliving medium. Thus, Pasteur had shown that life could not develop "spontaneously"; the living organisms in the flasks must have come from the outside. In Pasteur's own words:

> But with wads of asbestos previously calcined and not filled with dust, or filled with dust but heated afterwards, no turbidity, nor Infusoria, nor plants of any kind were ever produced. The liquids remained perfectly clear . . . never once did any of my blank experiments show any growth, just as the sowing of dusts has always furnished living organisms.[33]

However, Pouchet, whose inquiries Pasteur respected greatly, did not accept Pasteur's conclusions about spontaneous generation. After some acrimonious debate, both scientists repeated their experiments before the French Academy of Sciences. Immediately, Pasteur convinced the Academy that he was right.

Pasteur was also concerned with the process of fermentation. For some years scientists had insisted that fermentation was primarily a chemical, rather than a biological, process. The chemist Liebig, for instance, believed that there was some yet undiscovered substance which, though it did not change itself, could cause other types of matter to ferment. The imperial cytologist Schwann believed in a similar theory and argued that alcoholic fermentation depended on yeast, but, since yeast was living material rather than a simple chemical, nearly everyone else rebuffed him. Pasteur, however, was not quite so ready to dismiss Schwann. In fact, his own findings proved that Schwann was right. In 1856, Pasteur published his *Researches on Fermentation,* in which he first described his deduction that it is not chemical reactions but rather microorganisms that cause fermentation. Specifically, yeast converted ordinary sugar into alcohol and carbonic acid. By 1857, he further discovered that still other living organisms caused sugar to break down into lactic acid. For this reason he speculated that many—perhaps all—kinds of fermentation depended on some strain of microorganism or other. He then designed a number of experimental trials to test his general supposition and found an enormous variety of living microorganisms that could cause several different types of fermentation. He believed, for instance, that organisms similar to those which brought about the fermentation of alcohol caused the formation of lactic acid. However, biologists later discovered that these organisms were not quite as similar as Pasteur had believed.

Scarcely resting on any laurel leaves, he upped his pace and, in 1863, discovered the microbe that causes wine to sour. He continued this work for thirteen years, later publishing his *Studies of Beer, Its Diseases and the Causes That Provoke Them.* In this tract Pasteur revealed that there are organisms which can live despite the absence of oxygen. Biologist now customarily call these "anaerobic" bacteria, meaning "without air." In 1877, Pasteur resumed his analyses of bacteria, noting that certain groups of microorganisms were not congenial to others. If he placed a culture of one variety together with a culture of another, one variety would not survive, though he understood little of the significance of this. Today we know that what Pasteur observed was an antibiotic reaction, which would later become a major branch of epidemiology and modern medicine. It would lead to monumental improvements in public health and sanitation and to the now-common process of ridding food of bacteria through "pasteurization."

By 1880, Pasteur was ready to publish his monograph "On the Extension of the Germ Theory to the Etiology of Certain Common Diseases." The germ hypothesis was actually quite a natural extension of Pasteur's preceding research on fermentation. For he theorized, quite fittingly, that living organisms produced illness in men and animals in much the same way as the other organisms produced fermentation and decay. Pasteur was not, of course, the first to suggest something like a "germ theory" of sickness. In fact the sixteenth-century scientist Hieronymus Fracastorius, in his essay "On Contagion," had already anticipated this notion. In Fracastorius's view, seedlike organisms he dubbed "seminaria" could carry a malady from one individual to another, either via clothing, air, or bodily contact.

By the closing decade of the nineteenth century, Pasteur had devised the first convincing vaccine directed at the lethal scourge of both animals and men—anthrax. Within weeks, he had proven success, because among farm animals exposed to the heinous germ, those he had not vaccinated expired, while those he had injected did not contract anthrax.

Pasteur was a humanitarian before he was a scientist and could not rest as long as adversity threatened humanity. So it was that in 1886, by unveiling the world of viruses, he developed a vaccine against rabies, even saving the life of a boy named Joseph Meister. Eventually, his immunological work led to antiseptic surgery by Joseph Lister.

OTHER SOLDIERS IN THE WAR ON DISEASE

In Pasteur's impeccable struggles, he conceived what has come to be known as the "germ theory of disease." The wonderful force of Pasteur's legacy and the enchantment of his name inspired others. One of these was Joseph Lister, who, in 1865, suggested using carbolic acid to cleanse open incisions during surgery on the grounds that it would prevent infection by precisely the sort of bacteria Pasteur had discovered. Indeed, Lister paid tribute to Pasteur in a speech he delivered at the Sorbonne in 1892 on Pasteur's seventieth birthday:

> Medicine owes not less than surgery to your profound and philosophic studies. You have raised the veil which for centuries has covered infectious diseases; you have discovered and demonstrated their microbic nature . . .[34]

Also, the London physician John Snow, working at the Killingworth Colliery, found that some wells harbored bacteria which were causing massive outbreaks of cholera. When a cesspit that was leaking into the well was sealed, new infections all but ceased. Pasteur continued his work and, in 1879, found quite serendipitously that he could prevent chickens from contracting the dreaded ailment cholera if he allowed them to be infected with the weakened cholera bacteria.

At approximately the same time, others were also fighting the war on disease. By the closing decade of the nineteenth century the Russian bacteriologist Waldemar Haffkine and Jaime Ferrar of Spain added astonishing weapons to the war on this plague. With steady and wearisome testing while in India, he managed to create a weakened strain of the cholera bacterium, even daring to use himself as a test subject. He tested both oral and injected vaccines as well. Soon physicians would use this vaccine to reduce the mortality rate in India by nearly 80 percent. Pasteur himself proselytized in favor of new cleanliness procedures, persuading army surgeons to thoroughly sterilize all surgical tools.

One cannot forget the work of Rudolf Virchow. Although he is perhaps better known for his work in cytology, Virchow was undeniably one of the remarkable humanitarians of science; he was engrossed in biology and disease not solely for their own intrinsic interest, but to improve the quality of existence and reduce the anguish of humankind. For that reason, he lobbied to have better sanitation laws passed and toiled constantly in the field of public health. It can scarcely be an exaggeration to say that he saved thousands of lives.

EMBRYOLOGY IN THE VICTORIAN ERA

By 1881, the biologist Wilhelm Roux had given some of the earliest and most tentative hints as to the nature of theoretical, as opposed to observational, embryology. Born in 1850, he was another distinguished pupil of Haeckel, so his fabled contributions to embryology were hardly a shock to anyone. Not all of his puttering immediately bettered the science, however. Indeed, in some sense he set it back for a while, though for understandable reasons. For Roux suspected that he had found some evidence to support the ancient "preformation" concept.

His interest in this folk tale began in 1881 with the idea of "functional adaption," the axiom that virtually every part of the animal body was flexible enough to undergo whatever changes were necessary to perform a function the body charged it to perform. Like any able scientist he did attempt to test the principle by observing the development of several embryos. Roux was not content merely to describe, say, the stages of a developing chick. Instead, he proposed that one could split its development into two categories, the "embryonic" period and the period of "functional development." So in many ways he viewed the developing embryo as something of a "machine," with all of its parts integrally related to the other parts.

To test his assumption he devised, in 1888, his most renowned experiment: First, using a needle, he killed one of the first two cells that issued from a frog's egg just starting to divide. Since the egg ultimately formed only half an embryo, Roux concluded, erroneously, that the theoretical "machine" had already existed in the cell before division, containing a "blueprint" of the whole organism. He further argued that only half of the machine existed in each daughter cell, causing the half-embryo to form. For this machine to survive cell division—a process that was incompletely understood at the time—it had to somehow disassemble itself when the cell divided and "recombine" in the final products—an ingenious idea for the day.

However, in more careful experiments in 1891, the biologist Hertwig, initially a proponent of Roux's teachings, showed that each daughter cell could mature into an entire embryo. Consequently, Roux's theoretical machine did not exist preformed as he had contended, and Roux's hypothesis had to be jettisoned.

Although his beliefs are chiefly of historical concern now, Roux was the first scientist to at least try to go beyond mere description and actually dare postulate hypothetical mechanisms to explain embryonic development. For that reason history has rightly named him one of the founders of the modern science of embryology. By 1896, the biologist Edmund Wilson of Bryn Mawr and Columbia had looked still more deeply into the recondite science of embryology when he surveyed the process whereby eggs develop into embryos, which he discusses in his book, *The Cell in Development and Heredity*. He also first realized the existence of both X and Y chromosomes, although he did not fully grasp their significance in sex determination. Beyond this, he also actively promoted Mendel's ideas.

EMBRYOLOGY IN PLANTS

No one was neglecting plant embryology either, even if science was confessedly paying more attention to animal embryology. In the field of plant reproduction a find of the first importance transpired when the entirely self-taught German botanist Wilhelm Hofmeister discovered the phenomenon of "alternation of generations" in nonflowering plants such as mosses, horsetails, ferns, and liverworts. A primitive kind of reproduction, in these plants a sexual generation would alternate with an asexual one—a process no one before Hofmeister had even conceived of. In 1847, Hofmeister gave a painstaking account of the development of a fertilized ovule into a plant embryo, and also traced the entire life history of ferns.

However, the Danish biologist John Steenstrup (1813–1897) quite properly receives credit for spotting the phenomenon of alternation of generations in animals, and, in particular, in jellyfish. Steenstrup was born in Denmark, the son of a clergyman. His initial training was at the university in Copenhagen where he eventually became professor of biology.

He studied, early in his career, both botany and zoology, exploring, among other things, the then poorly understood sphaghum mosses. In zoology he found that jellyfish offspring, although they looked like their grandparents, looked nothing like their parents. It was at this point that Steenstrup began to realize that there was indeed such a phenomenon as alternation of generations. He examined it in a multitude of worms, in medusae, a stage in the life cycle of jellyfish, and in *Salpa*, a free-swimming planktonic invertebrate with a barrel-shaped body. He noticed in these forms that the adolescent, or "nurse," stage, as he called it, alternates with a more sexually mature second stage. More secrets about this phenomenon revealed themselves in 1894 in the investigations of Eduard Strasburger. Born in 1844, Strasburger was a precocious undergraduate who ultimately became professor of botany at Bonn. Undeniably, his most noted effort was his analysis of mitosis and its various stages, which he both described and illustrated in massive detail. He maintained that in the so-called non-flowering plants—mosses, ferns, and others—the spore-bearing generation has paired chromosomes, while the sexual generation has only a single chromosome of each type. This is analogous to meiosis in animal reproduction, whereby the chromosome number is reduced by half in sperm and egg cells. By 1879, he had proven that the nucleus of a cell could come only from a preexisting nucleus.

CYTOLOGY, SCHLEIDEN, AND SCHWANN

Closely related to the study of the cell nucleus was, of course, the field of cytology, or the study of the entire cell. The biologist Robert Brown made perhaps the first principal advance in this field when he discovered the cell nucleus in 1831. A few years afterward, the Czech biologist Jan Purkinje would discover that both plant and animal substances were composed of cells, although neither he nor anyone else yet appreciated the myriad differences that existed between the cells.

Very soon, one of the legends in this field would appear on the biological landscape—Theodor Schwann of Germany. He studied biology at both the universities of Würzburg and Berlin, conducting innumerable microscopic inquiries into both animal and plant tissue. After several years he became professor of biology at the University of Louvain in Belgium. His early studies were in physiology. In 1836, he showed that an enzyme, pepsin, aids in the process of digestion in animal digestive tracts. But, by far his greatest work was the in the field of cytology.

In 1839, Schwann's tract "Microscopical Researches on the Similarity in Structure and Growth of Animals and Plants" appeared. Here he gave a complete histological analysis of both plants and animals. He discusses the universal cellular nature of all living organisms in the opening passages of the above work:

> During development, also, these cells manifest phenomena analogous to those of plants. The great barrier between the animal and vegetable kingdoms, viz. diversity of ultimate structure, thus vanishes. Cells, cell-membrane, cell-contents, nuclei, in the former are analogous to the parts with similar names in plants.[35]

Schwann then further refined the cell hypothesis by disclosing that, in fact, it was false that living organisms were composed entirely of cells. Quite serendipitously he found that bone and connective tissue comprised few cells, but greater amounts of cell by-products. He amended the cell thesis to claim that "all living things are composed of cells and cell products."

Schwann also coined the term "cytoblastema" to apply to what is now routinely called "protoplasm," the living material within cells, as well as the word "metabolism" to apply to all biological processes going on within the cell. With these concepts, he firmly entrenched cytology as a unique branch of biology.

PURKINJE'S CONTRIBUTIONS TO CYTOLOGY

As of that time, however, scientists understood exceedingly little of the internal parts of the cell. Soon, Schwann would prove that yeast is actually composed of a myriad of living microscopic organisms, although only Pasteur accepted Schwann's premise. At least a small step toward understanding the internal workings of a cell occurred in 1839, when Purkinje began a probe of the gellike internal content of a cell, which he dubbed "protoplasm."

J. E. Purkinje was born in 1787 in Lobkositz in Bohemia. His father worked in the legal profession, and, though he died prematurely, his father had saved enough so that with his mother's backing, Purkinje was able to matriculate at a nearby theological school. For some years he fully intended to become a cleric, but he shortly began having theological doubts and began studying medicine and philosophy at Prague.

Purkinje has an interesting place in the story of biology. For one thing, he built his own plant physiology laboratory at his home in 1824—the first time anyone had done anything like that. Realizing the importance of his experiments, the Prussian government eventually gave him his own institute with extensive laboratory facilities. His preoccupations ranged far and wide, from cytology to physiology, medicine and plant physiology. After completing his university education in science he joined the faculty of the university at Breslau in 1823 and became professor of physiology upon a decree by King Frederick William III of Prussia. Unlike many noted biologists, he also earned high marks as a teacher.

Purkinje, nevertheless, understood little of the traits and function of protoplasm—an enigma that would dissolve only with the findings of the German biologist Hugo von Mohl in the mid-nineteenth century. What Purkinje left to biology was, though excellent, a rather unsystematic collection of data and experimental inferences. Still, he was the first to show that animal as well as plant tissue is composed of cells and introduced the term "protoplasm" into the scientific lexicon. (Von Mohl would later disclose that protoplasm is the main living element in the cell and the medium in which most metabolic activities take place. He, too, claimed to have invented the term "protoplasm".) Also, Purkinje was the first to describe accurately the movement of cilia in vertebrates. Finally, according to his biographer, in 1835, J. E. Purkinje first showed that the mass of the nervous system consists of both nerve fibers and nerve cells.

SCHLEIDEN

During the same era, another giant in cytology would surface, the eminent Matthias Schleiden. He early on displayed an acute interest in biology in the fields around Hamburg, the town in which he was born in 1804. He was the scion of a well-known physician and biologist. Schleiden originally trained for the bar but steadily lost interest in the legal profession and returned to graduate school to study medicine. At this stage his major interest was botany, and he subsequently became professor of botany at the University of Jena. By 1837, building on Robert Brown's heritage, he began a series of systematic reflections on the development and organization of plants. He recognized, in 1838, that cells were the basic unit of composition in plants, something that Brown had failed to appreciate. Even so, Schleiden comments respectfully on Brown's views in his essay "On Phytogenesis":

> Robert Brown . . . with his comprehensive native genius, first realized the importance of a phenomenon which, though previously observed, had remained neglected. In many cells in the outer layers of Orchids he found an opaque spot, named by him the *nucleus of the cell.* He traced this phenomenon in the earlier stages of the pollen-cells. . . . The constant presence of this nucleus in the cells of very young embryos struck me also [and] . . . led to the thought that it must hold some close relation to the development of the cell itself.[36]

Schleiden's most compelling contribution, therefore, was to fully appreciate the significance of the cell's nucleus, or the "cytoblast," as he later called it. He started by looking at cells in embryonic tissue, making careful notes on their nuclei. This led to his famous disclosure of still another entity, the "nucleolus"—a diminutive object within the nucleus, which biologists now know is critically important in protein manufacture. There were errors in his studies; he believed incorrectly, to suggest one example, that as the development of the cell progresses, the nucleus vanishes, which, of course, is not the case. Nevertheless, his appreciation of the significance of the nucleus makes him worthy of a place among the immortals of biology.

Oddly, Schleiden's attainments received a mixed reaction during his lifetime. Possibly because he cloaked his discussions in obscure, almost mystical philosophical terminology, coupled with the fact that he, like his countryman Justus von Liebig, could be quite abusive toward anyone whose opinions he disliked, he was not a popular figure in his day.

Strangely, no one at this point understood the relation of plant to animal cell protoplasm. In his essay "The Threefold Unity of Life," for example, T. H. Huxley was able to say only:

> Under these circumstances it may well be asked, how is one mass of non-nucleated protoplasm to be distinguished from another. Why call one "plant" and the other "animal?" The only reply is that, so far as form is concerned, plants and animals are not separable, and that, in many cases, it is a mere matter of convention whether we call a given organism an animal or a plant.[37]

This all changed with the German botanist Julius Cohn, who, in 1850, proved beyond any doubt that there was no important difference between the protoplasm of animal and plant cells, though he did not deny that plant and animal cells differed in many other ways. The history of cell theory is a topic of enormous scope by itself. Ever since ancient times biologists had regarded the tissues of both plants and animals as having no meaningful structure or organization. Many felt that below the level of what was possible to see with the naked eye was nothing of interest. This began to change in the seventeenth century with the work of Robert Hooke, as noted earlier.

BOTANICAL CYTOLOGY TAKES A SURGE FORWARD

Then, a curious reversal took place in biological progress. Zoological knowledge had previously moved faster than botanical science, maybe because man felt more of a kinship with animals or that a key to human health lay in first demystifying animal biology. Whatever the case, the usual pattern reversed itself in cellular investigations. The chief reason for this was that plant cells, since they alone have cell walls, were easier to see than animal cells, though both have cell membranes. Despite subsequent quests by Robert Brown and Purkinje, science made only halting steps forward in cell theory, in large part because exceptionally few really believed in the existence of cells.

VON NÄGELI

Toward the middle of the nineteenth century, Schleiden would publish his resplendent *Principles of Scientific Botany.* That biology today recognizes both Schleiden and Schwann as the founders of cell doctrine is absolutely proper. It is also true that later scientists would make emendations and additions to the basic theory. The German biologist Karl von Nägeli determined, in 1844, that Schwann was wrong in saying that new cells "budded" off of already extant cells. Credibly, Nägeli had the root idea of mitosis, or cell division, though it would be years before science disentangled the details of that process.

Beyond this, Nägeli conducted extensive chemical tests of cells and demonstrated that the nucleus consisted mainly of nitrogen compounds while the cell wall consisted primarily of carbohydrates. Nägeli also applied some of Darwin's suggestions to plant life when he argued that plants, as well as animals, "compete" with one another in myriad ways.

VON SIEBOLD AND CELL THEORY IN ZOOLOGY

Another contributor to the cell postulate was Karl von Siebold, the son of a professor at Würzburg. Born in 1804, the year of the passing of the eminent philosopher Immanuel Kant, he began his instruction at the University of Berlin, later getting his doctoral degree at the University of Göttingen. But again the familiar pattern appeared. He soon gravitated toward "pure" biology rather than medicine. He began looking at primitive marine life and finally grew sufficiently notable that the University of Erlangen appointed him professor, as did the university at Breslau, to shoulder Purkinje's duties. He passed away in 1885 after many years of failing health.

He collaborated with his close friend, Friedreich Stannius, a German biologist of the University of Berlin, and the two produced their great *Comparative Anatomy,* dividing the labor between them. Siebold discussed invertebrate zoology and Stannius discussed vertebrate zoology. This work was a thorough and systematic discussion of almost every form of life known at the time. Their method was to focus on a specific organ and then compare it with similar organs within that group of animals. They had used the microscope extensively in their research and were able to apply cell precepts to many kinds of life-forms, such as single-celled animals like protozoa and paramecia, taking special notice of the cellular structure of all of these primitive forms.

Siebold also made invaluable contributions to the study of parasitic worms and the theory of parasitism. He unconditionally rejected, for instance, the notion that parasites, such as intestinal worms, could arise through "spontaneous generation." His proof of this was simplicity itself. Meticulous studies revealed eggs in these worms. Accordingly, he considered it probable that parasites could reproduce the same way many other forms of life reproduced.

MIESCHER AND CELL THEORY

Still, the cell dogged biologists. To date the cell had yielded precious few of its mysteries. So it was that biologists pressed on. Johann Miescher, for instance, found an acidic, phosphorus substance "nuclein," that part of the cell that contains the reproductive element, in sperm cells. Oddly, many of his fellow scientists discouraged him from publishing these results because they ran so contrary to prevailing dogma that there was a risk the scientific community would ostracize him. Wilhelm Roux, for example, was known to be an advocate of the old preformationist doctrine and, about decade after Miescher's work, would perform his most famous experiment intended to show that preformation was the correct account of reproduction.

VON MOHL, BOTANY, AND CELL THEORY

More insights emerged from the laboratory of the German botanist Hugo von Mohl. Born in 1805 at Stuttgart, he was the heir of an influential and politically active family. He obtained a medical degree and forthwith became professor of physiology at the University of Bern, finally moving to Tübingen as professor of botany. In deportment, he was apparently a man of stately severity and quiet temperament. He never married, and his life was entirely given over to the pursuit of science.

By this period in the maturation of biology, scientists had already begun to understand the living cell in considerable detail. Robert Brown, as already noted, demonstrated in 1831 that each cell had a nucleus, which he had discovered in his perusal of orchids. Schleiden, too, had done thoughtful experimentation in this field, as discussed earlier.

Von Mohl's approach was methodical and thorough. As is so often the case, this was both a strength and a hindrance. Because he so preoccupied himself with his "hands-on" indulgences, he never came up with a biological theory of any great importance. However, in his *Principles of the Anatomy and Physiology of the Vegetable Cell,* von Mohl argued for extending the cell hypothesis to plants. He argued that cells exist in algae and higher plants. He revealed further that these cells arise from previously existing cells through a variety of cell division he called "partitioning." He proved that there was a cellular structure in bark, spiral vessels, leaves, flowers, and many other parts of plants. He also devised some helpful, albeit minor, improvements on the microscope lens. He details all of this in his book *Micrographia.*

VIRCHOW, CELL THEORY, AND DISEASE

In the mid-nineteenth century, the German Rudolf Virchow appeared on the scientific landscape. He was born in 1821 in Pomerania, the son of a merchant. As a youth he mastered medicine and pathology under Müller. He completed his internship at the Charite's Hospital in Berlin and was already publishing a great deal on pathology, much of which culminated his great humanitarian achievements in disease noted earlier. Soon he opened his own laboratory for the study of both pathology and cytology. He began conducting extremely accurate observations of a variety of animal tissues, including connective tissue and bone. Given his early interest in pathology, it was understandable that he studied such tissues in both healthy and ill animals. In his paper "On the Evolution of Cancer," he argued convincingly that disease always followed deviant cell behavior. Later, in 1858, he formulated his concept of "cellular pathology." Accepting the by now well-founded belief that all cells arise from cells, Virchow applied cell tenets to pathology. He believed, rightly enough, that cells can alter their normal behavior and attack the very organism of which they are a part. In a word, disease occurs when the body attacks itself—an idea that is remarkably close to the truth in many cases. Naturally, there were other causes of illness, such as bacteria and viruses, causes that others, such as Pasteur and the French bacteriologist Charles Chill-le-Vignoble, had already begun to explore.

Virchow was a versatile scientist as well. Beyond his cell studies he surveyed an assortment of major and minor complaints and the significance of infection in causing human ailments, as noted earlier. Ultimately, the scientific community of Germany rewarded him with a professorship at the University of Berlin. They also appointed him editor of the prestigious *Archives for Anatomical and Physiological Pathology,* a post he held for over fifty-five years.

BERNARD, DIGESTION, AND PHYSIOLOGY

Creative insights came from the studies of the Frenchman Claude Bernard, the peerless Victorian physician. Born in 1813 at Saint-Julien in France, he became one of the most distinguished pupils of the French surgeon and anatomist Magendie. His family were indigent peasants, and for some years it looked as if he might never acquire an education. He began reading omnivorously in science while still a child. His mentor was a local administrator who was probably more proficient in theology than science but who quickly recognized Bernard's talent. In 1832, his family was virtually impoverished, and he had to move to Lyon to continue his education as well as to earn some money. There he apprenticed to a knowledgeable local pharmacist named M. Millet. Although much of the job was witless drudgery, including sweeping and cleaning, it did give him a chance to learn a little about medicine firsthand.

At a nearby veterinary school he regularly sat in on classes, and, after much coaxing on his part, school officials allowed him to sit in on laboratory demonstrations and vivisections. When his skills had improved to the point that his pharmacist employer allowed him to mix uncomplicated preparations, Bernard exclaimed, "Now I could make something; I was a man."

For a while, Bernard tried to make a living as a writer, but, as is usually the case, he failed. But when Magendie met him, he recognized Bernard's extraordinary ability and began to tutor him. Together they performed innumerable dissections, and Bernard stayed with Magendie for several years. The relationship was a fortunate one for Bernard. Magendie, born in 1785 in Bordeaux, came from an powerful medical family. Again, according to custom, Magendie received his initial education in medicine, though he ultimately became a professor at the Collège de France. He became an enormously successful scholar and teacher. Although he caused some revulsion among colleagues because of his willingness to perform vivisections, his contributions to the technique of experimentation were sizable. Among other things he contributed greatly to biological erudition about the circulatory system and respiration.

His pupil Bernard, on the other hand, hardly fared as well, at least not initially. Finally, despite his reputation as a pitiable speaker, Bernard did get a job as professor of medicine at the prestigious Collège de France, even replacing his teacher Magendie. Among his many books is the *Introduction to the Study of Experimental Medicine,* which first articulated some of the most basic laws of physiology and medicine.

Bernard's Physiology and Anatomy of the Nervous System

In the mid-nineteenth century, Bernard found that control of body temperature in warm-blooded animals rests in the nervous system. He pursued this theme again in his *Lessons in Experimental Psychology,* in which he proposed the thesis that despite changes in temperature in its environment, an animal can sustain a stable body temperature, a concept now called "homeostasis." Later he learned more about energy in the body when he realized that the liver stores glucose, which the body converts to glycogen and uses for energy when the body needs it. One of his other notable discoveries involved a detailed description of the vertebrate ear, including his discovery of the cranial nerve that extended from the brain to the front of the tongue—the so-called chorda tympani nerve. Because of his initial training with Magendie, Bernard had acquired considerable virtuosity in dissection and attained his goal of exhaustively tracing the paths of even the longest nerves in the body. Relying on this ability, he traced the long vagus nerve, which travels from the throat through the torso and he mapped out the connections between the vagus nerve and the spinal accessory nerve. Searching relentlessly, he began to discover the function of the "vasomotor" nerves, the nerves that control both constriction and dilation of blood vessels. He also found a collection of chemicals that were critical in controlling dilation and constriction and, therefore, blood pressure. He was one of the first physiologists to pay close attention to the scientific method. In his words,

> I believe, in a word, that the true scientific method confines the mind without suffocating it, leaves it as far as possible face to face with itself, and guides it. . . . Science goes forward only through new ideas and through creative or original power of thought.[38]

Bernard on Digestion

Bernard carried out some profound investigations of digestion as well. In the late 1830s, he began looking at gastric juice and its function in digestion. Again relying on his skills in dissection, he traced the route of food from ingestion through the various parts of the digestive tract. Beyond this he analyzed the unique chemical processes involved in metabolizing food in vertebrates, even discovering the enzyme steapsin, as well as a number of other digestive enzymes. Long after this, the biologist Ernst Hoppe-Seyler would add to physiological findings about enzyme action, as well as about glucose, when he discovered invertase, an enzyme that accelerates the transformation of sucrose into glucose and fructose. The German physiologist Wilhelm Friedrich would add, as well, another enzyme to what would eventually become an enormous number. He found trypsin in the pancreas; in fact, he also coined the term "enzyme," to refer to any chemical substance that functions just as well outside a living cell as within it. Following a similar path, in 1897 the German physiologist Eduard Buchner of the University of Bern quite serendipitously discovered zymase—an enzyme which catalyzes the conversion of sugar into alcohol. Zymase was the first enzyme ever isolated. Also, Buchner showed that enzymes cause alcohol to ferment.

Somewhat after Hoppe-Seyler's research, the German physiologist Karl von Voit proved beyond any doubt that only certain components of food would provide a living organism with energy. Proteins, to cite an example, apparently did not produce energy since they metabolized at the same rate whether the body was at rest or engaging in strenuous activity.

BEAUMONT ON DIGESTION

Around 1833, the American surgeon William Beaumont began to understand the role of gastric acids in digestion—principally by the admittedly grisly method of studying a man who had been so gravely maimed by gunshot that a permanent hole had remained in his stomach, even after he had otherwise recovered. In 1888, the Russian biologist Ivan Pavlov broke still more new ground. Pavlov, like so many prominent biologists, first studied medicine. By 1891 he had began looking into digestion at the St. Petersburg Institute of Experimental Medicine. In the theory he devised, digestion consists of three phases: nervous, pyloric, and intestinal, though physiologists now recognize that digestion is more complex than this. However, Pavlov would eventually achieve world renown for his stimulus–response theory, or the doctrine of the "conditioned reflex." He demonstrated that when a stimulus caused the secretion of gastric juices, the nervous system controlled the entire reaction.

Shortly after the turn of the century Pavlov conceived the idea of a "conditioned reflex." According to this belief, still favored by an occasional psychologist, if one trains an animal by presenting a stimulus (such as a ringing bell—used in Pavlov's actual trials) at the same time you feed it, then the animal will start to salivate when the stimulus occurs, even if there is no food in front of it. In 1904, this work would earn Pavlov the Nobel prize. In 1927, his succinct summary of these novel concepts would appear as the book *Conditioned Reflexes.*

SUMMARY

All in all, this period saw perhaps the greatest progress in the annals of biology. With the work of Pasteur on disease; Beaumont and

Bernard on digestion; Purkinje, Schleiden, Schwann, Virchow, and von Mohl in cell theory; and Roux in embryology, it is difficult to point to one scientist who towered above the rest. As organizations flourished and the scientific method became more widely used, it is evident that science was preparing for the twentieth century.

CHAPTER 21

Biology in the Twentieth Century

It was, almost certainly, only in the twentieth century that science truly became a profession. It was the era of Einstein and the theory of relativity as well as the era of quantum mechanics and the computer. Germany and Italy would retain their places as the leaders in science for decades to come. Throughout an impressive part of the twentieth century it was *de rigueur* for newly graduated American scientists to head for European capitals to complete their education. Irving Langmuir, Robert Millikan, and Linus Pauling, among others, did precisely this.

With the advent of huge foundations, such as the Rockefeller and Sears foundations, which offered grants to scientists, as well as the opportunity for paid academic posts appearing on a regular basis, science escaped the amateur status it had held throughout most of the nineteenth century and before. Academic journals, too, were proliferating. With the invention of the telephone and the airplane and modernization of printing techniques, travel and communication with other thinkers became much easier. In addition to this, the public was becoming more educated about science in general and biology in particular. Modern biology was becoming more theoretical and even mathematical, rather than merely descriptive and concerned with classification, as in the halcyon days of classical biology. There were other kinds of changes as well. As the twentieth century progressed, one saw less and less of the intrepid "tinkerer" battling alone in his backyard or field. The uniquely twentieth-century research team began to replace the lone individual.

It is by no means evident that all of this was a boon to civilization. This was the era of the Manhattan Project and the atomic bomb, as well as genetic "engineering," with its doomsday potential for creating lethal new scourges. In 1942, the legendary Italian physicist Enrico Fermi created the first controlled chain reaction under the abandoned squash courts at the University of Chicago, the ultimate result of which was used to subdue Japan, thus ending World War II.

Even so, the war would not be without its benefits. Many technological achievements came about primarily through war ventures. Among these were penicillin, DDT, radar, accelerated progress in computer technology, and so forth.

PHYSIOLOGY AT THE TURN OF THE CENTURY

In physiology, the name of the biologist Otto Meyerhof stands out prominently in the dawning decades of the twentieth century. Building on earlier observations by Bernard in 1884, Meyerhof realized that when an animal system is under stress, the body reaches into glycogen "storehouses" in the muscles. It then converts that into lactic acid which, in turn, combines chemically with oxygen to rebuild the "storehouse" of glycogen. For this contribution and for Cambridge physiologist Archibald Hill's related find that a muscle produces heat during exertion, Meyerhof and Hill of England shared the 1922 Nobel prize. Hill, in addition to his fine work on muscle physiology, was an early and most outspoken champion of a controlled, disciplined scientific method and, most controversially, of vivisection. Hill explained his findings in an address to the British Medical Association at Manchester:

> I am speaking this evening to those, and the friends and relations of those, who spend their lives in mitigating the results of the experiments which Nature makes upon suffering mankind. Some of these experiments

> involve bacterial infection and, therefore, are . . . to some degree avoidable. . . . Nature, however, is an extremely bad experimenter; she is, in fact, the imaginary vivisector of anti-vivisectionist literature whose experiments are made without mercy and without apparent cause. So badly and so casually performed are they, so ill-controlled, that it is often impossible for you to reason accurate from them at all. . . . The only way in which the confusion may be avoided is by comparing the results of Nature's casual, random and complex experiments on human beings with those of simple, properly controlled experiments on living animals.[39]

Meyerhof would, after barely more than a decade, write his fine treatise *The Chemical Dynamics of Life,* which summarized most of his own inquiries in that field.

By 1904, the Englishman Arthur Harden of the Lister Institute had discovered the first coenzyme, a factor that allows a polypeptide chain—the strings of amino acids constituting a protein molecule—to act as an enzyme. This he did as part of a study of the fermentation enzymes.

ENDOCRINOLOGY

Another field that was progressing in a prolonged rise to importance and independence was endocrinology, or the science of hormones. An interesting early account of hormones is offered by the British biochemist Philip Eggleston of the University of Edinburgh in his essay "What Can the Chemist Tell Us about the Living Cell?"

> Hormones are another set of materials that interest the biochemist a lot today. A hormone is a chemical compound—often a fairly simple one—which is known to be manufactured in the body in special organs called glands. These hormones travel all over the body, dissolved in the blood, and produce the most remarkable effects . . .[40]

Arguably, the field of endocrinology began to take on its own identity in 1909, when the Italian physician Nicole Pende coined the term "endocrinology." The next luminary in this still-new field was undoubtedly the brilliant Philip Edward Smith of South Dakota, who spent most of his career studying the structure and function of the pituitary gland. Among other things, he proved, in 1884, that this gland was a sort of "master switch"—removing it will shut down the endocrine system entirely. Beyond this, he developed new surgical techniques for pituitary surgery. Smith's discoveries opened up the floodgates in pituitary scholarship, with innumerable books and articles being written about it. One of the first and most preeminent of these was the American physician Harvey Cushing's *The Pituitary Body and Its Disorders,* which became widely known very rapidly. His book was, at the time, the best and most comprehensive summary of pituitary functioning.

Cushing was an American neurosurgeon and one of the early movers in this field. In the first decade of the new century he collaborated with the Swiss surgeon E. T. Cohen and then joined C. S. Sherrington at Oxford University. He gave fresh vitality to brain surgery when he became one of the first to perform operations lasting up to nine hours. His studies of the pituitary gland led to a greater comprehension of what medicine today calls Cushing's syndrome, a malady of the pituitary arising from a tumor.

Soon, the English physiologist William Bayliss made his mark in this field. Bayliss was the son of an iron manufacturer and was educated at University College in London and at Oxford. Among his accomplishments are a study, with his colleague Ernest Starling, of the hormone secretin, which causes the pancreas to secrete digestive juices. They published these results in their classic paper, "The Mechanism of Pancreatic Secretion," which appeared in the *Journal of Physiology* in 1902. Later in life, Bayliss published *Principles of General Physiology* as well as *The Nature of Enzyme Action,* which together offered the most sweeping description of the mechanisms of endocrine activity as well as most other physiological actions in the body up to that point.

A few years later, in 1913, the German-American biochemist Leonor Michaelis of the University of Berlin and his team constructed a mathematical equation that explained and predicted the rate at which enzymes speed up any kind of chemical activity, which soon became known as the Michaelis–Menten equation. (In a more domestic vein, his research on keratin soon led to the development of home permanents!)

SUMNER AND UREASE

After 1913 there was something of a drought in the quest for the truth about enzymes. It would not be until 1926 that the Massachusetts biologist James Sumner would, by a process of crystallization, isolate the enzyme urease—the first enzyme anyone had ever found using crystallization techniques. Sumner received his degree in chemistry in 1910 from Harvard, ultimately becoming professor of chemistry at the Cornell Medical School. It was at Cornell that he began his investigations of enzymes. He attacked the problem of isolating urease, an enzyme that speeds up the conversion of urea into ammonium carbonate, among other things. The biologist J. H. Northrop fully confirmed Sumner's find in 1930. Northrop also proved that Sumner was right in his speculation that urease was a protein as well, as in fact are all enzymes. In 1905, Lister Institute chemist Arthur Harden discovered biochemical "catalysts," or compounds that the body creates when converting one material into another during meta bolic processes. Such catalysts, while their production is not the body's ultimate aim, do serve to assist and speed up the production of the "target" element. Specifically, he showed that phosphates accelerate the rate of fermentation.

The biochemist D. W. Ewer of Rhodes University in England, in his lecture "What Are Enzymes and Why Are They So Important?" describes catalysts and enzymes as follows:

> This acceleration of chemical reactions which normally proceed slowly is a well-known phenomenon in chemistry and has great industrial importance. The chemists call such substances *catalysts*. Enzymes are catalysts elaborated by living organisms. There is an enormous variety of enzymes in an animal's body and each is capable of accelerating only one type of chemical reaction.[41]

By the 1940s, this insight had led to the work of Britton Chance. This University of Pennsylvania biophysicist would embark on his epochal attacks on the function of mitrochondria and the enzyme peroxidase, an enzyme that catalyzes the breakdown of peroxide compounds in the body, thereby facilitating oxidation processes. This discovery, in turn, led to the theory that enzymes function by actually combining temporarily with the constituents they catalyze. A versatile intellectual, Chance also contributed to the development of automatic steering as used in ships, for example, and to better bombsights.

THE ANATOMY OF THE EAR AND THE BRAIN

By 1914, anatomy and physiology had advanced far enough for the Swedish-Hungarian physician Robert Bárány to acquire basically a complete command of both the anatomy and function, as well as infirmities, of the ear. He developed, for example, the "indication test" for studying the relationship of the brain and spinal cord to the ear. For all of this work he received the Nobel prize.

One of the earliest studies in electrical stimulation of the brain was in 1870 when the Berlin scientists Gustave Fritsch, a student of insanity, and Julius Hitzig began their famed experiments on the brain. They found that it was feasible to "map" the cerebral cortex to correlate different regions of the brain with diverse kinds of biological activity. For example, they showed that the control of respiration is located in the "lower" brain, or brain stem, and that stimulation of the brain causes muscles to contract.

Insight into muscle physiology took a surge forward in 1913, when the British biologist Archibald Hill found that muscle cells require oxygen immediately after any contraction—not while they contract, as many had previously supposed. Closely related to this and doubtless even more pacesetting was the experimentation of the German biochemist Fritz Albert Lipmann of the Kaiser Wilhelm Institute, who first began to understand the role of adenosine triphosphate, or ATP, in muscle contraction and energy production generally. As Lipmann pointed out, this molecule is a "storehouse" of energy which an individual cell can use as circumstances require. As it turned out, this was not all there was to oxygen metabolism in cells. In 1929, Lipmann actually separated out ATP from muscle cells. He also discovered coenzyme A and probed the biosynthesis of proteins. In 1921, the biochemist Sir Frederick Hopkins of Cambridge University discovered the compound glutathione. Consisting of three separate building blocks of protein, he proved that without this substance, cells could not utilize oxygen, no matter how much was immediately available. He also studied the role of lactic acid in muscle contraction. Then, in 1923, the German biologist Otto Heinrich Warburg of the Kaiser-Wilhelm Institute first showed that cancer cells can grow without oxygen and that they derive energy from lactic acid. The Nobel Committee awarded him, in 1931, its highest honor for these findings. Not long afterwards, the German physiologist Max Ruber proved that the ultimate source of energy for the body lay in carbohydrates and fats.

A closely related find was the discovery of cytochrome, a pigment, in yeast cells, by the Russian biologist David Keilen of Cambridge University. This enzyme also turned out to play a critical role in cellular respiration. Another important step forward occurred in 1929, when Sir Arthur Harden of the Lister Institute and Hans von Euler-Chelpin of Sweden captured the Nobel prize for clarifying the process by which enzymes cause sugar to ferment. In particular, Harden showed that inorganic phosphates will accelerate fermentation.

KREBS AND HIS CYCLES

By the mid-1920s, the legendary biochemist Hans Krebs, who was at the University of Berlin, was beginning to understand metabolism. He received his medical degree in 1925, and then headed to Berlin and Freiburg. Later, during the Nazi regime, he found it necessary to emigrate to England, where he cooperated with F. G. Hopkins at Cambridge. Finally, he became professor of biochemistry at Oxford. Subsequently, in 1932, Krebs outlined the "urea cycle," the process by which mammalian bodies convert ammonia to urea. In 1937, Hans Krebs again distinguished himself with his most famous and justly applauded conceptions. He described what is today called the "Krebs cycle," a sequence of metabolic and biochemical changes which is charged with the production of energy in the cells of all organisms. Also called the "tricarboxylic acid cycle," the Krebs cycle is the process whereby the body oxidizes pyruvic acid into water and carbon dioxide, a complex series of chemical reactions. It goes on in the body constantly as the body metabolizes the fats, proteins, and carbohydrates taken in as food. Krebs also illuminated the so-called glyoxalate cycle, a modification of the Krebs cycle that is active in plant cells. For all of this work, he deservedly received the 1953 Nobel prize with Fritz Lipmann.

THE STORY OF INSULIN

Very soon, still more enzymes were discovered. In 1930, John Northrop of the Rockefeller Institute, building on the enzyme crystallization work of J. B. Sumner, isolated the digestive enzyme pepsin. At this stage, nonetheless, most biologists had a far too naive view of the metabolism of foodstuffs into energy. The prevailing canon held that the body "burned" food in one step and transformed it immediately into energy. Soon, Karl von Voit, a student of Leibig, would show that the sequence of metabolic processes the body uses to convert food into energy is far more complicated than anyone had realized. Voit developed the "basal metabolism" test, which measures the rate of physiological activity of an organism at rest. He also first determined the average amounts of protein used by the body, which led to the discovery of the essential amino acids.

The story of insulin was entering its modern phase by this time. Seminal inquiries into the pancreas, the body's insulin "factory," had begun through the persistence of Oskar Minkowski of the University of Göttingen, who, in 1889, in addition to important studies on acromegalia, discovered the fundamental function of the pancreas after he surgically removed one from a dog. When he saw that the patient's urine attracted flies, he realized that there must be something in the pancreas that metabolizes sugar.

He, along with Joseph von Mering, had found insulin. By 1901, the physiologist Eugene Opie of the Rockefeller Institute had begun to grasp the relationship between insulin and the islets of Langerhans, the endocrine cells in the pancreas that secrete insulin. Opie summarized his work in his 1902 book, *Diseases of the Pancreas*. Step by step, scientists learned more about the disease diabetes mellitus. Soon, in 1902, the British physicians William Bayliss and Ernest Starling (who also proved that chemical coordination in the body did not require nervous activity) discovered the hormone secretin. Emitted from the mucosa of the small intestine, this critical hormone controls the functioning of the pancreas and generally aids the digestive process, although it has no direct relevance to the most common forms of diabetes. After being secreted, secretin enters the blood, causing the production of pancreatic juice as well as bile.

It would not be until much later that Frederick Grant Banting of Ontario and his student Charles Best of Maine, aided by J. B. Collip and J. R. Macleod, would first extract human insulin, using it to experiment on dogs in the hope of finding a weapon against diabetes. These physiologists accomplished the isolation of insulin by using excretions of a pancreas the tissue of which had disintegrated because they had tied off and sealed the pancreatic duct. From this work came the first efficient, large-scale technique for preparing human insulin from an animal pancreas. In 1922, drug companies were able to commit to the manufacture of insulin en masse. For this work, Banting and J. R. Macleod of the University of Toronto won the 1923 Nobel prize. Much later, in 1953, the English biochemist Frederick Sanger of the Medical Research Council would first decode the structure of insulin by revealing the sequence of amino acids.

THE THYROID AND ITS FUNCTION

Thyroid experimentation had an analogous history to that of insulin experimentation. In 1896, the German Eugen Württemberg found "iodothyrin," which was secreted by the thyroid gland and contained the iodine so essential to physiological functioning. Ultimately, this would lead physicians to use iodine to treat endocrine disorders such as goiter. Swiss physician Emil Kocher, a student of Lister, probed the thyroid gland as well, winning the Nobel prize in 1909 for his contributions. A versatile scientist, he did important work in neurosurgery as well. This investigation would turn out to have immense practical benefits as well. In 1910, Major Frank Woodbury of the United States Army Medical Corps, realizing the medicinal possibilities of iodine, began using it as a disinfectant.

A more theoretical phase in iodine explorations came in 1914, when Edwin Kendall of South Norwalk, Connecticut, later to gain more fame with his research on Addison's disease, identified the hormone thyroxin, which the thyroid standardly secrets in a healthy organism. Others immediately added to scientific particulars regarding the thyroid gland. In 1926, University of Toronto biochemist James Collip isolated parathormone, which he realized was a secretion of the parathyroid gland. He then used it to treat tetany successfully. These intrepid observers were the very first to validate the notion that a properly functioning nervous system was definitive beyond any doubt for the coordination of all activities within the animal body.

ADRENALIN AND CORTISONE

Just after the turn of the century the American-Japanese biologist Jokichi Takamine, head chemist of the Imperial Department of Agricultural and Commerce, and Thomas Bell would both discover and synthesize adrenalin. Takamine is also known for devising an efficient system of production of starch-digesting enzymes. Then, in the 1940s, the University of Montreal physician Hans Selye would amplify this discovery by showing that not only did various changes within the body affect hormonal levels, but external, or "exogenous," stress would do so as well. He summarized his work in his 1956 book, *The Stress of Life*. In 1927, endocrinologists were searching for the secrets of this new science in other organs. The biologist Frank Hartman managed to pinpoint cortisone in his reflections on the adrenal glands. He also inferred, correctly, that a lack of this hormone would result in Addison's disease. This disease results from malfunctioning adrenal glands and is characterized by weight loss, darkened skin and sometimes anxiety. By 1929, the Yale anatomist W. W. Swingle and J. J. Pfiffuer had prepared extracts from the adrenal glands and very soon used them successfully in the treatment of Addison's disease. Closely related to this was the research of biochemist Edward Calvin Kendall of the University of Minnesota who meticulously conducted a prolonged study of the adrenal gland. He discovered many hormones secreted by this gland, the most important of which was cortisone, which he also studied as a possible treatment for arthritis and rheumatic fever. For this, Kendall, along with Philip Hench and Tadeusz Reichstein of the University of Basel, would receive the 1950 Nobel prize in medicine or physiology. Within ten years, Reichstein would conclusively demonstrate the structure of this indispensable hormone, as well as probe the structure of vitamin C, steroids, and various sugars.

HORMONES OF SEXUALITY

After the work on adrenalin, the next critical phase involved the sex hormones. Interest in this phase of physiology was not new. The German biochemist Adolf Butenandt had been immersed in this area for many years. He had first begun to study biology and biochemistry at the universities of Marburg and Göttingen and in 1936 became director of the Max Planck Institute in Germany. In this period, he began his experimentation on sex hormones. He isolated estrogen in 1929 and testosterone in 1931. Finally, in 1934, he isolated the first pure sample of the female hormone progesterone. The next logical step was, of course, to analyze the chemical structure of these hormones.

His analysis of sex hormones was not limited to vertebrates; he looked into the sex hormones of insects as well—the pheromones. He attained the Nobel prize in 1939, although his government prevented him from accepting it because of the portentous political turmoil taking place in Germany. By 1935, the University of Zürich chemist Leopold Ruzicka had convincingly deciphered the structure of the male sex hormones testosterone and androsterone, in addition to becoming the first scientist to synthesize musk.

BIOLOGY-RELATED GEOLOGIC WORK

During the middle part of the nineteenth century, epochal investigations were going on in areas far removed from hormone research. At this point a breakthrough occurred in geology that would have the most profound significance for biology. The Scottish astronomer Johann von Lamont of the University of Münich published his *Handbook of Terrestrial Magnetism*. In addition to many other feats, including determining the mass of Uranus and cataloging 34,674 stars, he confirmed that the earth has a magnetic field and that this field transmutes itself in mysterious ways. His own opinion was that such modifications were related to sunspot activities, although this is questionable today. Remarkably, his discovery connects to some seemingly unrelated inquests. One hundred years after Lamont, some zoologists began looking at the "mechanisms" by which animals navigate. In 1960, to give one representative case, Kenneth Norris of UCLA (in addition to his exotic work in *fish* dentistry) found that the bottle-nosed dolphin uses echoes to locate objects in the water, very similar to the way bats use radar. Then, in the 1970s, ornithologists in Ithaca, New York, would start studying a bizarre phenomenon—the peculiar and inexplicable loss of guidance/navigational systems in birds. Through experiments they found that certain birds, such as homing pigeons and migratory birds, although able to pilot their way from one part of the globe to another under ordinary circumstances, lost this ability completely at specific points. In Ithaca again, ornithologists released homing pigeons that could ordinarily return to their "home" at Hornell, New York, without difficulty. But any birds released in Ithaca could not do so. So far, this is unexplained.

Although this is a comparatively minor issue in zoology, especially since there is no single accepted hypothesis of bird navigation, the phenomenon is interesting. The connection with Lamont's research is the fact that one speculation publicized to explain this peculiar behavior in birds holds that it is precisely the magnetic field of the earth that causes such disturbances in navigational ability.

EVOLUTION IN THE TWENTIETH CENTURY

In evolutionary doctrine during the twentieth century, one sees names like Hugo De Vries as well as a variety of significant preliminary developments in genetic theory. In 1895, for example, various groups of biologists showed that chromosomes do not lose their identity when cells divide—precisely what one would expect if chromosomes were the carriers of the critically significant genetic information from one generation to the next. However, one of the earliest problems biologists had to solve was the problem of gaps in parts of Darwin's theory. Despite the comprehensiveness of Darwin's picture, he had no notion as to how alterations in an organism could pass to future generations. That is the inquiry that Mendel at least began to answer. In fact, De Vries was one of the first to recognize the significance of Mendel's studies, which the biological world had neglected for so many years, and he helped reconfirm many of Mendel's findings. Mendel had discovered his laws of genetics roughly simultaneous with Darwin's epic investigations. But while Darwin achieved immediate notoriety, Mendel had to wait forty years. Part of the explanation for this was that he could not locate any backers of his convictions at first, nor even a publisher. Finally, an obscure natural history society agreed to publish his thoughts, though Mendel still did not achieve any great recognition. Even Darwin did not live to really appreciate him, although Mendel's work was critical to establishing Darwin's own ideas.

DE VRIES AND MUTATIONS

Hugo De Vries added to Darwin's hypothesis by introducing the idea of a "mutation," or a radical emendation in the genetic structure of an organism. De Vries acquired a medical degree in Holland, and subsequently taught in Amsterdam. His earliest preoccupations were with problems in genetics or, more accurately, botanical genetics. With his colleague J. von Sachs, he explored the ways in which plants use water and how water functions in plant metabolism. This, in turn, led him to an even more marked obsession with heredity. Hence, De Vries, along with Karl Correns of Germany and Erick Tschermak von Seysenegg of Austria, returned to and revived Mendel's theories through sheer coincidence, at about the same time.

Beyond rediscovering Mendel, De Vries was also interested in a puzzle that had plagued Darwinism since its inception—the fact that the geological evidence suggested that the earth simply had not existed long enough to produce all of the transformations Darwin spoke about. From the point of view of Darwinism, the single most attractive aspect of the mutation hypothesis was that it eliminated this one conspicuous weakness in Darwinian thinking. In Darwin's worldview, the tiny changes he envisioned would take so long that vast periods of time would be required to create a new species. Yet the fossil evidence showed that the earth had not existed long enough to bring about all of the species that currently existed.

De Vries's basic concept, on the other hand, was that evolution materializes from radical transformations in an organism, rather than from small, gradual changes, as Darwin believed. When the mutations enhanced an organism's chances of surviving a given environment, individuals with these mutations would win the struggle for existence. With enough mutations, an entirely new species might ultimately come into being. He further suggested that alterations between generations can occur precipitously, via these mutations. Immediately, other astute scientists also realized that this 1901 "mutation" postulate supported Darwin's scientific opinions. An important aspect of De Vries's view was that mutations were directly observable. As he says in his *Species and Varieties, Their Origin by Mutation,* the plant that remained "constant and distinct from its allies in the garden" was an elementary species.[42]

It is worth saying a bit about Karl Correns since, despite the fact that he deserves equal credit for rediscovering Mendel, he often gets less than his fair share of attention. In the 1890s, he conducted experiments similar to Mendel's at the University of Tübingen. Among the many aspects of heredity he scanned were sex, sterility, leaf variegation, and so forth. The plant he used most in his work was the so-called four-o'clock of the genus *Mirabilis.* Aside from his work at the University of Tübingen, he also held a post at the University of Leipzig, and, ultimately, he was head of the Kaiser Wilhelm Institute in Berlin.

OTHER PIONEERS IN MENDELISM

In 1905, more scientists jumped on the bandwagon of Mendelism. William Bateson of England proved that not all inheritances are independent of one another; some traits or groups of traits must be inherited together. Bateson would subsequently add to the library of genetics when he published his *Mendel's Principles of Heredity,* which applied Mendel's laws to animals other than man. Others also helped, including Cuenot in France and Harvard biologist W. E. Castle. In Europe, the invertebrate zoologist Lucien Cuenot also recognized the importance of Mendel's thoughts. In 1902, he proved that the inherited proportions of characteristics followed Mendel's mathematical ratios in white mice, and by 1903, he had realized that a more complex pattern of inherited colors also follow Mendelian patterns. Cuenot was not content, nevertheless, simply to confirm the conjectures of others. He was an intrepid watcher in his own right and conducted a rather clever series of investigations of cancer in mice.

EARLY STAGES OF DNA AND VIRAL RESEARCH

Simultaneously, W. E. Castle certified that Mendel's laws applied to a great variety of traits in many animals. He paid attention to inheritance in humans, cats, guinea pigs, mice, and so forth. One of the more interesting inherited characteristics he noticed was albinism in humans.

Before long, scientists would start to discover, at ever-deeper levels within the organism, what, exactly, happens when a "mutation" occurs. That, in turn, led to more interest in genes and chromosomes and, soon, to the incomparable discovery of the structure of DNA. Genetics moved ahead even more in 1902, when Walter Sutton of Utica, New York, in his paper "The Chromosomes in Heredity," supported the proposal that paired chromosomes are, in fact, the transmitters of genetic data from one generation to the next. Also, Wendell Stanley, known for his achievements on the tobacco mosaic virus, as mentioned earlier, was another pioneer in DNA experimentation. He acquired a degree in chemistry in 1929, later joining the Rockefeller Institute. Building on the conclusions of J. B. Sumner and J. H. Northrop on enzymes, he began applying many of their techniques and conceptions to the tobacco mosaic virus and, during World War II, the flu virus. In essence, he discovered viruses, those microscopic organisms which are midway between living and nonliving organisms. They reproduce, for example, but only in living cells. In 1946, he received the Nobel prize for his studies on viruses.

In 1937, British botanist Frederick Charles Bawden of the Rothamsted Experimental Station, beyond his sucess in isolating a number of plant viruses, was able to unearth an atypical form of nucleic acid that exists in the tobacco mosaic virus—RNA, or ribonucleic acid. Building on this work, later scientists discovered that it was exactly these nucleic acids that were responsible for the viral activity. By 1939, the Russian biologist Andrei Nikolaevitch Belozersky would discover that bacteria contain both RNA and DNA. As history shows, the tobacco mosaic virus turned out to be of the very greatest importance in the history of biology. For it was the study of this virus in the twentieth century by researchers like Max Delbrück of Vanderbilt University, Watson, and others that would eventually lead to discovery of the structure of the "molecule of life"—DNA.

The story of viruses began in the late nineteenth century. In 1892, the Russian Dmitri Ivanovsky became the first person in the history of science to prove that viruses, heretofore merely a speculation, actually did exist. Using Pasteur's porcelain filter candles first used to trap bacteria, he found that sap from diseased tobacco plants could infect healthy plants. Soon, scientists knew that viruses differed from bacteria and even had some idea of how they differed. One of the things they did not know, however, was the ecological relationship between the two at the time.

Frederick Twort of England and Félix D'Hérelle of Montreal are the key names in this field. Twort was born in Britain in 1877. He studied medicine in London and later served as head of the Brown Institution. From his student days he was drawn to the study of bacteriology, particularly the study of Johne's disease, a chronic and often fatal disease found in farm animals. In 1915, he and Félix D'Hérelle found a remarkable type of virus, one that consumes bacteria. They gave these viruses the name "bacteriophages," since such viruses and their bacteria were related very much as *predator and prey.* Bacteria, however, did not give up easily; in 1968, the Swiss biologist Werner Arber found that bacteria could fight off such "bacteriophages" by producing so-called restriction enzymes that would decapitate the viral DNA. Later, in the late sixties and early seventies, scientists would learn to deliberately manipulate these new enzymes to break new ground in recombinant DNA research.

PENICILLIN, LURIA, DNA, AND BACTERIA

In 1940, the Italian microbiologist Salvador Luria of Columbia University, using an electron microscope, took the first usable photomicrograph of a bacteriophage. By 1945, Luria had discovered that the same mutations occur spontaneously in bacteriophages as in their host, the bacterium itself. From this he theorized that DNA from the bacteriophage had some way or other found its way into the host cell's DNA.

Scientists today realize that this was a critical step in the eventual solution of the structure of DNA. Eventually, of course, all of these theoretical finds resulted in practical applications, though often quite by chance. Chance is often extremely critical in scientific progress. In 1921, for instance, Alexander Fleming had a cold. When he sneezed on some bacteria in a culture dish, the bacteria dissolved. Fleming later realized that he had found the enzyme lysozyme, which is at home in mucous and saliva and acts as an antibiotic. Similar good luck, as well as the knowledge emerging from understanding the nature of DNA, would lead Fleming to penicillin. This find occurred in 1928, while he was studying molds. Fleming reflected on his discovery of penicillin in his Linacre lecture of 1946:

> The story of penicillin has often been told in the last few years. How, in 1928, a mould spore contaminating one of my culture plates at St. Mary's Hospital produced an effect which called for investigation; how I would that this mould—a Penicillium—made its growth a diffusible and very selective antibacterial agent which I christened Penicillin; how this substance, unlike the older antiseptics, killed the bacteria but was non-toxic to animals or to human leucocytes . . .[43]

However, Fleming is not the only revered name associated with penicillin. The German-American chemist Ernst Chain, born in 1906, also did superb work in this area. He left Germany for England in 1933 to study and research at Cambridge. He then teamed up with the Australian physician Sir Howard Florey to begin studying a variety of microorganisms.

Florey's role in introducing the use of penicillin cannot be overestimated, despite the publicity given to Fleming. Born in Australia in 1898, he studied at Oxford under a Rhodes scholarship and eventually became the provost of Queens College, Oxford. His first important work was a study of the enzyme lysozyme mentioned above, which scientists knew could kill bacteria. That led him to further studies in antibacterial substances. In 1939, he and E. B. Chain began to study penicillin, finally proving beyond question that penicillin was a powerful and useful antibacterial drug. Although Fleming had, by this time, already moved on to other studies, Chain and Florey felt that the drug held still more mysteries. Their studies soon led them, after the war, to the development of a wide range of synthetic antibiotics. Florey particularly was the chief mover in persuading pharmaceutical companies to manufacture penicillin for the military. For their work, the world scientific community would honor Chain, Fleming, and Florey with the 1945 Nobel prize.

Knowledge of bacteria and antibiotic agents increased steadily after this. In 1939, for instance, the French virologist René Dubos, a colleague and collaborator with O. Avery, began a search for natural organic chemicals that would destroy each other. He succeeded marvelously. This was the first time anyone had deliberately sought and found antibiotics, since Fleming's discoveries occurred through sheer chance. However, no one began using antibiotics in regular medical practice until the 1940s, mainly because it was impossible to manufacture them in any impressive quantity. Indeed, the name "antibiotic" did not even exist until 1941, when the Russianborn American microbiologist Selman A. Waksman coined that term as a convenient label for anything that killed bacteria while leaving other life-forms alone.

Oddly, no one had yet authored a particularly authoritative book on bacteriology. That changed with the publication, in 1926, of Paul de Kruif's *The Microbe Hunters,* which essentially achieved best-seller status. As their knowledge of bacteria and the mechanisms of antibiotic reactions widened, microbiologists began discovering a host of other antibiotics. In 1943, for example, at Rutgers, Waksman, in a study of the soil microbe *Streptomyces griseus,* found the antibiotic streptomycin. This drug had its broadest application in the treatment of tuberculosis, though it also proved to be effective in other pestilences that did not respond to penicillin. In 1944 the physician Benjamin Duggar of the University of Wisconsin and an authority on plant disease, in a study of *Streptomyces aureofaciens,* discovered aureomycin, the first of the class of antibiotics known as the tetracyclines, which physicians use extensively today and which are effective against staphylococci, rickettsiae, and other invaders.

TECHNOLOGY AIDS BIOLOGY IN THE TWENTIETH CENTURY

The computer turned out to have applications to biology that no one had anticipated. One of the elite and earliest pioneers in computer-assisted biological experimentation was undoubtedly the celebrated Dorothy Hodgkin. She was arguably the first to recognize the scope of the computer in biology. After receiving her degree in chemistry at Oxford University in 1932, she collaborated with the British biochemist J. D. Bernal, eventually returning to Oxford to pursue X-ray crystallography. With X-ray techniques, she thoroughly clarified the structure of penicillin, as well as the structure of vitamin B 12, in 1956. For this, she received the 1964 Nobel prize. Not resting on her laurels, she continued using the computer for research, and in 1972 she solved the structure of insulin.

In 1956, the Chinese-American chemist Choh Hao Li, a collaborator with Frederick Koch, and his team at the University of California followed the lead of Dorothy Hodgkin and used the still-new computer to unravel the structure of ACTH, or adrenocorticotrophic hormone, a hormone secreted by the pituitary gland that stimulates the adrenal gland and which is essential to the health of the adrenal cortex, finding that it consisted of thirty-nine amino acids in a usual polypeptide chain. He also isolated five hormones of the pituitary gland and worked out the architecture of human growth hormone.

Structural mysteries continued to yield before the unrelenting probing of science. In 1960, the English biologist John Cowdery Kendrew, editor of the Journal of Molecular Biology, located the position of every single atom in a molecule of myoglobin—a molecule similar to hemoglobin. Kendrew, like Rosalind Franklin, Linus Pauling, and Richard Tolman, expanded the field of X-ray crystallography, focusing that technique on the structure of the giant biological molecules such as ordinary protein and the nucleic acids RNA and DNA. (Kendrew confirmed Pauling's alpha-helix in this way.) Kendrew got his degree in biology from Cambridge University, joining the biologist Max Perutz at the Cavendish Laboratory in England. Together they tackled chemical and structural problems in biological molecules such as protein, often using proteins in a crystalline state.

Perutz was born in 1914 and mastered chemistry as an undergraduate at the University of Vienna. He later moved to Cambridge University (where he received his doctorate) to join forces with J. D.

Bernal in the use of X-ray crystallography. In 1937, he began, as did many biologists, his own inquiries into the hemoglobin molecule, given its sway over health and sickness. It was comparatively easy to inspect the relatively "photogenic" hemoglobin molecule with X-ray techniques. Not long after this, Perutz joined forces with J. C. Kendrew to begin observing other large biological molecules. Eventually this group would found the world-renowned Medical Research Council Laboratory of Molecular Biology. In 1947, Perutz and Kendrew, then at Cambridge, tackled protein structure with X-ray techniques, focusing on the hemoglobin molecule as well as its near relative, the protein myoglobin, pirated from sperm whales. They had scrupulously probed the structure of both molecules by 1953, thereby capturing the 1962 Nobel prize.

CHEMISTRY IN THE TWENTIETH CENTURY

In chemistry, many pundits began to fully appreciate that the chemical properties of the different elements in the universe were a function of the electronic structure of the various "shells" surrounding the nucleus of the atom. According to the "shell" theory that Irving Langmuir, Mulliken, and G. N. Lewis developed, electrons orbit the nucleus of an atom at fixed distances, just as the various planets orbit the sun—Pluto being further from the sun than earth, for example. By 1916, Lewis had finished his classic tome, *The Atom and the Molecule,* in which he stated his famous "even-number" rule. According to this proposition, electrons exist in compounds in even numbers, almost without exception. Scarcely pausing, the prolific Lewis produced another treatise of considerable distinction when he published his *Thermodynamics and the Free Energy of Chemical Substances,* coauthored with the American Merle Randall, in which he applied some of proposals of thermodynamics elucidated by the American scientist J. Willard Gibbs and others to chemistry.

In the 1930s, Linus Pauling offered the most thorough explanation of chemical bonding to date and distinguished so-called ionic bonding—whereby two atoms bind by exchanging electrons—from covalent bonding—whereby two atoms bond together by *sharing* electrons. In accomplishing this work, he became one of the pioneers in applying the recommendations of the new science of quantum mechanics to the chemical bond. He embodied all of this in his timeless classic, *The Nature of the Chemical Bond,* written in the 1930s. Shortly after this, Pauling again added to the sum total of biological truths when he examined and carefully untangled the "polypeptide chain" hypothesis of protein structure. In this conjecture, he suggested that amino acids hooked together like links in a chain to form the backbone of a protein's structure.

By 1907, the eminent German organic chemist Emil Fischer made his presence known in protein science. Born in 1852, he taught at Erlangen and the University of Berlin. In Berlin he attracted a number of important chemists and successfully encouraged extensive explorations in that field. The most portentous parts of his work had to do with the chemistry of purines, nitrogenous compounds that include the DNA components adenine and guanine. From this work it was a natural step to protein analysis, since the DNA and protein molecules share many common features. Beyond being indispensable to life, both are gigantic organic molecules, and scientists can use the same techniques to study both of them, such as X-ray crystallography. Fischer's primary contributions to protein theory consisted of artificially synthesizing the first protein. This he did by developing a combination of laboratory techniques for "hooking" amino acids together in the polypeptide chain that constitutes the basic structure of protein molecules.

In 1935, William Cumming Rose, of the University of Illinois, after seminal research on pepsin and uric acid metabolism, found threonine, the last of the so-called "essential amino acids." Still, neither the number nor the function of amino acids in nutrition was unequivocally clear at this point. Barely two years after this, Cumming again distinguished himself by proving that of the twenty amino acids known to exist in the protein molecule, less than half were necessary for normal health in humans.

SUMMARY

The twentieth century was, of course, the era of the Manhattan Project and the atomic bomb, relativity and quantum mechanics. This was one of those rare moments in history when physics actually eclipsed biology and chemistry. Still, Meyerhof broke new ground in physiology when he learned that an animal system under stress pulls from its glycogen warehouse. In endocrinology, Harden discovered the first coenzyme, while Cushing did his epochal work on the pituitary gland. Similarly, Bayliss and Starling did their research on secretin while Banting and Best accomplished marvelous things in the study of diabetes and insulin. Kocher soon grabbed the Nobel prize for his work on the thyroid gland. Jokichi Takamine and Thomas Bell would both discover and artificially produce adrenalin. In geology, the physicist Johann von Lamont first showed that the Earth has a magnetic field and that this field varies in bizarre ways.

In evolution, perhaps most significant was the work of De Vries in rescuing Mendel's ideas from oblivion.

Primus inter pares was the monumental work on the structure of DNA. Walter Sutton, for example, showed that chromosomes are paired and do carry genetic information to the next generation. Soon, Italian microbiologist Salvador Luria, using the electron microscope, took the first serviceable photo of a bacteriophage. In medicine, Selman A. Waksman found the antibiotic streptomycin and added to our comprehension of actinomycin and neomycin. One of the very greatest scientists of this period was surely Dorothy Hodgkin, who pioneered the use of the computer in biology. With it, she solved the structure of penicillin, as well as the structure of vitamin B_{12}. In chemistry and protein studies, scientists like Kendrew, Rosalind Franklin, Pauling, and Richard Tolman expanded the field of X-ray crystallography to study the giant molecules of the body. In chemistry proper, we witnessed the evolution of the "shell" theory of chemical bonding with the toils of men like Irving Langmuir, Robert Mulliken, and G. N. Lewis.

CHAPTER 22

T. H. Morgan and the Rise of Genetics

It is inevitable that insights into the elusive secrets of the protein molecule, which is so integral to life, would lead to speculations about the origins of life on earth. Sidney Walter Fox, for instance, now at Miami University of Ohio, would shortly (1912) argue that a sufficiently hot environment eons ago caused amino acids to form polymers, or long chains of organic molecules linked together—the first step in "manufacturing" a total organism. He has even managed to get proteins to coalesce into quasi-living entities he calls 'protenoids.' Polymerization, as it turned out, would have enormously far-reaching practical applications. In 1926, the German Hermann Staudinger of the University of Freiburg synthesized the first plastics, basing this work on his huge knowledge of polymer chemistry. In 1928, building on this work, the German scientists Paul Diels and Kurt Alder clarified the so-called Diels–Alder reaction, which allowed for rapid chemical combining of atoms into molecules. Ultimately, it turned out that this reaction led to a consummately efficient way of making plastics and virtually started the plastics industry all by itself.

THE ORIGINS OF LIFE

Shortly afterwards, the Russian biologist Alexander Oparin, known for his strong rejection of the idea that a living organism is a 'machine,' took up the question of the origins of life in his book *The Origins of Life on Earth.* Harkening back to the crude beginnings of western science in the ruminations of the Greek philosopher Democritus and others, he postulated that life may have evolved solely through random processes. In the oceans—which he termed a biochemical "soup"—molecules had been moving around and changing quite aimlessly until the earliest life-form appeared by happenstance. This idea has many adherents even today.

Oddly, this concept is, in a way, a revival of the antique notion of spontaneous generation. For the theory suggests that given the primordial soup, with the right combination of amino acids and nucleic acids, and perchance a lightning bolt or two, life might in fact have begun "spontaneously." The major difference is that according to what biologists customarily called spontaneous generation, life supposedly began this way all of the time. According to the "soup" suggestion, by contrast, it began this way only once in the immeasurably distant past. In a similar vein, in 1962, the Cornell physicist Carl Sagan used his adeptness in biology to probe issues relating to the emergence of life on earth. He first created a blend of chemicals thought to duplicate the primordial biochemical soup and conditions that might have existed eons ago on earth. By scrutinizing this odd brew, he found the chemical DNA—a strong indication that life might well have emerged out of such circumstances. Presumably, on account of this, humankind was one step closer to counterfeit life, since we now knew the "recipe" for the life-giving soup.

THE NATURE OF MIND

Beyond the matter of the beginnings of life, many biologists were interested also in the character of life, especially the much-pummeled conundrum regarding the nature of the human mind. Conceivably feeling omnipotent on the heels of ongoing psychophysiological research, science went off on a most peculiar, almost "metaphysical" philosophical tangent. The British physician C. S. Sherrington began to make his influence felt in discussions about the nature of the human mind. Sherrington studied biology in both London and Cambridge, later becoming professor of physiology at Liverpool. Finally, Oxford appointed him Waynflete professor of physiology. He devoted his career predominantly to contemplating the brain and nervous system, describing it in encyclopedic detail. In 1904, he published and summarized much of his lifetime work in the book *The Integrative Action of the Nervous System*. It was in 1906 that he urged a division of the nervous system into the "mechanical," "thought," and "mind."

LOEB AND MECHANISM

The eminent Jacques Loeb would follow this work with *The Mechanistic Conception of Life*, where, like so many others, he made it plain that he regarded the venerated theological and Cartesian doctrine of the "soul" as so much excess cargo in the science of life. In this book he tried to explain the very notion of a living organism using solely the principles of chemistry, physics, and biology. While this may be possible for the lower animals, philosophers have mounted strong arguments against this plan. The issues here are complex and go way beyond the scope of this book. However, the general line of attack involves pointing out that our concept of a "person" really has little to do with the brain or nervous system. For example, if a person were sincerely crying out and behaving in a way typical of someone in great pain, then no information about the "state of his brain" could possibly show he was not, in fact, in pain. The main point and the universal problem with physiological analyses of "mind" is that the critical issues are logical, not physiological. The philosopher Ludwig Wittgenstein, along with many others, has discussed these issues in great depth, and the reader is referred to these writings.[44] (On the other hand, I do not mean to imply that the traditional religious conception of the "soul" is necessarily the right approach to such issues either.)

NERVOUS SYSTEM PHYSIOLOGY

E. D. Adrian's research also followed these paths, probing the functioning of the nervous system and nerve cells, or neurons. Adrian was born in 1889 and graduated from Cambridge University in 1908. From his university days, he directed his ardor at neurology, particularly the relationship between muscles and nerve stimulation. During World War I he got the chance to attend to these issues from another angle, via an appraisal of injured tissue. In 1925, he began to contemplate memory and information storage in the brain. It was for this work that he and C. S. Sherrington received the 1932 Nobel prize.

Still, although biologists had done indispensable work in discerning the structure of nerve cells and nerve fibers, they knew less about the chemical and metabolic conditions necessary for transmission of nervous system impulses. In 1904 the eminent Spanish physiologist Santiago Ramón y Cajal had finished writing up his results of many years of research on the brain and nervous system. Following a long-established tradition among scientists in his country, he first began his training in medicine at Madrid, ultimately becoming professor of histology and anatomy there. His work demonstrates conclusively that the ultimate "building blocks" of the nervous system are nerve cells, and the axons and dendrites—fibers that carry nerve impulses from the nerve cells throughout the body—that emanate from them. His classic tome is *Histology of the Nervous System*.

He and Camillo Golgi also proved the existence of the "synapse," a gap between nerve cells, and they also added much to our perceptions of how nerve fibers can regenerate. In 1906, he shared the Nobel prize with the Italian biologist.

Then, in 1914, the physiologist Henry Hallet Dale of England advanced our understanding of nerve impulse transmission still further when he suggested that acetylcholine, a neurotransmitter that acts to stimulate the nervous system, is basic in the transmission of nerve impulses. Although it took many more years, Dale ultimately isolated this chemical in 1929. In his honor, medical science named the 'Dale Reaction,' a test of muscle contractibility. For this breakthrough, Sweden awarded Dale and Loewi the 1935 Nobel prize. Among those who carried this endeavor further in more recent periods was the American husband-and-wife team Mabel and Edward Hokin of McGill. In 1953, they found that acetylcholine causes pancreatic cells to take in phosphorus, which then becomes part of the cell membrane. This was interpreted, somewhat metaphorically, as a mode of "communication" between cells. Phosphorus is of great physiological importance; among other things it is known to be important in converting glycogen into the sugar glucose, which in turn provides the body with energy.

In the 1920s the American physiologists Joseph Erlanger of Washington University, also known for his work in the physiology of the circulatory system, and Herbert Gasser had begun measuring the velocity of the propagation of a single impulse along a nerve fiber using the oscillograph. For this they eventually received the 1944 Nobel prize. In 1921, this biological front advanced still more when the German biologist Otto Loewi of the University at Graz, experimenting with acetyl choline on a frog's heart, found that there existed a group of chemicals that could cause nerves to fire; that is, these chemicals caused nerve cells to produce electrical impulses which then traveled to the next nerve cell, resulting in the general transmission of nerve impulses throughout the body. Science was starting to realize that the functioning of the nervous system had a chemical basis.

Another pioneer in the general area of nervous system physiology was Ulf von Euler of Sweden. Ulf Svante von Euler got his medical degree at the Karolinska Institute in Stockholm, later winning a Rockefeller Fellowship to continue his research in Britain and Germany. It was in London that he began his historic collaboration with Julius Axelrod on the transmission of nerve impulses. Others, including Dale, had already verified the existence of a group of chemicals called neurotransmitters, which acted to move impulses from one nerve cell to another. What they did not yet know was how such chemicals were stored and, in general, what their mechanism of action was. Euler found that one of these transmitters, noradrenaline, is stored in diminutive granules in the neuron. Then, in 1935, Euler discovered the first of a class of chemicals called prostaglandins—a compound similar to hormone that science knew were necessary for proper functioning of the immune and nervous systems. Their search disclosed that prostaglandins caused muscles to contract and blood pressure to rise. For this Axelrod, Euler, and Bernard Katz received the 1970 Nobel prize.

By 1982, the English biologist John Vane and the Swedes Sune Bergstrom and Bengt Samuelsson were conducting similar inquiries.

Soon, they too captured the Nobel prize for their findings on the structure and internal architecture of prostaglandins. Due to complex political dissension within the Nobel committee, Bergstrom was well along in his career before he received this honor. He was born in 1916 and received his medical degree from the Karolinska Institute in Stockholm in 1943. In 1947, he became professor of biochemistry at the University of Lund and began probing the mechanism of blood coagulation. From there he went on to pioneering work on bile acids and cholesterol. His finest gift to biology, for which he won the Nobel Prize, was his research on prostaglandins. He determined that chemically they were merely fatty acids with five-member carbon rings—but they had a wider spectrum of physiological activity than anyone had previously realized. More specifically, they include any kind of unsaturated fatty acids that are critical for the standard contraction of smooth muscles, as in the walls of the intestines. They also control body temperature, electrical activity in the nervous system, and a myriad of other physiological activities.

Future physicians would put these insights to enormously beneficial use with the development of many varieties of drugs to treat depression. As scientists know today, the so-called MAO inhibitors and tricyclic antidepressant drugs function on precisely the chemical basis that Otto Loewi explained. Still another convincing practical advance in therapeutic application came in 1972 when David Janowsky found that "manic–depressive psychoses," a disorder characterized by wildly fluctuating mood swings, befalls a person because of an imbalance in neurotransmitters—in this case adrenergic, or nerve fibers that release epinephrine, and cholinergic, nerve fibers that release acetylcholine. At this point many scientists made an assumption that seemed reasonable. They believed that any particular neuron would contain only one neurotransmitter. In 1977, however, the Scandinavian biologist Tomas Hokfelt found that most neurons actually contain an assorted class of neurotransmitters.

T. H. MORGAN AND THE BEGINNINGS OF GENETIC RESEARCH

The "holy grail" of biology, DNA, was beginning to seduce more and more biologists in this era. Among other pioneers in this area, Phoebus Levene of the Rockefeller Institute, also known for an early theory of DNA structure, found, in 1929, that there was a sugar, deoxyribose, in the DNA molecule which no one had ever seen before. By 1936, progress in this field had reached the point where the Moscow University biochemist Andrei Nikolaevitch Belozerskii was able to isolate pure DNA "in vitro," or in the laboratory. He then went on to break new ground on the effects of antibiotics on DNA content in bacteria. By 1944, the Canadian biologist Ostwald Avery, of the Rockefeller Institute, along with Maclyn McCarthy and Colin MacLeod, proved decisively that DNA is the molecule that carries genetic and hereditary data to the next generation.

Building on this and other feats, Watson and Crick would finally unravel the structure of the DNA molecule—the substance making up the gene, which holds the genetic details in an organism. Roughly simultaneously, Archibald Garrod of St. Bartholomew's Hospital in England found that genes do not all have the same modus operandi: some function merely "defensively." That is, they block biological changes, preventing them from occurring. Garrod was also one of the first to directly study albinism.

Further down the road, the gifted T. H. Morgan of Caltech would undertake a spirited reassessment of William Bateson's past experimentation. He would prove Bateson right and carry his campaign even further, developing the theory of chromosomes as the carriers of genetic information.

Morgan spent most of his professional career at the California Institute of Technology, as well as at Columbia. It was at Columbia that Bateson, the chief advocate of Mendelian genetics at the time, visited Morgan in his laboratory. Although Bateson had had some doubts about the chromosome assumption, this visit all but eradicated them. As he said in a talk before the American Association for the Advancement of Science in Toronto in 1921,

> "For the doubts—which I trust may be pardoned in one who had never seen the marvels of cytology, save as through a glass darkly—cannot as regards the main thesis of *Drosophila* [the fly Morgan used in his experiments], be any longer maintained. . . ."[45]

Morgan was inquisitive almost from boyhood about developmental embryology. From that beginning he turned full time to genetic exploration. His initial work, *The Development of the Frog's Egg*, published in 1887, was perhaps less experimental and more hypothetical and speculative than much of his ensuing writings. He then began a

DR. THOMAS HUNT MORGAN

PRESIDENT OF THE AMERICAN ASSOCIATION FOR THE ADVANCEMENT OF SCIENCE.

T. H. Morgan (1866–1945). (Courtesy of the Library of Congress.)

more experimental approach to embryology and genetics, culminating in a comparatively minor book titled *Regeneration,* which appeared in 1901. Still another preparatory manual was his *Evolution and Adaption.* Published in 1911, this work convincingly showed that Morgan was a rather orthodox Darwinian—which, of course, is what one would expect by this time. Morgan elaborated further on evolution in his 1919 book, *The Physical Basis of Heredity.*

By 1911, he and his colleagues had published the first sustained description of a chromosome, displaying more than 2,000 genes. During the first decade of the twentieth century he discovered the existence of "sex-linked" characteristics in his *Drosophila* experiments. In observing fly mutations, he found, among other things, that if a male fly with white eyes appeared as a mutant in a colony of red-eyed flies, the next generation of flies exhibited white eyes only among male flies. That demonstrated that the male and female chromosomes also carry genetic information. Morgan discusses this in the very beginning of his 1927 paper in *Science,* "Sex-Limited Inheritance in *Drosophila*":

> In a pedigree culture of *Drosophila* which had been running for nearly a year through a considerable number of generations, a male appeared with white eyes. The normal flies have brilliant red dyes. . . ."

After displaying the mathematical tallies of the crosses Morgan concludes:

> *No white-eyed females appeared.* The new character showed itself therefore to be sex limited in the sense that it was transmitted only to the grandsons.[46]

By 1926, Morgan had completed *The Theory of the Gene,* which covered all of his previous accomplishments in the physiology of genetics. It was here that Morgan successfully began to demonstrate how traits passed from parent to offspring. In this picture, he postulated that paired genes in the germ plasm carried the inherited traits. He displayed in some detail how the germ cells contain only half the customary complement of chromosomes which, during fertilization, combine with the other parent's genetic material to produce an organism with the entire complement of chromosomes for that species.

In 1927, still another classic came from Morgan's pen—*Experimental Embryology.* At this point some were starting to conjecture that genetic studies had gone as far as they could due to the limits of the light microscope typical of the times. Prophetically and perhaps heedlessly, new technology would strikingly increase the tempo of genetic inquiry. Epic advances would come from X-ray techniques and the still gestating field of molecular biology. In 1932, yet another classic of Morgan's appeared, called *The Scientific Basis of Evolution.*

Somewhat after this, Morgan opened new windows into the gene when he was able to interpret the separation of certain kinds of inherited characteristics. There are some qualities that are usually linked together during hereditary transmission. A given eye color, for instance, might normally pass on to offspring along with a certain skin color. Occasionally, however, such "pairing" breaks down and possibly only the eye color would appear in the next generation. Morgan suggested that, occasionally, during cell division, a chromosome might fracture into two pieces, thereby explaining the collapse of the "pairing" phenomenon. Inspired by this, Morgan immediately began "mapping" the locations of genes on chromosomes of *Drosophila.* It was no surprise that Morgan captured the 1933 Nobel prize for physiology or medicine. Among numerous other attainments, he discovered that Mendel's laws were roughly dependable, which, in turn, ultimately led to his view that it was the chromosome that carried hereditary knowledge.

PLOUGH'S CONTRIBUTIONS TO GENETICS

Still another previously unknown phenomenon surfaced in 1917. That year, the geneticist Matthew Plough confirmed that during cell division, chromosomes not only can break, but can also rearrange themselves. Today, scientists call this phenomenon "crossing over." That same year, R. C. Punnet's (who worked closely with W. Bateson) *Mendelism* appeared, presenting his widely used diagram of hereditary processes. Genetics was fast becoming practically a uniquely American science. Then came the finding of University of Kansas zoologist Clarence McClung that males of any mammalian species have an X and a Y chromosome, while females have two X chromosomes. McClung then went on to do extensive chromosomal studies of the *Orthoptera* (crickets, e.g.) and other insects.

THE MULLER ERA

The same year Morgan completed his *Experimental Embryology,* another well-known geneticist Hermann J. Muller, found that he could induce mutations by bombarding *Drosophila* with X-rays. Muller described his findings with characteristic modesty at the end of his 1927 paper, "Artificial Transmutation of the Gene":

> In conclusion, the attention of those working along classical genetic lines may be drawn to the opportunity afforded them by the use of X-rays, of creating in their chosen organisms a series of artificial races for use in the study of genetic . . . phenomena.[47]

Muller majored in biology as an undergraduate at Columbia University and then in genetics at Rice, although he spent most of his career at the University of Indiana. Like his Columbia colleague T. H. Morgan, Muller was interested in genetics, especially the factors that might affect mutations. In his mutation work, he found that X-rays would increase the mutation rate by a factor of 150. That finding alone would have solidified Muller's place in the legacy of science, for after this work the progress of mutation theory accelerated considerably.

MULLER'S SOCIAL CAMPAIGNS

Lamentably, like academics such as Pauling and Teller, Muller found the lure of political proselytizing irresistible. For many years he tried to "educate" the public with his controversial convictions about the menace of radiation for future generations. The low point of his political career was a nasty public as well as private debate with Linus Pauling. The litigious Pauling, believing that Muller had libeled him various times in print, even threatened at one point to sue him.

His political activities aside, Muller's contributions to genetics were immense. For these fine efforts, the biological community honored him with the Nobel prize in 1946.

PLANT GENETICS

In 1937, Albert Blakeslee, a relatively unknown botanist from Geneseo, New York, discovered that the chemical compound colchicine, found in the simple crocus plant, could induce mutations by interfering with regular mitotic division in cells; more precisely, colchicine caused chromosomes to divide, while preventing the cell from dividing along with the chromosomes, as is typically the case. The next principal revolution in plant reproduction had to await Watson and Crick's illustrious unraveling of the structure of DNA in the 1950s.

It is important to point out that not all advances in genetics came from fruit flies. In 1928, for instance, the American biologist Fred Griffith of the British Ministry of Health, in experiments with laboratory mice, noticed that the progeny of certain species of bacteria will acquire the qualities of another species of bacteria if grown in a chemical extract from the other species. This told him that chemical compounds transmitted genetic instructions, though no one at this time completely understood what genes or chromosomes were. His findings were soon confirmed in Berlin and at the Rockefeller Institute.

Wilhelm Johanssen was still another biologist who made substantial additions to genetics. Born in Denmark in 1857, his major preoccupation early in his career was plant physiology, although his most acclaimed contributions were to genetics. When his reputation had grown sufficiently, he became the head of the Institute of Plant Physiology at the University of Copenhagen. His classic paper is undeniably the 1896 essay "On Heredity and Variation," where he noted Mendelian laws again in relation to plants, particularly corn and barley. The Danish biologist and geneticist then coined the terms "gene," to mark the basic unit of hereditary tidings, "genotype," as a standard descriptive term for the genetic profile of a given life-form, and "phenotype," to designate those aspects of an organism's outward appearance that were a direct consequence of its genetic makeup.

Johanssen's engrossment in the teaching of biology continued throughout his days. In fact, he produced what is indisputably one of the first trustworthy genetics text, his 1905 classic *Elements of Genetics,* which reflected a substantial amount of his own exertions, but also covered the toils of others fairly comprehensively. The art of teaching biology would advance still further with the talent of the British physiologist Ernest Henry Starling (who also coined the term "hormone"). His *Principles of Human Physiology* would establish itself as the definitive text on physiology for decades to come.

PROGRESS IN STATISTICAL GENETICS: THE HARDY–WEINBERG LAW

The huge advances in genetics notwithstanding, many still did not consider genetics a full-blown science since no thorough mathematical treatment of it existed. In some ways this attitude is surprising since, as early as 1915, R. C. Punnet had already begun to fill in some of these gaps when he wrote his *Mimicry in Butterflies*. He later reminisced on this work in his 1950 essay, "Early Days of Genetics." This was the first mathematical analysis of Darwin's proposals. Specifically, Punnet demonstrated mathematically how Darwin's hypothesis of natural selection operated on the genetic recasting in populations over several generations.

There was, however, still one important question in the science of genetics that no one had yet succeeded in answering: given Mendel's laws, how frequently could an observer expect designated traits to materialize in upcoming generations of offspring? If a dominant gene governed white fur in a polar bear, for instance, what percentage of its offspring could be expected to have white fur? Here the mathematician-turned-biologist G. H. Hardy and W. Weinberg pioneered what genetics today calls the Hardy–Weinberg law. The law states that assuming random mating in a given generation, the frequency of a specific gene will stay roughly the same in upcoming generations.

As usually occurs, this mathematical approach to genetics started others down the same road. By 1925, statistician Ronald Aymler Fisher of University College, London, had written his *Statistical Methods for Research Workers,* arguably the finest research up to that point on mathematical statistics as it relates to genetic inquiries. He followed this with his 1935 classic, *The Design of Experiments*. In 1925, A. James Lotka, a mathematician at the *Metropolitan Life Company,* wrote his *Elements of Physical Biology,* which, like Fisher's publication, also applied mathematics to biology. Lotka tried to understand the Darwinian view of survival of the fittest as well as predator–prey relationships mathematically. Another classic of Lotka's is his 1930 work, *The Money Value of a Man*.

Biology moved ahead even more when the English logician and biologist Joseph H. Woodger attempted to systematize biology in *The Axiomatic Method in Biology,* of 1937, and in *Biological Principles* of 1939, though his "family-tree" conception of evolution is passé today.

CALVIN BRIDGES

In 1938, still another fabled geneticist would make his presence known. Calvin Bridges of the Carnegie Institute, American geneticist and student of T. H. Morgan, constructed, in that year, a detailed blueprint of the X chromosome of the fruit fly—the organism universally used in genetic trials. Among his finest work is his study of the locations of genes on the giant salivary chromosomes.

FISHER AND MIMICRY

Another milestone in genetics came a few years later with the publication of the British geneticist and statistician Ronald Aymler Fisher's 1927 paper on the genetic basis for mimicry—a fascinating phenomenon. Especially common in the insect world, mimicry can confer protection to an organism that resembles a different species or even a natural object. Certain species of palatable butterflies, for example, benefit by resembling unpalatable species, thereby warding off potential predators. Beyond this, Fisher was one of the founders of evolutionary theory based on Mendel's principles.

BEADLE AND HIS TIME: LONG-STANDING MISCONCEPTIONS

The American geneticist George Wells Beadle continued the tradition of seeking the genetic basis for various anatomical and physiological traits of organisms. Beadle was surely one of the founders of molecular biology. He received his postdoctoral instruction at Cornell after 1925 and then began to tackle the genetic factors controlling eye color. He became professor of biology at Stanford in the late 1930s and, together with his colleague, the celebrated Stanford biochemist E. L. Tatum, began scrutinizing the genetics of a common fungus, *Neurospora crassa*.

Tatum is one on the true geniuses of genetics. He majored in biology at both Stanford and Yale, and in 1957 he joined the Rockefeller Institute. In 1946, he began his collaboration with New Jersey geneticist Joshua Lederberg on mutations in bacterial cultures using the mold *Neurospora*, proving that the earlier theories that genes did not exist in bacteria were untrue. In fact, he helped prove that all biochemical processes in bacteria were controlled by genes. After a while he teamed up with George Beadle. A sinister feature of Tatum and Beadle's technique was the deliberate inducement of mutations by using X-rays. Beadle and Tatum shortly devised a theory to the effect that genes monitor and modulate all intracellular chemical reactions. They quickly proved this supposition to be legitimate, and biology has since come to know it as the "one gene–one enzyme theory." That is, nature charges every gene with the production of a unique enzyme. In 1958, Tatum, Beadle, and Lederberg of Berkeley would split the Nobel prize. Lederberg, specifically, won the prize for his exploits in genetic recombination and decoding the organization of nucleic acids in bacteria, while Beadle and Tatum earned it for their struggle with the phenomenon of gene and enzyme interactions.

LYSENKOISM

Worse than mere error is ideological dogmatism. The twentieth century also saw some unfortunate ideological developments that actually set back the progress of science. In Russia, between the 1930s and the 1960s, Trofim Lysenko took over, with Stalin's support, all Soviet biological laboratories. Although Lysenko was an otherwise competent and well-trained scientist, he held to the long-refuted belief of Lamarck that one generation can inherit characteristics acquired by a previous generation. Why he believed such an outdated notion has never been clear, save for the fact that the philosophical reflections of Marxism–Leninism could be read as supporting Lamarckianism. More rational scientists opposed him valiantly and often at significant peril to themselves. In 1940, for instance, the Russian geneticist Nikolai I. Vavilov, who had had a distinguished career, courageously and openly opposed Lysenko and everything he stood for.

That, in turn, got him arrested. For his "crimes," the Soviet government sentenced him to death. Although Soviet authorities subsequently commuted his sentence to incarceration because of particularly visible and sustained outcries from a number of western intellectuals, he eventually succumbed to maltreatment by prison authorities. Nevertheless, in 1951, the world would see the publication of his book *The Origin, Variation, Immunity and Breeding of Cultivated Plants*.

Yet Russia was hardly the only nation to find itself awash in a heap of irrational ideology. In the United States, attacks and counterattacks over evolution had escalated as the date of the famed "Scopes monkey trial" approached. In 1925, zealous fundamentalists, with the support of other cult groups, successfully prosecuted John T. Scopes for teaching evolution in high school biology. Undaunted, courageous biologists continued to proclaim the truth of Darwin's teachings. A watershed, these events showed that public understanding of science and the scientific method had had precious little impact in some corners of society.

But scientists were not merely interested in how life forms perpetuated themselves. They wanted also to understand how they behaved.

SOCIAL AND ECOLOGICAL ISSUES IN ETHOLOGY

In the first decade of the twentieth century, a novel field, ethology—the science of animal behavior—started to take shape. It would not fully assume a definite identity until the work of the German biologist Konrad Lorenz in the 1930s, and not much of any immense importance appeared for a number of years. However, in 1913 a high school teacher named Johann Regan used the telephone to prove that the cricket's characteristic sound is in fact a mating call. Regan allowed a male cricket to chirp into the telephone and found that a female cricket instantly headed for the receiver.

The Austrian ethologist Karl von Frisch did seminal studies on insect communication. In 1919, Von Frisch found that while bees did not customarily use the telephone, they did communicate through a kind of "morse code" of bodily movements. Years after, in 1965, scientists would build on von Frisch's classic achievements as well on that of Adolf Butenandt in a marvelous way. W. A. Jones, Morton Beroza, and Martin Jacobson created ersatz "love potients," or synthetic pheromones, for several different species of insects. In moths, e.g., a pheromone known as *disparlure* can attract male moths several miles away.

Fortunately, interest in animal behavior led others in Russia to proceed along more rational and productive lines than the ill-fated Lysenkoism discussed above. In 1934, the Russian ecologist G. F. Gause devised what is now known as "Gause's principle," which states that if two species are sufficiently similar to one another, they cannot survive in too close proximity to each another, primarily because they will be competing for a limited fund of similar resources. Gause proved this in his classic experiments with *paramecia*.

Konrad Lorenz had done similar research when he published a general analysis of social behavior in animals in 1935. It was undeniably Lorenz who founded ethology, the science of animal behavior. Born in 1903, he started his training in medicine at the University of Vienna and, in 1940, became professor of psychology at the University of Königsberg, the hometown of the imperial German philosopher Immanuel Kant. By the 1960s he had moved to the Max Planck Institute in Germany, where he first began to learn about the behavior of animals in their regional habitats. Among other things, he found that bird behavior is "deterministic" to the extent that either genetics or the environment imprints definite instinctual behavior on birds almost from birth—behaviors that limit and direct their impending behavior, such as courtship, mating, nesting, and the like. Lorenz summed up much of his work in ethology when he published *King Solomon's Ring,* a general account of social behavior in animals. In 1966, he completed *On Aggression,* arguing that among all of the world's species only man deliberately butchers members of his own species—a hypothesis that drew its share of controversy, to say the least. For his work he received the 1973 Nobel prize.

Robert Ardrey's 1966 book *The Territorial Imperative* counters Lorenz's controversial views to some degree by arguing that man in many ways is similar to lower animals. Like the songbird, for instance, man is naturally territorial. In his *African Genesis* he argues that because man is naturally aggressive, moral responsibility may be a vacuous notion.

During the 1950s and 1960s came the investigations of zoologist Nikolaas Tinbergen of Holland. He authored *The Study of Instinct,* perhaps the most comprehensive and exacting treatise on animal behavior to date. He mastered biology at Leiden and Yale, eventually returning to Leiden to take the post of professor of experimental biology. By 1966, he had moved on to Oxford to investigate animal behavior. He did most of this seminal experimentation on an assortment of birds, including the herring gull. He found that animal behavior often follows highly definite patterns; that is, it is more "deterministic" than zoologists had previously realized. In 1953, his classic book *The Herring Gull's World* appeared. Tinbergen also surveyed autism in human beings, contributing much to medical erudition regarding this disease. He captured the Nobel prize in 1973, along with Karl von Frisch and Konrad Lorenz.

LOEB AND PARTHENOGENESIS

More progress came in other areas. The American physiologist Jacques Loeb also known for extending H. Jenning's work on stimulus responses to higher animals, unmistakably proved that there was such a thing as "parthenogenesis," or the development of an egg without fertilization. This became evident in his observation of unfertilized sea urchin eggs when, by altering the environment in various ways, he managed to prod the eggs into dividing and eventually to reaching maturity. Some scientists who remained skeptical were to receive a nasty shock when, in 1958, the Armenian biologist Ilya Darevsky would find a species of lizard in his country that had no males. Instead the females reproduced parthenogenically, though other species, such as aphids and rotifers also do so.

PHYSIOLOGY AND DISEASE IN THE TWENTIETH CENTURY

Physiology in the twentieth century entered a fresh phase with the efforts of Karl Landsteiner. Landsteiner was one of those exceptional scientists with a broad range of interests. At various periods he did important research in anatomy, virology, and immunology. Born in Austria, he graduated in medicine from the University of Vienna in 1891 and spent several years in further education in Europe. Ultimately, he accepted a post at the Rockefeller Institute of Medical Research in the United States. In 1909, he first determined that there were at least three main types of blood, though a fourth was found sometime afterwards. He also confirmed that only certain types could match other types. With this discovery, the harrowing practice of blood transfusion suddenly became routine and safe.

Yet Landsteiner's fame did not rest only with this phase of his immunological studies. In 1940, Landsteiner, combining his resourcefulness with Alexander Wiener's, found the Rh factor and proposed a relationship between blood cells in humans and the rhesus monkey. The Rh factor refers to a group of antigens—any substance that can stimulate the production of antibodies—found in certain types of blood. The name "Rh" comes from the fact that scientists first found this group of antigens in the blood of the rhesus monkey. Blood having this factor is called "Rh positive." It is known today that if blood having this factor is transfused to persons with blood not having this factor—Rh negative blood—a severe reaction can occur. Today we know the Rh factor is critical in reckoning blood compatibility.

In the biology of disease, Jules Bordet of the University of Brussels ranks high. In 1901, he first began to devise the doctrine that antibodies fight invaders (antigens) of the body by "complementing" them—combining with them—and rendering them harmless. Beyond this, he researched the physiology of bacteria and developed a precursor of the Wasserman test for syphilis. For this work, the Nobel committee would award him their highest prize in 1919.

From the 1940s to the 1960s the findings of Sir Frank McFarlane Burnet were becoming increasingly compelling. Born in Australia, he got his medical degree from Melbourne University and then went to study and conduct research at the Walter and Eliza Hall Institute of the Melbourne Hospital. Before 1957, his interests centered primarily around influenza viral infections. After this he became increasingly consumed by the enigmas of immunology and how host organisms accept or reject alien tissue as well as various autoimmune diseases and normal aging. Burnet, along with the British biologist Peter Medawar of University College, London, would forthwith capture the Nobel prize in 1960 for his investigations into immune reactions in tissue transplants.

In 1961, physiologists realized that the thymus helps regulate the activity of the immune system. For the immune system to become functional, the body manufactures the necessary white blood cells in the thymus gland even before the organism is born. In 1962, the French physician Jacques F. Miller of the National Cancer Institute confirmed and expanded on this fact when he found that if he cut out the thymus gland from neonate mice, they easily accepted tissue graphs that thymus-intact mice would routinely reject. In this way he confirmed the supposition that the thymus gland is an essential part of the immune system. He then further showed that tinkering with the thymus affected cancer rates in mice.

THE POLIO VACCINE VALIDATES THE SCIENCE OF IMMUNOLOGY

In 1930, the American biologist Ernest Goodpasture of Vanderbilt University demonstrated a process for culturing, or growing, viruses in eggs. That done, viruses could now be produced practically at will. With the emergence of this artificial technique for growing viruses, scientists could cultivate viruses for any number of viral diseases. Once grown, minute amounts of a virus could then be injected directly into a person. The presence of the virus in the body would then cause the body to produce antibodies, thereby making the person immune to the disease. This technique, in turn, made the polio vaccine possible in the 1950s. By 1952, Jonas Salk of New York had developed a polio vaccine using dead polio viruses. Salk is one of the true immortals in the chronicles of biology and medicine. Born in 1914, he received his M.D. from New York University in 1939. Subsequently, he held positions both at the University of Michigan and the University of Pittsburgh. At Michigan he began his attack on the flu virus, ultimately recommencing this line of research at Pittsburgh. In the 1940s, Salk started his historical pilgrimage to create a polio vaccine. He succeeded masterfully and almost immediately mass inoculations began. By 1954, physicians were using the vaccine nationwide.

Although this initial vaccine was unmistakably impressive, another "live-virus" vaccine that Albert Sabin developed in 1957 eventually replaced Salk's original formula, since the live virus produced a stronger antibody reaction in the body and, therefore, a higher level of immunity. This was a pivotal event, for at about this time science escalated into high gear on the riddle of the relationship between viruses and human ailments. In 1936, for instance, the Pennsylvania biologist John Bittner of the Jackson Memorial Laboratory proved that a female mouse can transmit cancer to her young in her milk via the "Bittner milk factor"—the first solid evidence that viruses may cause cancer since it was well-known that mammalian milk can carry viruses.

REPRODUCTIVE BIOLOGY

Reproductive biology would come into its own as well in this era. In 1901, the Russian biologist Ilya Ivanov founded the first institution for artificial insemination. Later, artificial insemination would be perceived either as a boon or a loathsome evil, depending on one's point of view. Childless couples would turn desperately to newly developed reproductive technologies in the 1970s to have children by a variety of techniques involving artificial insemination. In so-called *in vitro* fertilization, physicians would pluck an egg from a woman and fertilize it outside the body. That, in turn, would lead to a storm of controversy about "tampering with nature," led by legions of theologians, philosophers, legislators, and even some physicians themselves.

Still, reproductive biology was not always such serious stuff, laden with moral austerity. Other scientists, pretensions to "research for its own sake" notwithstanding, were finding newer and better toys to play with.

TECHNOLOGY IN THE SERVICE OF BIOLOGY

Technology would develop at a faster rate than ever before in the twentieth century.

By the end of the first decade of the twentieth century the Dutch biologist Willem Einthoven of the University of Leiden created a 'string galvanometer' that was, for all realistic purposes, the precursor of the electrocardiograph, since it could measure the electrical current emanating from the heart. Fourteen years later, in 1924, Einthoven would capture the Nobel prize in physiology for this technological feat. He passed away two years later in Leiden.

Theodor Svedberg of Sweden created the "ultracentrifuge"—a device that spins so quickly that it can separate particles of practically anything from the liquid in which they were suspended. Svedberg noted that the time it required to separate out such particles correlated reasonably well with their molecular weight. That observation allowed upcoming biochemists to "weigh" gigantic organic molecules, an essential step for future growth in biochemistry. This work got the Nobel prize for Svedberg in 1926. The eminent British biochemist J. D. Bernal commented on Svedberg's work:

> The work of the physico-chemists, particularly of Svedberg, has shown that active proteins exist in the form of molecules of definite molecular weight, and more recently X-ray structure analysis has shown that they are perfectly definite chemical compounds with identical molecules . . .[48]

Thus it is no wonder that theorists such as Roscoe Dickinson and Linus Pauling of Caltech introduced the use of X-ray diffraction into the United States from Europe. In this technique one shines an X-ray beam on a substance, and the reflection pattern from the beam enables one to divine the structure of what one is observing. Although Max von Laue and Sir William Bragg had introduced this technique to probe the structure of inorganic crystals, which one routinely sees in any tourist rock shop, some scientists are beginning to apply it to the giant biological molecules in the expectation that they might similarly unravel the secrets of these molecules' architecture.

One of the most important of these X-ray crystallographers was the English biologist W. T. Astbury of the University of Leeds. While inspecting wool, he found that the X-ray picture of stretched wool differed markedly from that of its relaxed state. The bearing of this on biology consisted in the fact that X-ray diffraction could yield data about the structure of proteins. He then applied these finds to the study of textiles, as well as to crystal structure. By 1934, the British biologist John Desmond Bernal of Birbeck College, who all but invented X-ray studies of protein, was able to use the new technique to take the first photograph of the protein crystal pepsin.

The technology of X-ray crystallography took another leap forward in 1953. In that year the Austrian-born Cambridge biologist Max Perutz devised a technique with the assistance of John Kendrew, for extraordinarily enhancing the quality of X-ray photos. To do this, he added an atom of some heavy element, such as lead, to the molecule under consideration. Then, in 1960, Max Perutz again distinguished himself by specifying the structure of hemoglobin in imposing detail by adding such heavy elements. In succeeding years, many others would use this technique advantageously.

Probably the most far-reaching technological innovation of this period was the analog computer. In 1930, the physicist Vannevar Bush would design the forebear of such a machine. Barely six years afterwards, Konrad Suze nurtured the growth of computer technology still further when he designed a digital computer, using relays instead of vacuum tubes. Lest anyone suffer from the delusion that computers first appeared during the twentieth century, the anthropologist Derek Prince, in 1959, found a sort of primitive computer in the Mediterranean. The most interesting aspect of the find was the soon-to-come realization that it dated back to about 65 B.C.!

In 1939, the American theoretical physicist John Atanasoff developed the first electronic computer, intended to solve mathematical equations. He expanded this technology when he and Clifford Berry designed the so-called Atanasoff-Berry computer, on which most recent computers are based. By 1944, the mathematician Alan M. Turing of the University of Manchester began working on computers. Today, Turing has some, perhaps unfair, notoriety because of his infamous "Turing test." Allegedly, it was a way of determining whether machines could think. However, thanks to the reflections of savants like John Searle of the University of California, many today regard it as wrongheaded. Echoing sentiments expressed earlier by the philosopher Ludwig Wittgenstein, Searle says in his book *Minds, Brains and Science:*

> The reason that no computer program can ever be a mind is simply that a computer program is only syntactical, and minds are more than syntactical. Minds are semantical, in the sense that they have more than a formal structure, they have a content.[49]

These issues are enormously complex and go beyond the scope of this book. Roughly, however, Searle's idea is that all any computing machine could ever do is manipulate strings of symbols without being able to understand the *meanings* of the symbols. In the above book, he describes his now famous "Chinese room" thought experiment, where groups of computer programmers—who understand no Chinese whatsoever—by blindly following certain rules, correctly match up English words and sentences with their Chinese counterparts. By doing this, correctly formed Chinese sentences emerge. As Searle argues, no one would claim that the programmers in the room therefore understand Chinese—even though they would pass the Turing test. Although not everyone accepts Searle's ideas, they have been enormously influential and have provoked considerable discussion.

In any case, the Turing test was a brilliant conception. On a more practical level, Turing made real progress in computers. He with the aid of F. C. Williams and T. Kilburn, in 1943, developed *Colossus,* the first full-scale electronic computer. Doubtless the most vital use of it consisted in decoding German war messages. Intriguingly, Turing did little work in biology as well. In 1952, he proposed the notion that there is a select class of chemicals, which Turing called "morphogens," that control a person's biological appearance. An example would be the idea that two morphogens allow zebras to have stripes.

Despite these achievements it would be several decades before computers would really filter into the American mainstream, primarily because computer science had to await more efficient "languages" with which to program the computer. It would not be until 1956 that John Backus of the T. J. Watson Research Center would develop FORTRAN, which was the first computer programming language that even slightly resembled human language, making it much easier to handle. No surprise, since Backus had earlier developed the IBM *70H* as well as *ALGOL*.

In 1945, the American computer engineers John P. Eckert of Philadelphia and physicist John Mauchly both of the Eckert–Mauchly Division of Sperry Rand, had constructed ENIAC, more powerful and with incomparably greater applications than any previous machine. But it was still exhausting to program (as well as being an electricity hog; whenever anybody turned it on residents bombarded them with complaints, since the enormous appetite of the machine drained electricity all over town.) In 1951, the Remington Rand Corporation would begin selling BINAC and UNIVAC I, also designed by Eckert, the first computer sold to the masses. It was also the first machine that stored data on magnetic tape. From that point onwards, the computer would increase in brawn, complexity, and affordability, until *Apple,* in 1984, brought an efficient and affordable computer into the American home. (Mauchly, incedentally, first proved, using computers, that the moon affects rainfall.)

Soon enough, the computer would all but dominate American life and would figure prominently in, for instance, mathematical genetics. Still they came—one technological marvel after another. In 1930, the Swedish scientist Arne Tiselius added considerably to the stock of biological tools. Tiselius received a degree in science from Uppsala, then served as professor of biochemistry there. His first scientific struggles were with his colleague Svedberg on the study of protein structure, using the technique of centrifugation, or spinning samples at high speeds to separate out the protein from other unwanted materials. He then invented electrophoresis, a technique that allowed laboratory separation of proteins with the use of electric currents. More specifically, this technique relied on the fact that large molecules, such as proteins, move at different speeds in an electric field depending on their size and structure, thereby providing a method for effectively separating dissimilar proteins. Tiselius was active also in the politics of science; he became vice-president of the

Nobel Institute in 1947, and in 1948 the scientific elite awarded him the Nobel prize in chemistry.

In the ensuing years, Linus Pauling and his associates would use electrophoretic techniques to isolate the error in the hemoglobin molecule that caused sickle-cell anemia, a disease primarily affecting blacks. In 1959, still other "genetic" illnesses would yield their secrets. To mention one, the biologist C. E. Ford proved that Turner's syndrome, a chromosomal abnormality resulting in failure of sexual maturation, also results from the lack of one X chromosome. The very same year, P. A. Jacobs and J. A. Strong showed that Klinefelter's syndrome, another chromosomal abnormality also resulting in incomplete sexual development and occasionally reduced intelligence, occurs when an individual has one sex chromosome too many and a genotype of XXY.

Perhaps surprisingly, chemists contributed to biological technology as well. In 1944, for instance, the English chemists Archer Porter Martin of the Books Pure Drug Co. and Richard Millington developed a technique known as "partition chromatography" as well as new techniques in gas chromatography. These ingenious tools allowed researchers to more easily analyze the chemical elements making up organic compounds, leading to discovery of new antibiotics.

In 1902, the Austrian chemist Richard Zsigmondy and H. Siedentopf dramatically advanced microscopic vistas when he devised the ultramicroscope. Far more powerful than any previous microscope, it allowed the scientist to see particles in colloidal and other sorts of suspensions (a "colloidal suspension" consists of particles uniformly dispersed throughout a liquid, gas, or even a solid medium, without actually dissolving in the medium). With the new device, one can see objects down to 100 Å in size, about 20 times stronger than ordinary light microscopes. Further progress in microscope technology appeared in 1908, when the Berkeley biophysicist Robley C. Williams devised a "shadowing" technique for taking electron microphotographs, a technique involving spraying the object to be viewed with special opaque metals. Using similar ideas, he also greatly improved telescope mirrors. In 1945, the German physicist Hans Geiger invented the Geiger counter. Although it had no immediate applications to biology, it later would have the most profound relevance as scientists and the public began to worry about the effects of radiation on living tissue—fears which would escalate after the incidents at Three Mile Island and Chernobyl.

Arguably, the predominant technological breakthrough for biologists came in 1937, when the American physicist James Hillier and Albert Prebus drastically redesigned the electron microscope, improving its efficiency and resolving power so much that their design eventually became standard. After this, the next leap ahead would come when Russian engineer Vladimir Zworykin of Westinghouse reached his goal of lifting the magnification of the electron microscope to the point that it could expand things fifty times more than the mightiest light microscope. Zworykin also invented the iconoscope and was a pioneer in television.

Soon electron microscopy was advanced enough that the Belgian biologist Albert Claude was able to use it to reveal actual structures in the endoplasmic reticulum—the network of tubules distributed throughout the cytoplasm of cells. Claude studied medicine at Bucharest University, finally becoming professor of anatomy at the same institution. Subsequent to the Second World War he left for the United States to assume a position at the Rockefeller Institute and then at Yale. He began using the electron microscope to survey the minutest details of cells, particularly the mitochondria, or "energy manufacturers" in the cell. He and others found that the mitochondria are in fact miniature "organelles" within the cytoplasm of the cell. They differ from other elements found in cytoplasm in that they have their own DNA, taken directly from the maternal line. The primary function of the mitochondria is to manufacture the enzymes necessary for the production of energy in the cell. After the discovery of mitochondria, Claude turned his attention to ribosomes, tiny, mitten-shaped organelles existing in huge numbers in cells. As he and others found, they are the sites where a cell manufactures protein molecules. For their superb exploits in the field of cytology, as well as electron microscopy, Albert Claude, the American George Palade of the Rockefeller Institute, and the Belgian cytologist René de Duve shared the 1974 Nobel prize.

Yet despite such exploits scientists kept trying to improve the weapons of microscopy still further. They succeeded wonderfully, for in 1955, the German physicist Erwin Mueller of the Max Planck Institute devised the field-ion microscope, the first microscope that afforded a direct view of individual atoms in a molecule, which led to further knowledge about the physical effects of field ionization.

TRACER TECHNOLOGY AND MODERN PHYSIOLOGY

While the above were arguably the most dramatic finds of the twentieth century, scientists were conducting research no less important in the field of physiology. There were also some amazing results in scientists' efforts to understand the circulatory system. In the 1950s, the American biochemist David Shemin of Columbia University used carbon-14 as a "tracer." By tagging a certain chemical with carbon-14, he could follow the route it took through the body, thereby allowing him to attend to chemical reactions as they happen, so to speak. In this way, he learned how a cell synthesizes both heme and vitamin B_{12}. Not too many years afterward, the British biologist Melvin Calvin was launching inquiries that would lead to more progress in tracer technology and botany.

Born in 1911, Calvin began his schooling at Manchester University in 1935. He later joined the Manhattan Project—the secret World War II project that built the first atomic bomb. From there he went to the University of California at Berkeley to direct the Laboratory of Chemical Dynamics. From the outset he devoted most of his time to studying photosynthesis, that process in plants whereby, using chlorophyll and sunlight, they convert carbon dioxide from the air to glucose and oxygen. Following the lead of the Hungarian chemist G. von Hevesey, Calvin began applying radioisotopes, one of several forms of an atom with different atomic weights which is also radioactive, to follow chemical changes in plants. By 1945, using carbon-14, he discovered the process whereby plants reduce carbon dioxide in the atmosphere via a particular enzyme, as a step in manufacturing sugar, later called the Calvin cycle. For this, he gained the 1961 Nobel prize. "Tracer" technology surged ahead again in 1935 when the German biologist Rudolf Shoenheimer first used isotopes of hydrogen to trace the metabolic pathway of fats. Later, in 1964, the German scientist Konrad Emil Bloch of Harvard along with German colleague, Feodor Lynen, captured the Nobel prize for their discoveries in unsaturated fatty acid and the role of acetic acid in cholesterol metabolism.

Still another milestone in tracer technology came with the exploits of the Canadian biochemist Martin David Kamen of the University of California, a specialist in photosynthesis. In 1940 Kamen found carbon-14, a radioactive isotope of carbon with a colossally long half-life—the time it requires for a given amount of an element to decay by radioactivity to half its initial amount. Unquestionably, carbon-14 is among the most extensively used tracer in science. Scientists have applied it to everything from physiology to dating of ancient artifacts. The distinguished American chemist Willard Libby of Berkeley was the first to use radiocarbon dating to divine the age of such artifacts relying on the fact that ordinary carbon has a minute amount of radioactive carbon in it.

However, imaginative academics hardly limited the use of tracers to zoological and archeological surveys. By 1934, tracers were weaving their way into physiological experimentation on plants. In that year, the Hungarian chemist Gyorgy Hevesey while working with Rutherford at the University of Manchester, used a radioactive isotope of phosphorus to investigate plant and human metabolism. The scientific aristocracy gave him the 1943 Nobel prize for this research. After that, a number of scientific thinkers became interested in the role of phosphorus in cell metabolism.

The first real insight into the mechanism of action of insulin on carbohydrate metabolism came from the studies of Jacob Sacks at the University of Michigan with radioactive phosphorus as a tracer. Sacks and his group studied how the hormone insulin affected the turnover of phosphorus compounds in muscle as well as the role of insulin in metabolism generally. In 1940, the American anatomist Herbert Evans of Berkeley used radioactive iodine to prove what had previously been only a conjecture—that the thyroid gland requires iodine for healthy functioning. Beyond this, he introduced a variety of dyes to study blood volume and studied a variety of pituitary hormones.

CELLULAR PHYSIOLOGY

Continuing the survey of mitochondria, the American biologist James Bonner of the California Institute of Technology, in 1951, proved that the process of oxidative phosphorylation—the process whereby the body converts glycogen into the sugar needed to provide energy for the body—takes place in mitochondria. Subsequent to this, in 1955, James Bonner again took a step forward when he and Paul Tso actually isolated the mitochondria inside cells. Two years later, Bonner and Tso summed up much of their work in their epochal book, *The Next Hundred Years*. In 1960, Bonner added still more to his resume by discovering that DNA not only directs the formation of proteins but also orchestrates the manufacture of RNA. He then developed a technique for studying nucleic acids in test tubes. That same year the physician and physiologist Murray Lewellyn Barr of the University of Western Ontario realized that the two X chromosomes in the female were not identical; one of them carried an unusual entity. This "Barr body" as it was later called, causes the X chromosome having it to be genetically inactive—the other X chromosome alone carries on all genetic activity. Barr also did extensive research in mental retardation and sex chromosome defects. Still another advance in grasping intracellular metabolic activities emerged in 1964, when the American cytologist Thomas Roth and the Canadian Keith Roberts Porter of Harvard found that the parts of the cell that gather environmental elements in the cell's cytoplasm soon use them to manufacture yolk, nutritive material in the cytoplasm of an ovum. They also found that some vesicles that carry the yolk, once formed, stay deep inside the cell to convert it into usable energy for maturation and reproduction. Porter subsequently authored the book, *Fine Structure of Cells and Tissues*.

PHYSIOLOGY OF THE BLOOD

Other biologists were also briskly adding to our knowledge of the constituents of blood. In 1925, George Whipple of the University of Rochester Medical School, discovered iron as a fundamental element in blood and went on to write over 200 articles on anemia alone. Then, in 1928, the physiologist Hans Fischer of the University of Münich Germany unerringly described the atomic structure of heme, the one segment of the hemoglobin molecule not constructed out of proteins.

INVERTEBRATE PHYSIOLOGY

At the other end of the biological cosmos, disciples of classical biology were continuing the grand tradition of invertebrate biological inquiry. In 1944, the American biologists R. B. Cowles and C. M. Bogert invested an huge amount of energy and dollars in scanning the mechanism by which desert reptiles control their body temperature. Such "cold-blooded" animals cannot support a constant body temperature as humans can; instead, they seek out cooler or warmer places in their environment to stabilize their body temperature. Today biologists call such animals "poikilotherms," or cold-blooded, as opposed to "homoiotherms," or warm-blooded animals.

THE LANGUAGE OF BIOLOGY

Interestingly, even by this late date—so long after Linnaeus, some were still doing extraordinary things in the science of classification. In 1904, Cambridge biologist George Nuttall in addition to showing that bacteria in the digestive tract are needed in digestion, wrote his *Blood Immunity and Blood Relationship,* which proved that biologists could utilize blood-testing to order an unknown organism from phylum through class, order, and genus. Subsequent to this, the scientific community published its *International Rules of Zoological Nomenclature.* This at least eliminated one obstacle preventing the solution of the most intractable problems in biology and one that had persisted for eons—the difficulty in communicating, one scientist with another. At least, with a universal standard classification, scholars could talk with one another in the same language.

MIRACLES IN PALEONTOLOGY

Classification was not the only branch of biology where biologists would have to confess that they did not yet know all there was to be known. In 1938, a catch occurred in the Indian Ocean that generated dumbfounded incredulity in every biological laboratory on the planet. For some years already, evolutionists and paleontologists felt fairly sure that they knew which species were extinct and which were not. They "knew," for instance, that the extraordinarily ancient animal, *Latimeria chalumnae* or the coelacanth, had been extinct for over sixty million years. Paleontologists knew from fossil finds that it had lived on the ocean floor, had limblike fins, and a heavy body with hollow spines. The scientific community therefore received a monstrous shock when Captain Hendrik Goosen caught one in 1938 off the coast of Africa. It was, defensibly, the nearest thing to a miracle that had ever occurred in biology.

In just under twenty years, the ichthyologist James L. B. Smith, who had first examined the coelacanth and who had also named it in honor of Courtenay Latimer, a colleague who first drew his attention to it, managed to get hold of still another one through advertising a reward for anyone who could bring him such a creature. Smith finally realized that they were numerous near the Comoro Islands, at depths close to 1,000 feet. Comoro are a group of islands located between Africa and Madagascar in the Indian Ocean, and it turned out that fishermen had been seeing them there for generations. After this, biologists left these bottom-heavy, quasi-fish alone for a few years, until 1987. In that year Hans Fricke, using a submarine, began following them in their normal environment, deep in the Indian Ocean. He found that they periodically swim upside down, among other oddities. Sadly, such happy events as finding "extinct" animals were to be few and extremely far between. Still, in 1981, zoologists did discover another animal they had believed was extinct—the black-footed ferret, a prairie-weasel (*Mustela nigripes*), surviving in minuscule numbers in Wyoming.

Very nearly as spectacular was the discovery, in 1954, of seeds of the genus *Lupinus*—a plant in the legume family having thick clusters of leaves—found preserved in the Yukon territory of Canada since the last ice age. By 1967, the American biologists Alf Porsild and Charles Arington were able to make a stunning announcement—they had planted and successfully cultivated arctic lupines from these seeds, estimated to be over ten thousand years old. Apparently, there is virtually no limit to how long seeds can remain functional when frozen as the lupine seeds were. Still the tundra did not cease to yield her frozen wonders. In 1954, again, American biologists Elso Barghoorn a Harvard botanist specializing in decaying organic matter, and Stanley Tyler found minuscule fossil remains in Canada which closely resembled modern-day algae and bacteria. Scientists estimated that these organisms were about one and one-half billion years old—more ancient than anything ever found before. Absolutely the most climactic find of all came in 1977, when explorers dug out a baby mammoth that had been, by all estimates, frozen for over forty thousand years.

VITAMINS: THE GROWTH OF NUTRITION

While many scientists were focusing on dead animals, others were focusing on "substances of life"—the vitamins. In biology, the once-tainted science of nutrition was starting its rise to prominence. By 1901, the German biologist Adolf Windaus of the University of Göttingen proved that ultravilet rays can accelerate the production of one of the "fat-soluble" vitamins, vitamin D. Windaus also contributed to our understanding of bile acids and the B vitamins. Within two decades the American biologist Harry Steenbock proved that it was only the ultraviolet part of the spectrum that speeded up vitamin D production in the body, a discovery he subsequently patented. By 1913, the Johns Hopkins physiologist Elmer McCollum, along with his colleague Marguerite Davis, first drew the vital distinction between fat-soluble vitamins such as A and D (both of which he also discovered), and water-soluble vitamins such as the B-vitamin group. McCollum also broke new ground in creating a variety of new methods for analyzing food.

By 1906, the English physiologist Frederick Hopkins began to suspect that fats, proteins, and carbohydrates did not exhaust the list of substances critical to conserving life. Hopkins had begun his work soon after completing his education. Born in 1861 in Eastbourne, England, he studied chemistry at Guy's Hospital in London, and later Cambridge University appointed him to the post of lecturer in chemical physiology. Among other things, he isolated the amino acids glutathione and tryptophan, two of the "building blocks" of the protein molecule. He showed too how essential milk was in the diet. In a series of explorations in the 1920s, he proved that milk dramatically accelerated development in rats who were on otherwise spotty diets. The Nobel committee deemed this effort definitive enough to give him the 1929 Nobel prize, though he also received the Copley Medal and the Order of Merit.

In 1912, the Polish biologist Casimir Funk would coin the neologism "vitamin" for the broad category of food elements Hopkins had analyzed. That was not the end of Funk's contributions, however. After a number of posts at the Pasteur Institute, the Cancer Hospital of London, and the State Institute in Warsaw, he began his most perceptive work in nutrition at an institute he founded, called the Funk Foundation. Among many other things, he ascertained that yeast (which is high in thiamine) was effective in curing the nutritional disorder beriberi.

Still, despite such progress, no one yet realized that a deficiency of vitamins could actually result in illness. Then, in 1915, the Austrian physician Joseph Goldberger of Washington, D.C. broke through this barrier when he proved conclusively that the disease pellagra would result if a person did not absorb enough vitamin B. Also, his research interests extended to diptheria, yellow fever, and typhus. By 1922, knowledge of nutrition had advanced to the point that Herbert McLean Evans of the University of California at Berkeley and his colleague K. J. Scott were able to propose the existence of, and consequently reveal, vitamin E. Three years after that, the American biologists George Minot and William Murphy would prove that liver was an effective treatment for anemia.

In 1929, the Nobel committee offered their grand prize to bacteriologist Christiaan Eijkman of the University of Utrecht and Sir Frederick Hopkins of Great Britain for their many years of struggle that culminated in significant insights into both the metabolic and nutritional role of vitamins for diseases such as beriberi and polyneuritis. By 1931, the Zürich chemist Paul Karrer had worked out the structure of vitamin A as well as B_2 and E, and had learned a substantial amount about their function, as well as about the B vitamins generally. Six years later, the University of Wisconsin biochemist Conrad Elvehjem found vitamin A and first isolated niacin. In 1934, still another valuable step occurred when Robert Williams successfully isolated thiamine from rice.

Then, in 1928, a discovery took place which would generate of the most acerbic controversies in the history of nutrition. The chemist Albert Szent-Györgyi of Hungary discovered vitamin C. Subsequently, the American chemist Linus Pauling would take up the banner for this vitamin, touting it as an effective treatment for the common cold, cancer, and several other ailments. Pauling has also intimated that vitamin C may be useful in the treatment of AIDS.

Biologists probed still another dimension of physiology in the twentieth century when Szent-Györgi found the four acids required by a muscle cell for respiration to proceed productively. In 1937, Szent-Györgyi captured the Nobel prize for his discovery of vitamin C and for furthering the understanding of respiratory physiology. It was plainly this work which ultimately led to the discovery of the

"Krebs cycle," the process whereby the cell oxidizes food in order to provide it with energy. That same year, Teodor Axel Theorell of the Nobel Institute of Medicine proved that certain coenzymes used in cell respiration are similar in structure to vitamin B_2. Theorell, a Swedish biochemist, took a medical degree at the Karolinska Institute in Stockholm, as had Euler years before. He then headed for the Pasteur Institute in Paris, returning to Sweden in the ensuing years to become Director of the department of biochemistry at the Nobel Medical Institute. Most of his initial prospecting was in the field of hematology, but after some years he devoted himself completely to the enzymes invòlved in respiration in cells. He was also the first to isolate myoglobin. It was during this phase that he did the influential inquiries sketched above. By 1936, the American chemist Robert Williams of Bell Labs had worked through the structural riddles of B_1, a vitamin physicians already knew was indispensable in preventing beriberi, a disease characterized by a variety of neurologic disorders such as sleep and memory loss. Williams later campaigned for the addition of B_1 to flour, rice, and bread.

In 1940, the American biochemist Vincent du Vigneaud also known for his work on oxytocin and vasopressin, first identified biotin, part of the B-vitamin complex, previously and erroneously known as "vitamin H." That accomplished, Vigneaud began scrutinizing the structure of the biotin molecule itself, finding that it had a two-ring architecture.

By 1943, other scientists, building on Vigneaud's conclusions, synthesized biotin. Vigneaud himself followed in his own footsteps with his 1953 efforts on amino acids. In that year he finished demarcating the exact sequence of amino acids in the polypeptide chain of the hormone oxytocin, a hormone that causes the uterus to contract and the mammary glands to produce milk. Subsequently, in 1954, Vigneaud would become the first to produce oxytocin. It was the first time anyone had artificially synthesized any hormone. In 1955, the biological community would honor him with the Nobel prize. Inspired by Vigneaud's toils, the American investigators Rosalyn Yalow of Mt. Sinai, who probed peptide and growth hormones, Roger Guillemin of Baylor University, and Andrew Schalley of Tulane Medical School, who also studied peptide hormones, proceeded with still more successful attempts to synthesize several new hormones, most notable being Guillemin's work on the hormones of the hypothalmus, eventually capturing the Nobel prize as well.

Then came the research of the American physiologist Edward Doisy. In 1935, he became the first to pinpoint and isolate the class of hormones known as estrogens. This was followed almost immediately by the feats of D. David, who isolated testosterone from testicular tissue and, in the act, coined the term "testosterone." In 1943, the Danish biochemist Henrik Dam and Edward Doisy captured the Nobel prize—Dam for finding vitamin K and Doisy for analyzing it into its constituent elements. The vitamin saga was not over.

Henrik Carl Dam graduated from the Copenhagen Polytechnic Institute in 1920 and subsequently lectured at the physiological laboratory of Copenhagen University, eventually becoming senior research associate at the University of Rochester. By 1929, he was studying chickens on low-fat diets. He quickly found that their blood rapidly lost the ability to clot. After ruling out possibilities such as scurvy, he theorized that there was a still-unknown vitamin crucial to clotting. He called this vitamin K, and demonstrated that it existed naturally in a wide variety of plants as well as in the liver of many animals. Some years later, Edward Doisy managed to isolate vitamin K in the laboratory, though he also studied the role of vitamin E, fat, and cholesterol in health.

The onslaught on vitamins went on unabated, and in 1948 Karl August Folkers of Merck and Co. found that vitamin B_{12} was an cardinal element in the continuing vigor of bacteria. The less B_{12} a culture had, the slower it would multiply, and vice versa. Thus, by a comparative analysis of the developmental patterns of dissimilar bacterial cultures, he was able to track down and inevitably isolate B_{12}, which later proved so effective in treating pernicious anemia. Not satisfied with this, Folkers went on to do pioneering research on antibiotics.

After many years of time-consuming labor and numerous failed attempts, Robert Woodward offered his legacy to science. Born in 1917, he finished his Ph.D. in chemistry at MIT, immediately taking a position at Harvard. In 1963, the Woodward Research Institute in Basel, Switzerland, appointed him its head. Continuing his career-long interest in the structure of biological molecules, he was able, in 1948, to exhume the structure of strychnine, which others fully confirmed by 1954. Tarrying not a bit, he began working on the structure of protein steroids and a miscellany of other biological molecules. In the end, he synthesized B_{12} in the laboratory. His lifetime of labor justifiably garnered him the 1965 Nobel prize. After this time, because of the claims of people like Linus Pauling, Adelle Davis, and others, vitamins and their alleged cures mushroomed into a multibillion-dollar industry, along with stupendous promotion of "herbal" cures from the Orient, anticancer claims for fiber, and so forth.

THE VITAMIN MANIA

But was there anything to these claims? Assuredly, there was some support for Pauling's claims in experiments conducted by Arthur Robinson and Ewan Cameron in Scotland. However, there was also evidence that was not so hospitable to the claims of the vitamin gurus. Two independent reports appeared in the *New England Journal of Medicine* in the 1980s, for instance, which militantly dismissed all of Pauling's claims, though it is by no means obvious that these experiments had precisely duplicated the vitamin-favorable work Pauling had been relying on. In a related vein, in 1988, Elias Corey, Myung-choi Kang, Manoj Desia, Arun Ghosh, Ioannis Houpis, and Wei-guo Su reported that they had manufactured ginkgolide B. Many had believed that this compound was the active element in herbal remedies based on ginkgo leaves. These researchers found that ginkgolide B did in fact have some effect on asthma and a number of allergies. They hypothesized that it functioned by suppressing the immune system.

EMBRYOLOGY

During the second decade of the twentieth century, biologists had begun to make further inroads into the enigmas of embryology. In a series of cleverly designed experiments, German zoologist Hans Spemann of the Kaiser Wilhelm Institute and his colleague Hilda Mangold found that if they grafted the "organizer" portion of an amphibian embryo to another embryo, the organizer would influence the subsequent development of the host embryo. The essential principle here is that cells do not act alone; in fact, they necessarily affect other cells in their vicinity. This "organizer" effect thus put the final nail in the coffin of preformationism. About two decades subsequent to these experiments, embryology would move further ahead when, in 1950, zoologists triumphantly performed an actual embryo transplant in cattle—a procedure that would find its way into the treatment of infertility in humans, whereby physicians fertilized an egg outside the body and then transplanted the embryo back into the mother.

In 1932, Oxford biologist Julian Huxley's epochal book *Problems of Relative Growth* appeared. In it he stated his now-famous rule that the ratio of the growth rates of all organs in the body to one another remains constant throughout an organism's lifetime. He returned to similar themes in his 1947 book, *Evolution and Ethics*. In 1936, the German embryologist Etienne Wolf consummated his epochal probings into defects of embryological development—an initiative that would shortly help physicians understand the makeup and causes of human birth.

Yet another discovery in endocrinology emerged in 1958 with the experimental work of Harris, Michael, and Scott, who helped "map" the routes of the endocrine system by showing that hormones do not act directly on the central nervous system, but indirectly via the hypothalamus, a region of the brain responsible for sleep cycles, body temperature, and the activity of the pituitary gland. By 1965, the sum total of intensive endocrinological research began to pay off in serviceable applications. That year Morris E. Davis of the University of Chicago published the reports of his trials, which revealed that administering estrogen can reduce both osteoporosis and atherosclerosis in women past menopause. He has also done extensive work in the physiology of aging generally. In a related enterprise, Michael Brown and Joseph Goldstein found the part of the cell membrane in human beings which "traps" fats, inevitably depositing them in the arteries to cause atherosclerosis.

DNA TARGETED

Scholars would return to the general question of sequences in biochemical "chains" in 1961. Charles Yanofsky and the South African molecular biologist Sydney Brenner, later at the Medical Research Council laboratories in Cambridge, had been looking at Vigneaud's legacy and wondered if it might be feasible to unmask sequences in DNA itself, given that the order of amino acids in oxytocin was already known. Brenner was a tough, brilliant, and implacable molecular biologist. He graduated from the University of the Witwatersrand in South Africa and then furthered his education at Oxford. At last he joined the Molecular Biology Laboratory of the British Medical Research Council and became director of the lab in 1980.

By 1967 he had disclosed the sequences in DNA above. From his work emerged the concept of the codon. A codon is a sequence of three genetic units in RNA whose job it is to direct the manufacture of a particular amino acid when the cell assembles protein chains. Brenner originally suggested the term 'adaptor,' for the RNA molecule that directs the synthesis of proteins. He ascertained that the sequence of nucleotides, or codons in DNA follows an order identical to the order of amino acids in protein. He further proved that the "triple codon" sequences did not overlap one another and that there were no gaps in the sequence of bases in DNA, still another critical step in understanding how a cell manufactures protein. That same year, the fabled British biologist and physicist Francis Crick went with Sydney Brenner on another genetic "detective" mission; they tried—and succeeded—in decoding another message on the DNA molecule, the message that shuts the machinery off or, in other words, tells the molecule to quit building protein. They found also that a given amino acid had to correlate with more than one codon, since there are more codons than amino acids.

MOVING TOWARD THE PRESENT

As science moved through the turbulent era of the 1960s and 1970s, it yielded, not surprisingly, some of the most electrifying and far-reaching accomplishments ever. In 1969 the first man walked on the moon. In cosmology, scientists would create two theories purporting to explain the launching of the cosmos—the so-called Big Bang and Steady State theories—only to abandon the latter when physicist Arno Penzias of the Bell Telephone Laboratories detected microwave radiation that conclusively established the big-bang theory as the correct one. One of the key notions in this hypothesis was that the universe was perpetually expanding—an expansion that had begun with the explosion of a single atom of hydrogen fifteen billion years ago. The microwave radiation was one of the prophecies of the hypothesis, and Penzias was the first one to find it, thus confirming that forecast.

In particle physics, swarms of previously undiscovered particles would surface, constituting, as Caltech scientist Robert Bacher told the author, a "bottomless pit." The Beatles would come and go, and Elvis sightings escalated. The cost of living soared, and Egypt signed an armistice with Israel.

Entire new fields would either open up or take on fresh vigor. With escalating global disquiet over pollution, damage to the environment, and population expansion, the old field of ecology recharged itself. Still, while it was gaining new vitality, it was also accruing eccentric and exaggerated verbiage. In 1957, biologists "defined" the ecological niche as "an abstract hypervolume in a space with axes for each of the environmental and biological variables that affect the organism whose niche it is." One supposes this means that an animal's surroundings can affect it, a proposition whose truth is self-evident and decidedly unremarkable—an example of the timeless academic principle that the more ten-dollar words one finds in a piece of writing, the fewer genuine concepts there are.

A little less metaphysical were the reflections of G. E. Hutchinson in his essay "Homage to Santa Rosalia, or "Why Are There so Many Kinds of Animals?" Intended more for the lay public then a scientific audience, the ideas were nonetheless credible. Hutchinson argued that two closely related species of fauna could not occupy the same environmental niche because the similarity would cause too much competition for the same resources. The newly energized field of ecology benefitted too from computers; with them, ecologists could construct mathematical "models" of the behavior of populations. By entering particular classes of data into computers, the computer would predict, for instance, when and how populations grew or diminished. Related to this was another new field called "sociobiology," in which biologists, occasionally using mathematical procedures, tried to appraise the behavior of communities of animals. Some even tried to apply these creative musings to man, hoping to elucidate such things as ethical behavior in humans.

SUMMARY

The first half of the twentieth century was, manifestly, a time of absolutely fantastic progress. Oparin took up the question of the origins of life in his book *The Origins of Life on Earth,* speculating that life may have evolved solely through random processes from a primordial biochemical "soup." Jacques Loeb a physiologist at the Rockefeller Institute, made fine contributions with his views on mechanism and parthenogenesis embodied in his 1912 book, *The Mechanistic Conception of Life*. It was also an era of vast and far-reaching insight into the nervous system, with such eminent scientists as Dale, Loewi, and the Hokins. Also, Euler and Hillarp did seminal work on neurotransmitters, which would become the foundation for modern biological psychiatry. They found that one of these transmitters, noradrenaline, is stored in diminutive granules in neurons. T. H. Morgan also published the first complete description of a chromosome, revealing more than 2,000 genes on it. Completing this was the great work of Muller on mutations, as well as the discovery of the Hardy–Weinberg Law governing gene frequencies in heredity.

There was a darker side to biology in this period. There was the infamous dominance of Lysenkoism in the USSR and the Scopes "monkey" trial over the teaching of evolution in the United States. In physiology, we see Landsteiner's work on blood groups and the Rh factor, while medicine surged forward with the work of Goodpasture on cultivating viruses and Salk and his polio vaccine.

There was tremendous technological innovation as well. The ultracentrifuge became standard hardware in biological laboratories. It was the age of X-ray crystallography, Pauling, Perutz, and others. With the work of Turing and others, the world saw the dawn of the computer age as well. The toolbox of biology grew with the ultra- and electron microscopes, enabling scientists to see deeper into the innards of the cell.

Communication among scientists became much easier when the scientific community published its *International Rules of Zoological Nomenclature,* which provided a universal standard classification.

Among the most dramatic finds of this period were those in paleontology, with the discovery of the extant coelacanth and arctic lupine seeds.

Most significant of all was, without question, the early work on DNA. Sydney Brenner of the Medical Research Council took one of the significant early steps when he was able to determine the sequences of amino acids in that important molecule. The world was getting ready for Watson and Crick.

CHAPTER 23

On the Trail of DNA

The most significant development in molecular biology after the war was, indubitably, the epochal discovery of Watson and Crick. They would complete their analysis of DNA, for which they, along with Maurice Wilkins, would receive the 1962 Nobel prize.

The road to DNA was not easy. Nor was its ensuing field, molecular biology, without controversy. Scientists such as the American geneticist T. H. Morgan, the Italian biologist Salvador Luria, the Canadian physician Ostwald Avery of the Rockefeller Institute and others had done much of the early spadework. By the 1930s Russian-American chemist Phoebus Levene of the Rockefeller Institute and his colleagues had identified the sugar ribose in nucleic acids, of which both RNA and DNA are illustrations. In the 1920s Levene had found that there were two forms of the sugar, ribose and deoxyribose. Hence there were two sorts of nucleic acids, RNA and DNA. Avery and his team then proved in 1944 that genes were the bearers of genetic information—not proteins as many had previously believed. Avery subsequently researched *pneumococcus* bacteria and pneumonia.

COMPETITION AND CONTROVERSY

By the 1950s, science, therefore, already knew something about DNA. In fact in this decade, several scientists were getting closer to the secrets of this mysterious molecule. A dazzling step toward the acclaimed conclusions of Watson and Crick occurred earlier in 1953, when Rosalind Franklin and the English physicist Maurice Wilkins took some of the first X-ray photographs of DNA. By getting hold of these photos, Watson and Crick would shortly unravel the structure of DNA. This was now the "holy grail" of biology, precisely as the unified field theory—a hypothesis that would unite all the forces of the universe in a single equation—would ripen into the holy grail of physics. Besides Watson and Crick, Linus Pauling of Caltech was wondering about the structure of DNA. Initially, many alleged that Pauling would solve the structure first. Pauling had been thinking for years about protein structure, and in 1951 he and his colleague Corey came up with their most elegant discovery, that the alpha helix was an essential part of protein structure. Additionally, he had successfully proven the polypeptide chain hypothesis of protein structure to be valid—such a "chain" being a string of amino acids hooked together like boxcars. Pauling had accurately surmised that DNA had a helical structure, and many already knew that it did in fact carry genetic information from one generation to the next. But Pauling was handicapped by a thoroughly anchored stubbornness even in the face of contrary evidence. In 1951, he erroneously speculated that DNA was a three-stranded helix, despite warnings by colleagues such as the American chemist Verner Schomaker, who told the author in his office at the time of his retirement from the University of Washington that something was "not right" about Pauling's structure.

Still, even others not so handicapped, such as Watson, Crick, and Franklin, had no real picture of DNA's molecular "architecture." Using X-ray pictures "borrowed" from the British scientist Rosalind Franklin and by imitating the enormously powerful "model-building" techniques of Pauling, Watson and Crick eventually succeeded in divining the structure of DNA in 1953. In their famous 1953 paper, "A Structure for Deoxyribose Nucleic Acid," they acknowledge their debt to Pauling:

> A structure for nucleic acid has already been proposed by Pauling and Corey (1953). They kindly made their manuscript available to us in advance of publication. . . . We wish to put forward a radically different

> structure for the salt of deoxyribose nucleic acid. This structure has two helical chains each coiled round the same axis.[50]

Thus the most significant feature of this search was the realization that DNA was a double helix—two intertwined strands—a proposal that the X-ray pictures fully supported. As things turned out, Rosalind Franklin's analysis was unquestionably the most decisive of all. Had Watson and Crick not borrowed her notes, it is exceedingly likely she would have received the prize as well. For she was the only one to realize the importance of studying and photographing both the "wet" and "dry" forms of DNA, which turned out to be the key to the riddle.

Shortly, Watson began colluding with Francis Crick, a thirty-five-year-old Englishman, originally trained as a physicist but now more captivated by DNA. Francis Harry Compton Crick was born in England in 1916. He early on majored in physics but then switched to molecular biology, soon joining the Medical Research Council's (MRC) laboratory of molecular biology. From there he went to the Salk Institute where he remains today. While at MRC he became interested in of the structure of gigantic biological molecules, a field that Linus Pauling of Caltech had pioneered. By 1951, the American biologist James Watson was at MRC, and the two straightaway joined forces. Watson was born in the United States, but majored in molecular biology and genetics at the University of Copenhagen in 1950. After that he added his talents to those already working at the prestigious Medical Research Council's Cavendish Laboratory in Cambridge, England.

A NEW FIELD OF BIOLOGY IS BORN

Together they continued their imitations of Pauling's "baby toy" methods, pasting together jumbo tinker-toy-like models of molecular structures. What they found was that DNA is made up of four kinds of still-smaller molecules. These then hook together in intertwining spiral chains. It is the arrangement of these molecules in every cell that constitutes a unique array of genetic "instructions" which delimit the character of the offspring. When they had finished, humankind knew how genetic "instructions" passed from one generation to the next. For this, Wilkins, Watson, and Crick were awarded the 1962 Nobel prize.

With the unveiling of the structure of DNA, the budding field of "molecular biology" would expand dramatically. In molecular biology, the target of scientific investigation was the individual molecule rather than the entire organism. As a consequence of this tremendous progress, DNA studies would unconditionally dominate biological research for the rest of the century. For the first time science had an explanation of how Mendel's laws of heredity operated at the deepest level of the biological organism. Diabolic results followed. In 1965, Cornell chemist Robert Holley who aided Vigneau in the first synthesis of penicillin, was able to duplicate Watson and Crick's feat for transfer RNA, the molecule responsible for "reading" the instructions from the DNA molecule and interpreting them to build polypeptide chains. After this, Holley became interested in cell division in mammals.

Of course other sciences had an integral role in the "story of life" that was barely starting to reveal itself. The Scottish population biologist and Nobel laureate J. B. H. Haldane who had, in the 1930s, helped put Darwin on a Mendelian basis, found that amino acids, which make up proteins, were themselves constructed out of methane, water, ammonia, and hydrogen. This, in turn, led many to consider whether life itself may have actually hatched from chemical reactions among inorganic materials. An "inorganic" material is anything not based on carbon. Water is inorganic, for instance, despite its sizable role in the lives of organic beings.

GENETIC ENGINEERING

After Watson and Crick solved the DNA riddle, the macabre field of "genetic engineering" evolved at a geometric rate. In essence, scientists in this field were curious about whether and how it was possible to actually alter an organism at the genetic level, so that the next generation would be radically modified. Edward Tatum of Stanford and Joshua Lederberg of the University of California had begun searching for examples of genetic recombination and actually located instances of it in the primitive life-form *Escherichia coli,* a bacterium found in animal digestive tracts. In 1946, the Vanderbilt University biologist Max Delbruck already known for his studies in chemical bonding and nuclear physics, and the American Alfred Hershey who would show later that DNA was the bearer of genetic information in the nucleus found independently that it was possible to merge chrososomal material from two different viruses to create a unique virus—one more ominous step into the new realm of genetic engineering. In 1969, Delbruck, Hershey, and Salvador Luria received the Nobel prize for this and other offensives on the problem of how viruses reproduce themselves.

By 1952, Lederberg would again make a contribution by finding that bacteria-eating viruses can, while they are eating the viruses, also transmit genetic material to them. That same year he coined the term "plasmid" to denote those parts of bacteria that contain genetic material outside of ordinary chromosomes. More specifically, a plasmid is a piece of the DNA molecule that exists independently of the DNA of the chromosome and which can therefore replicate itself. Then in 1959, a Japanese group found that one of the functions of plasmids was to carry the ability to resist antibodies from one bacterium to another. They in fact found this trait in the bacterium *Shigella dysenteriae.*

Lederberg, incidentally, helped to develop the field known as "exobiology," or the study of life on other worlds. In his essay "What Do We Seek in Space?" he poses the great question for exobiology:

> In the world we know, nucleic acids and proteins come about only as copies of what has evolved before them. Their blueprints are handed down from parent to offspring. But how did these complex substances come about originally—without pre-existing cells or brains to guide their production?[51]

PROGRESS IN IMMUNOLOGY

A frightening phenomenon connected with bacterial "defenses" against outside invaders surfaced in the late 1960s. In 1967, The National Academy of Sciences published an article reporting that the practice of adding antibiotics to animal food could allow some of the antibiotic to linger in the meat, thereby increasing the ability of bacteria to resist antibiotic drugs.

These inferences electrified the biological community, especially since the myriad practical possibilities of this labor seemed virtually unlimited. So it was that in 1965 the biochemists William Dreyer who had already studied mutations in DNA, and J. Claude Bennet began to think about the problem of how a single antibody can fight off so many individual kinds of antigens, the latter term referring to any external entity invading the body. They theorized that this was possible because the antibody, while it has one gene that never varies, also has many others that are infinitely "plastic." This means that they can transform their structure in any way necessary to fight off an invader. In 1967, starting from these assumptions, Gustave Nossal posed the hypothesis that antibodies, with their malleable disposition, operated by detecting the size and shape of the invading organisms or antigen and changing their own shape to "bind" with the antigen, thereby rendering it harmless. (This is sort of like the liquid metal Terminator in *Terminator II,* starring Arnold Schwarzenegger albeit with good and evil reversed.)

BIOCHEMISTRY AND VIROLOGY MOVE AHEAD

Science made still more headway in biochemistry with Imperial College chemist Derek Barton of England, who, in 1949, began his evaluation of the giant molecules constituting the class of compounds known as steroids. Building on Pauling's findings among others, he proved that the geometry of a molecule is as consequential in determining its properties as its actual constituents. Specifically, he found that reaction rates of ring-shaped steroids depend on how they are 'twisted.' Related to this work was the creation by the American microbiologists John Enders, Thomas Weller, and Fred Robbins of a faster and more efficient method of cultivating viruses. John Enders had perhaps the most exotic career; in his early days he majored in exotic languages at Harvard, but before long gravitated to biology. After majoring in science at the University of Missouri, Frederick Robbins acquired an M.D. degree from Harvard in 1940. He served with the United States Army during World War II and then returned to Harvard Medical School, where he began his collaboration with Weller and Enders.

After graduating from Harvard Medical School Enders stayed on at Children's Hospital and began his research into virology. Weller and Enders cultured both the mumps and polio viruses in cells extracted from a chicken embryo. Banding together with Robbins and Weller, Enders found that he could culture whole colonies of viruses in isolated tissues. That is, he did not need an entire organism as a host, as previous scientists had believed—pieces of tissue would suffice. With this technique he was able to set up viral cultures to grow the virus known to cause mumps, measles, and polio. Robbins afterward went on to investigate hepatitis and typhus. In 1951, Enders and the same colleagues created a vaccine which physicians used widely and efficaciously in the 1960s against measles.

After this, antigen–antibody studies had advanced sufficiently that biologist Cesar Milstein was ready to add his contribution to the inexorable progress of biological wisdom. Born in Argentina in 1927, he received his chemistry degree in Buenos Aires and then spent three years at Cambridge as a postdoctoral fellow. Eventually he became a permanent member of the prestigious Medical Research Council in Britain. From his earliest days he was concerned with immunology and enzyme activity. His most notable deed is the discovery of and his devising a method for manufacturing of monoclonal antibodies—an indefinitely large number of antibodies, all of which have a common origin in a single "ancestral" cell. All of these antibodies behave identically, as one might expect from clones; that is, all can bind with the same antigen. Since antigens are customarily illness-producing organisms, such identical antibodies were immensely useful in fighting any malady caused by a single antigen. By 1975, along with Goerges Kohler, he had devised a technique for manufacturing them in usefully large amounts, which involved fusing lymphocytes with marrow-cancer cells to yield a hybrid called a 'hybridoma.' For this, the Nobel committee gave him, along with Georges Kohler and Niels Jerne, their highest award in 1984.

In the 1950s, the Italian virologist Renato Dulbecco now at the Salk Institute, began his search for clues on how viruses interact with host cells. Dulbecco received his initiation in science at the University of Turin, finally joining the faculty of the University of Indiana in 1947 and then the California Institute of Technology in 1954. But it was during his tenure at Salk Institute in the 1960s and 1970s that he did his universally acclaimed analysis on phages—viruses that infect bacteria. He found first that ultraviolet light would eliminate the ability of a phage to react to any type of light. Subsequent to that he created the so-called patch test that allowed virologists to easily recognize mutations in animal viruses. This, in turn, led to tremendous progress in the understanding of viruses suspected of causing cancer. For all of this, he would share in the 1975 Nobel prize with Howard Temin and David Baltimore. Later, Dulbecco's findings would grow especially acute in the battle against AIDS. He explained much of this work in his 1987 book, *The Design of Life.*

BUILDING VIRUSES

Still, up to this point no one actually knew how a virus was "constructed." The problem dissolved in the hands of the Berkeley biochemist Heinz Fraenkel-Conrat, who, after brilliant work in protein chemistry in 1955, found that they consisted of a protein coat and an infection-producing center consisting of nucleic acids. This, in turn, allowed the assembly of an active tobacco mosaic virus. In the ensuing years, the presence of nucleic acids in viruses gave rise to the view that viruses were actually living organisms rather than mere inanimate matter, as some biologists were insisting. In an effort to break new ground the American biologist Sol Spiegelman certified that they were alive when he actually managed to synthesize a self-reproducing virus—the ability of an organism to reproduce being, arguably, a pretty fair test of whether or not it is "living." Shortly afterwards, in 1966, Spiegelman and his partner Iciro Haruna found an enzyme that causes RNA molecules to duplicate. Spiegelman also showed that the DNA helixes would separate if heated and that an RNA helix could fuse with a DNA helix to form a hybrid molecule. Not long after this, a parallel "architectural" problem dissolved when Gerald Edelman of Rockefeller University and Rodney Porter of Oxford captured the Nobel prize for exposing the chemical structure of antibodies. This they accomplished by studying patients with multiple myelomala (a form of cancer).

By 1976, the Japanese geneticist Susumu Tonegawa of the Basel Institute proved that the genes dedicated to making antibodies actually move close to one another on a chromosome to combine their potency, so to speak, to manufacture antibodies en masse.

Educated at Kyoto University, Tonegawa did his postdoctoral studies at the University of California at San Diego. He spent many years exploring immunology in Switzerland before returning to the United States in 1981 to accept a post at MIT. The basic question he wanted to answer was this: how could one explain the body's ability to produce the vast and bewilderingly varied number of antibodies needed to fight off so many different possible ailments? In the 1970s he began finally to answer that question when he showed that antibody-producing cells, or "T cells," contain huge numbers of distinct genetic "instructions." The enormous number of different realizable combinations of these instructions is what allowed the manufacture of the wide array of antibodies necessary to meet the body's defensive needs. Though he had to wait longer than one might commonly expect, the biological community was pleased to honor him with the Nobel prize in 1987 for his profound attainments in grasping the immune system.

Virologists were not, of course, targeting only antibodies; many wished to discover more about antigens as well. That was the trail that the American biologists Baruch Blumberg and D. C. Gajdusek followed. Daniel Gajdusek studied medicine at the Harvard Medical School and, in 1958, became the Head of a National Institutes of Health program to observe the progress of disease in underdeveloped nations. In Papua, New Guinea, he identified a fatal disorder called "kuru." By discovering the cause of the illness, he thereby also found an entirely new strain of virus, a kind that incubates so slowly that its effects ordinarily take at least a year to manifest themselves. He postulated that cannibalism transmitted the sickness throughout the Papua communities. Following this work, using similar techniques, he identified the hepatitis antigen. In 1976, both Gajdusek and Blumberg shared the Nobel prize for this feat.

By the 1970s, inspired by the distinguished results in the 1950s of Heinz Fraenkel-Conrat on the structure of the protein coating of viruses, other researchers applied their energies to clarifying the genetic structure of still another virus—SV40. Relying again on Fraenkel-Conrat's data, they unraveled the genetic structure of the coat of SV40.

Other inquiries into genetic physiology worth pondering are those of the legendary Italian biologist Rita Levi-Montalcini. After studying medicine in Italy, she moved, in 1947, to the United States to conduct physiological trials at Washington University. It was here that she became enthralled with the nervous system in embryos, eventually discovering a "redundancy" factor—the mechanism that causes the nervous system to produce far more cells than it genuinely needs, to allow for the yet unknown final size of the new infant. The factor she found to be accountable for this she called nerve growth factor, or NGF. She then returned to Rome to head the Laboratory for Cell Biology. For her historical pronouncements on cell physiology she and her colleague Stanley Cohen captured the 1986 Nobel prize.

BUILDING MORE GIANT MOLECULES

In 1954, geneticists finally unraveled a haunting mystery. They knew that the instructions for assembling proteins come from the DNA molecule, but their grasp of the intermediate steps remained aboriginal. But in 1955, the molecular biologist Mahlon Bush Hoagland of Harvard Medical School, in addition to seminal studies on cancer of the liver, claimed that a type of RNA called "transfer RNA" combined with amino acids and that still another never-before-seen version of RNA assembled the amino acids into phenomenally long protein chains. Hoagland promptly called the new RNA "messenger RNA." In 1961, Robert Holley of Illinois using chromatographic techniques, managed to extract samples of at least three unlike variants of transfer RNA, while, in the same year, the Cavendish chemist Howard Dintzis proved beyond any doubt that transfer RNA assembles the polypeptide chain one amino acid at a time, going systematically from one end of the polypeptide chain to the other. Before this, Dintzis had drastically improved techniques for preparing myoglobin for X-ray study.

Fascinated by Hoagland's proposals, the French biologists Jacque Monod and Francois Nancy began their own confrontation with messenger RNA. Monod majored in biology as an undergraduate studying with Ephrassi and Tessier, then went on to earn his doctorate in 1941, eventually joined the zoology department of the Sorbonne. By 1971 he had become the director of the Pasteur Institute. Soon Monod and Nancy confirmed Hoagland's conclusions. More explicitly, they devised the concept of the "operon," a group of genes that directed the manufacture of myriad enzymes. Another breakthrough of Monod's was the idea of a "repressor"—a part of a protein molecule that was also involved in regulating enzyme manufacture. Ultimately, Monod and his collaborators would be awarded the Nobel prize for their breakthroughs.

Ultimately, however, it began to dawn on scientists that enzymes might do other peculiar things to nucleic acids such as DNA. The Czechoslovakian biologist Martin Gellert performed some of the critical preliminary experiments in these areas. He began in the 1970s and by 1976 he had found the enzyme gyrase. The name was well chosen. For Gellert shortly discovered that gyrase would cause DNA to go into a "supercoil." DNA, like most proteins and nucleic acids is, in fact, already coiled in its natural state. That is, DNA is standardly a double helix—two spiral strands twisted and intertwined around one another. Amazingly, gyrase caused such helices to coil even further into supercoils, although to date, no one has found a serviceable application for this phenomenon. By contrast, there were, in some cases, ample practical—and profitable—uses for genetic inventions. On June 16, 1980, the Supreme Court by a 5–4 decision, ruled that researchers at General Electric had the right to patent an oil-eating microbe *Pseudomonas,* developed by A. Chakrabarty, expressly to eradicate oil spills.

ARTIFICIAL LIFE

Still further progress in genetics came in 1955, when the Spanish biochemist Severo Ochoa of New York University, beyond his great work in fatty acid metabolism and photosynthesis, created a facsimile of RNA in a test tube, while the next year the Stanford biochemist Arthur Kornberg concocted DNA in the laboratory using an enzyme he had discovered, *DNA Polymerase,* to alter the action of bases found in the DNA molecule. Actually, this was little more than a superficial imitation of the DNA molecule, as it could not actually modify the genetic structure of an organism. In a word, it was a DNA "dummy." Nevertheless, it was a striking step forward, and for it Kornberg received the 1959 Nobel prize, sharing it with Ochoa.

But the progress toward counterfeit life did not stop. Having successfully synthesized the DNA molecule, others wanted to go one step beyond this and synthesize an entire gene. This happened in 1976 with the epochal explorations of Gobind Khorana and his staff. Khorana mastered biology at the Punjab University as well as at Cambridge University. He then moved to Simon Fraser University in Canada and after that to the Institute for Enzyme Research at the University of

Wisconsin. For many years he probed the puzzles of the structure as well as the function of nucleic acids, all of which led him to be the very first to so construct a gene. Not only did it "look like" a natural gene, but, unlike Kornberg's dummy DNA, it *functioned* as one as well.

In 1956, the Rockefeller Institute cytologist George Emil Palade in addition to work on mitrochondrial microsomes found that ribosomes, long known to exist inside cells, contained mainly RNA. This was scarcely a shock, since they assist in directing protein construction. Scientists find them near the "manufacturing" site of protein synthesis in cells. By 1962, the downstate New York biologist Marshall Nirenberg of the National Institutes of Health began to decipher the mysteries of the genetic code. He showed that in RNA molecules, when the genetic code has the formation UUU, three uridylic acids in a row, the RNA "tells" the cell to manufacture the amino acid phenylalanine. In 1968, the Illinois chemist Robert Holley and Gobind Khorana of the University of Wisconsin merited the Nobel prize because of their success in further deciphering the genetic code. Some of the codes were stubborn. One in particular was the codon formation UGA—uracil, guanine, and adenine. Though researchers had untangled the sense of most of the other "words" in the genetic language, the meaning of this had so far eluded them. But in 1968, the University of California biologist David Zipser, also known for his interest in the neural basis of consciousness, found that it was a "terminator," a message to the cell to stop making protein.

Toward the end of the 1960s, therefore, science had learned much about DNA, the "messages" it sends, the role of "messenger" RNA, and so forth. Kornberg, his pioneer spirit not appeased by the creation of a mere dummy of the DNA molecule, was still trying to produce functional DNA. He relentlessly bore through all obstacles. In 1967, he successfully assembled a DNA molecule that was biologically active. By taking a strand of natural DNA he simply constructed the rest of the molecule from basic chemicals in his laboratory. A virtual repeat of this came, once again, with another illustrious performance by Gobind Khorana at the University of Wisconsin in 1970. He found ligase (from phage T_4), an enzyme that would join pieces of DNA together. Khorana now at M.I.T., was able to proclaim, after some years of exacting and tedious battling, that he had artificially created "aniline-transfer RNA." Beyond this, Khorana helped show that RNA is read in linear order, one codon at a time. By 1988, Ya-Ming Hou and Paul Schimmel had contrived a method for "translating" some of the genetic "language" found on the transfer RNA molecule, the molecule that, following DNA instructions, actually links amino acids together into a polypeptide chain.

THE LANGUAGE OF GENES

Yet there was one query, rather an obvious one in hindsight, that few had actually posed. When one speaks of a "message," the immediate dilemma is whether the message is "universal"—do all forms of life use it or not? That is, did all the myriad species of animals have their own genetic codes peculiar to their own species? Or was the language of genetics a universal one, a sort of Esperanto of the gene? In the sixties Charles Caskey, Richard Marshall, and Marshall Nirenberg of the National Institutes of Health set out to solve this problem. Along with Khorana, they quickly proved that there was indeed a universal language of the gene. They demonstrated beyond question that DNA is identical in reptiles, amphibians, vertebrates—even bacteria. They found also that identical messenger RNA molecules manufacture identical amino acids. The link between man and other categories of biological existence was strengthening again, much to the despair of religious extremists, who contended that there was something completely unique about *Homo sapiens*. In fact, as Nirenberg put it, all life forms "use essentially the same genetic language."

Even so, there were still many unanswered doubts. Although scientists had long been convinced that genes did, of course, exist and had even synthesized one, as noted above, no one had actually isolated a naturally occurring gene in the laboratory. In 1969, geneticist and recipient of the Eli Lilly Award Jonathan Beckwith of Harvard Medical School ended that problem when he isolated a gene found in bacteria that is part of the genetic "message" for the metabolism sugar.

A NEW GENETIC TOOL: THE RESTRICTION ENZYMES

In 1970, investigators were pursuing research on the new tools for recombinant DNA scholarship—the so-called restriction enzymes mentioned earlier. The successful scientist here was microbiologist Hamilton Smith of Johns Hopkins, who isolated the first of the restriction enzymes, the enzymes that will split a DNA molecule at a specific point (some luck was involved, since he happened to notice that certain bacterial cytoplasms will break up a DNA molecule). After this beginning, several new such enzymes began to appear. Much of the credit for this work goes to physiologist Daniel Nathans, who was Smith's colleague at Johns Hopkins in this chore. The laser found its way into genetic technology as well when, in 1986, Arthur Ashkin and his team were able to trap protozoans with lasers—a novel technique for manipulating and studying whole organisms.

Here the career of the American biochemist Herbert Boyer becomes critically significant. Boyer graduated from the University of Pittsburgh with high honors and swiftly became professor of biochemistry at the University of California at Berkeley. By 1970, the fact that DNA controls protein construction was an old story. Understandably enough, this led to speculation as to whether it was possible to "tell" an organism to produce alien proteins by somehow hooking it up to alien DNA. In 1973, Boyer and Stanley Cohen of Stanford did exactly that, thus proving that the restriction enzymes and other gene-splicing techniques used to place DNA inside bacteria had an even greater application. They could both cut DNA molecules to pieces and, together with other enzymes, join the pieces together in novel ways. Boyer, in 1973, spliced DNA from alien plasmids onto the DNA of *Escherichia coli* bacteria, thus making more and different *E. Coli* for further study, since it is easy to grow large amounts of bacteria.

BALTIMORE AND REVERSE TRANSCRIPTASE

Still, the genetic engineering "toolbox" was not yet complete. In 1970, the American virologists Howard Temin of Philadelphia and David Baltimore gave science another tool in their classic work on enzymes and DNA. Baltimore graduated from Swarthmore in chemistry and completed his postdoctoral education at MIT, Cold Spring Harbor, and the Rockefeller Institute. His first fulltime post was at the Salk Institute, and he became a full professor of biology at MIT in 1972. Some of his work emanated from his assaults on the polio virus, and he found out precisely how it reproduced as well as how it synthesized protein. By 1970 he was ready to overturn an absolutely fundamental cluster of axioms in biology.

Baltimore and Temin independently found an enzyme that Baltimore called "reverse transcriptase." This enzyme can cause RNA to be transcribed into DNA. In other words, the enzyme causes genetic information on the RNA molecule to be copied onto the DNA molecule. This was significant in that it is the reverse of the usual process and is a phenomenon that is seen occasionally in tumor viruses. Up to this time scientists had assumed that the steps to protein manufacture were one-way only. They believed, that is, that DNA is always transcribed into RNA which is then transcribed into protein. In short, Baltimore and Temin found that reverse transcriptase could convert protein back to DNA. Both earned the 1975 Nobel prize in physiology.

Reverse transcriptase turned out to be an immensely practical tool both for genetic engineering and for the understanding of disease. In 1982, for example, the American biologists William Mason and Jesse Summers showed it is precisely reverse transcriptase that allows the hepatitis B virus to replicate. (Since the HIV virus makes use of this phenomenon, the research was very timely.)

ESCALATING FEARS OVER SCIENCE

Much of the work described above presumably constitutes the earliest beginnings of what has since come to be called "recombinant DNA"—no wonder so many, such as Ruth Hubbard of Radcliffe, brooded over the possible inadvertent creation of some kind of doomsday virus. Soon, recombinant feats, along with Watson and Crick's advances, created profound disquiet about the threat of genetic experimentation that promptly seeped into the scientific community. Many were terrified that someone might inadvertently create an unstoppable new pandemic that would decimate the globe.

But DNA research as such was not the only concern. Fears about the consequences of developing technology had been surfacing since close to the turn of the century.

In an eerie foretelling of impending perils, the General Electric engineer Charles Steinmetz, inventor of the metallic electrode arc lamp, warned, in his *Future of Electricity,* that both burning coal and dumping waste into rivers would inevitably pose monstrous hazards for future civilizations. He wrote this in 1910, and concerned citizens from many fields would soon remember and take up Steinmetz's cause. Similar fears surfaced with the development of nuclear energy after the war. Still, the pace of the U.S. government in warning the public about the dangers of radiation proceeded slowly—all too slowly for many who feared the dangers of radiation. One of the papers that came out of the Genetics Conference on Atomic Casualties of the National Research Council, "Genetic Effects of the Atomic Bombs in Hiroshima and Nagasaki," states the following:

> Although there is every reason to infer that genetic effects can be produced by atomic radiation, nevertheless the conference wishes to make it clear that it cannot guarantee significant results from this or any other study on the Japanese material. . . . this material is too much influenced by extraneous variables. . . . In spite of these facts, the conference feels that this unique possibility for demonstrating genetic effects caused by atomic radiation should not be lost.[52]

But fears again escalated with the accidents at Three Mile Island and Chernobyl. At Three Mile Island, for example, the reactor lost the water protective shield and some radioactivity seeped out of the containment dome. By far the most ominous part of this frightening event, however, was the petrifying consequences of the total meltdown that many feared had occurred. Still, many commentators have charged that critics have blown such dangers all out of proportion to any real peril involved. Yet such criticism sounded hollow in the face of the even worse disaster at Chernobyl in April of 1986. In this calamity a Soviet nuclear reactor actually exploded, releasing a massive amount of radioactive material and generating contamination on an unprecedented scale.

It is true that there were catastrophes on the same scale, though less publicized, in nonnuclear fields. In Seveso, Italy, for instance, the Givaudan-La Roche Icmesa insecticide plant went awry, spewing a monstrous cloud of toxic gas over an immensely large area of land. In the process the accident exterminated thousands of farm animals. Much later, in the 1980s and in spite of the scope of this event, Italian physicians claimed that their observations on persons exposed to these chemicals showed no harmful effects. Then one has to wonder whether, ultimately, the depletion of the ozone layer by chlorofluorocarbons could be even worse than nuclear accidents. The ozone layer, after all, covers the entire earth, preventing harmful ultraviolet radiation from heating the planet's surface. In fact pressure from environmental groups became so severe that by 1987 twenty-four nations signed a pact limiting the use of chlorofluorocarbons, which scientists now know are systematically wrecking the ozone layer. Chlorofluorocarbons do so by forming various chlorine compounds that combine with and wreck the ozone.

Still other events plagued the environment. The Malaysian government had been using DDT extensively for many years to eradicate malaria-carrying mosquitoes. The problem was that it also killed cockroaches, which the village cats then ate. Of course, the cats then perished from DDT poisoning. That led to further imbalances. One event was tragicomic. Rats became so numerous that the government had to airlift a fresh supply of cats to restore the inherent balance of nature. Nevertheless, there lingers a quality of dread over nuclear power that does not exist with things like pesticides. The fact that nuclear energy can affect the biosphere at its core, at the level of the gene, is what scares us the most.

In 1962, Rachel Carson would sound an environmental warning alarm in her book *Silent Spring*. In this treatise she foretold the menace of unending contamination of rivers and oceans with chemical waste. As she so poetically and prophetically stated, speaking of the dangers of industrial chemicals:

> Residues of these chemicals linger in soil to which they may have been applied a dozen years before. They have entered and lodged in the bodies of fish, birds, reptiles, and domestic and wild animals so universally that scientists carrying on animal experiments find it almost impossible to locate subjects free from such contamination.[53]

While many challenged her, evidence surfaced in coming years revealing that she was not as far off the mark as many had supposed. In 1968, a Congressional report found that Lake Erie was so polluted that it might require half a millennium to restore it. By 1972, the United States, not having appreciated the moral of Malaysia, banned DDT only when ecologists realized that the insecticide was systematically thinning birds' eggshells in immeasurable numbers, since DDT destroys hormones that assemble the calcium needed in the egg.

This was hardly the end of catastrophes traceable to tinkering with the biosphere. In Egypt the Aswan High Dam, completed in 1968, has drastically cut down the flow of algae nutrients to the Mediterranean which, in turn, has severely damaged the fish population and, consequently, the fishing industry in the Mediterranean. But at least the flooding of the Nile is, after eons, now predictable. In 1978, a massive eruption of anthrax occurred in Russia. Though the final explanation never appeared, the Soviets have always maintained that the outbreak developed from tainted food. The United States insisted that it came from an industrial accident at a germ-warfare plant, something the Soviets, quite naturally, would not want to admit.

As more disasters occurred, more environmentalists tried to raise world consciousness on the issues as well as to halt these fiendish trends. In 1986, to mention one case, biologists managed to capture all known black-footed ferrets, a type of prairie-weasel, and place them in a program of controlled breeding. The principal cause of their reduced numbers was environmentally caused distemper. Similarly, in 1986, naturalists placed the last known wild California condors (of the family *Carthartidae*) in a breeding program.

Nevertheless, the lingering fears over genetic research remained throughout at the forefront of social and scientific concern—it was a kind of *leitmotif* in the unfolding of this area of research. Many continued to think not about conjectural benefits of this hair-raising technology, but about its darker side. Fears began to surface at a 1973 conference on nucleic acids. By 1974, numerous scientists and concerned laypersons were becoming worried about the dangers of DNA experimentation. In that year the scientist Paul Berg, himself a towering figure in cloning technology, leading a committee of 139 pedagogues from the ranks of the United States National Academy of Sciences, demanded an end to all genetic tinkering until it became manifest beyond any doubt that no real risk of creating "doomsday" organisms existed. He did this in the famous "Berg letter" of July 26, 1974. Not surprisingly then, in the midst of success came a rout.

Opponents of genetic fiddling at last scored at least a partial triumph. They managed, via a lawsuit, to block any more testing of genetically altered bacteria. Although such tests did resume in the years following, there is little question that the suit stonewalled the progress of genetics for several years. By 1976, however, the panic over genetic engineering had ebbed enough that entrepreneur Herbert Boyer, then on the faculty of the University of California, and 28-year-old venture capitalist Stephen Hall, felt secure enough to found Genentech, the first profit-making body dedicated to using the techniques of genetic engineering for commercial purposes. Their desire was to produce such things as insulin, human growth hormone, and other vital human materials. In less than five years the company was worth over 120 million dollars. They eventually produced the desired compounds, and many other life-giving substances.

Perhaps a counterweight in the controversy was the fact that DNA research, at least, had potentially promising life-saving medical applications. It appeared feasible, for instance, to tinker with the DNA of organisms like *E. coli* by grafting foreign DNA into it in such a way that the foreign DNA would "tell" the *E. coli* to manufacture, say, insulin—something *E. coli* does not do.

With all of this hardware and erudition at hand, scientists turned to an unusual sort of problem—a more advantageous medical application of genetic technology. As indicated already, there are several diseases "located" on genes. Pauling identified sickle-cell anemia, for example, in 1949, as a genetic malady. Others include certain categories of muscular dystrophy, hemophilia, Huntington's disease and many more. The great problem, of course, is that since there are over two million different genes, it is stupendously difficult to correlate a disorder with the gene that causes the infirmity.

This is when some of the restriction enzymes again come into play, the enzymes that snip DNA molecules into pieces. Using these enzymes, obtainable from bacteria, the biologist can control when the DNA is "cut," producing pieces of DNA that can then be sorted and "filed." Assume now that physicians have proven that there is a genetic disease in a given family, say sickle-cell anemia. While they do not know which piece of the DNA is liable for this condition, they do know or can easily reveal which family members suffer from it. They then extract pieces of DNA from family members who are afflicted with sickle-cell and pieces from those who are not. Typically, there is a

DNA piece in the afflicted relative that does not appear in the DNA of the unafflicted relative. That piece of DNA is then the "marker" for the complaint. Using it, the scientist can predict on which chromosome the gene typically resides.

The role of genetics in family biology became even clearer in 1984 when the British biologist Alec Jeffreys designed a procedure theoretically as credible as fingerprints for establishing identity. He found that there were individual sequences of DNA that could belong to one and only one person and, by extension of the same logic, that certain similarities between pieces of DNA from different persons meant that they had to be members of the same family. He speedily realized that it was possible to use this technique for learning family relationships, such as paternity. One such case occurred involving a Ghanian boy born in England who later went to Ghana to live with his father. Later, when he tried to return to Britain to live with his mother, immigration officials refused to admit him, claiming that he was not her son. Jeffreys, using the genetic fingerprinting techniques, was able, however, to prove that he was her son.

Using genetic techniques, Kay Davies and Robert Williamson of St. Mary's Hospital in London located, in 1983, the very first "marker" for the disease Duchenne muscular dystrophy, a degeneration of the brain stem and spinal cord, often affecting middle-aged men. (Williamson had also worked on cystic fibrosis.). Finding the marker, nonetheless, is not as unequivocal as locating the actual gene; the marker only bares a *region* of the chromosome where the suspect gene probably is—not its exact location. Sometimes, however, scientists can both find and duplicate, or "clone," the gene.

In 1987, the American physician Murray Bornstein, building on the above gains, told the world that the drug Cop 1 was immensely successful in checking the progress of muscular dystrophy. That same year, in related work, American physician Louis Kunkel, brilliant scientist and amateur gardener of Harvard Medical School and his team, found the protein "dystrophin." Their studies had partially confirmed the hypothesis that the lack of that protein could cause Duchenne muscular dystrophy. Comprehension of muscle physiology moved further ahead when Kevin Campbell and Roberto Coronado found the protein needed by the body to monitor the voyages of calcium into and out of muscle cells—such calcium interchange being a fundamental requirement in muscle contraction. They named the new protein the "calcium release channel."

Subsequently, innovative applications for gene markers were appearing. In 1978, the Stanford molecular biologists Mark Skolnick, Ronald Davis, and David Botstein, suggested while working on ways of 'mapping' the yeast genome, that a process called "DNA sequencing" would allow the creation of still other markers for genetic diseases. In 1984, James Gusella of the Massachusetts General Hospital found the gene marker for Huntington's disease, a find he described as "lucky." Then, in 1986, the Department of Agriculture granted the request of the Biologics Corporation for a license to sell a genetically engineered virus designed to function as a herpes virus vaccine for pigs. Possibly the most remarkable step ahead in the treatment of genetic illness occurred in 1988, when the American physician Rudolf Jaenisch and his personnel managed to pilfer a gene accountable for a hereditary malady and implant it in a mouse, preparing the way for "replacing" the defective gene in human beings with normal genes. Yet people paid the most attention, not surprisingly, to genetic onslaughts on cancer. There is gradually accumulating evidence, for example, that damage to genes can lead to cancer. For example, studies have shown that the majority of industrial chemicals linked to cancer cause the disease by damaging genes. In related work, many scientists have shown that certain RNA viruses will "turn on" cancer-producing processes in body cells. This has led, in turn, to a widening body of theory on "oncogenes" genes that directly cause cancer. (These originate in a benign and normal class of genes called 'proto-oncogenes.' When damaged, they can become active oncogenes.) Normally dormant, they may be activated by certain stimuli such as specific chemicals and pollutants. Once turned on, they begin sending out instructions that upset normal cell division in the body. However, there is a defense system of 'tumor suppressor' genes that help keep oncogenes in check.

Around 1988, the same year as Jaenisch's work on hereditary disorders, the Korean chemist Sung-Hou Kim of the Lawrence Berkeley Laboratory, already known for his research in crystallography, and the Japanese scientist Susumu Nishimura, diagnosed the molecular "architecture" of proteins they knew originated from the messages that a certain cancer-causing oncogene was sending. By knowing the construction of the faultily assembled proteins that emanate from oncogene "messages," there is hope that scientists will be able to reason backwards to the temperament of the malicious oncogene itself and, thereby, find a way to terminate such genes.

Another high point in the progress of treatment of genetic disease came in 1981 when scientists unsnarled the genetic code for the hepatitis B antigen. With that innovation came the possibility of a vaccine against this ancient scourge. In fact, the Merck Institute for Therapeutic Research engineered such a vaccine in the same year, which the Food and Drug Administration briskly approved for common use. In 1982, the Eli Lilly Company secured the approval of the FDA to market insulin fashioned from bacteria—the first really successful commercial application of recombinant DNA investigation. A few years later, the New York State Department of Health developed a more comprehensive vaccine that could protect not only against hepatitis B, but against herpes simplex and the commonplace flu as well. In 1985, the FDA approved for public use growth hormone that scientists had manufactured through genetic engineering techniques similar to those surveyed above. With this miraculous hormone, children who, in a previous era would have been doomed to abide anomalously small stature, could now develop to average or near-average size. There is now an interesting misapplication of the hormones; bodybuilders use them, along with steroids, to produce deviantly large muscles.

REDUNDANT GENES

In 1977, scientists found something that went one step beyond a flawed gene. The American geneticists Philip Sharp of MIT and Richard Roberts of Cold Spring Harbor, working independently of one another with an adenovirus (which causes upper respiratory infections), found that DNA contained huge sections of literally meaningless "information," which the cell discards during the manufacture of proteins. These impenetrable data the biologist Wally Gilbert and others call "introns," while the useful part of the gene—the part containing translatable information—they call an exon. In 1980, another even more peculiar oddity appeared. An American scientific group found what they dubbed "hypervariable" regions in genes, or short "messages" that repeat over and over again. Presumably this is a "backup" system in case something goes awry in the main coding.

TRANSFER OF GENES AND CLONING

In 1980, scientists took a phenomenal step in the war against cancer and, as scientists would soon realize, in the war against AIDS. Charles Weissmann of the Institute for Molecular Biology of the University of Zürich with the assistance of Biogen Corporation manufactured human interferon in bacterial cultures. He did so by placing cloned interferon genes into *E. Coli* bacteria. Acting exactly like natural interferon, the clones even prevented the growth of tumors. It came none too soon. A year later the Centers for Disease Control in Atlanta officially identified and publicly recognized AIDS as a new and lethal disease.

To date, scientists had found out how to alter genes, tear them to pieces, synthesize them, and duplicate them. Yet as science learned ever more about the makeup of the most fundamental units in a living organism, it was inevitable that scientists would dream about cloning, or using genetic material to duplicate a biological entity. Indeed, this phenomenon was beginning to fascinate both serious scientists and sci-fi hacks. Substantial progress in cloning came with the team of the American biologists Louise Clarke and John Carbon at University of California, who, having already succeeded in showing the size of the genetic "library" of *E. Coli* to be 200 "books," succeeded in 1980 in cloning a part of a yeast chromosome responsible for a key step in meiosis, the version of cell division that occurs in sperm and egg cells and which reduces by one-half the number of chromosomes in the daughter cells. Even so, no one had yet been able to actually transfer a gene from one organism to another. That impasse ended in April of 1980, when Martin Cline and his group at UCLA successfully transferred a gene from one mouse to another. In fact, he took DNA from cells resistant to methotrexate and put over 5 million cells worth of DNA into the mice, making the mice resistant to methotrexate. Even more amazing was the fact that the gene kept functioning even after the transfer.

After this the victories were both faster and scarier. In 1981, a group of Chinese biologists were able to artfully clone an entire fish—the golden carp. That same year, taking their cue from Martin Cline's conclusions, molecular biologists at Ohio University also succeeded in moving genes from one animal to another. This went one step beyond Cline's achievement, however, in that they transferred a gene to a mouse from an animal other than a mouse. Not surprisingly, the shifted gene was not functional. In 1982, nevertheless, scientists solved that problem when they took a gene involved in producing growth hormone from a rat and implanted it in a mouse. The mouse doubled in size because of the extra hormone produced by the new gene.

Quite naturally, research on cloning skyrocketed after these feats. By 1988, *Science* magazine reported that the American biologist Steen Willadsen was able to successfully clone an entire sheep. He achieved this by chopping up an embryo, placing a cell nucleus into a sheep ovum, and finally implanting this ad-libbed embryo into another sheep. No one stopped there. In fact the same year as Willadsen's probings, Allan Wilson and Russell Higuchi of Berkeley first, and rather amazingly, cloned genes from an extinct species. They did this by cloning genes from preserved tissue of the quagga—a species of wild ass related to the zebra but which was more like a horse in the rear, that had been extinct since the nineteenth century. They accomplished this by using DNA from quagga hides in museums. Although the feat had no immediate practical application at the time, other than providing another technique for the study of the history of evolution, it is not inconceivable that future developments with such techniques may make it a practical possibility to "rescue" extinct life-forms. Higuchi, for example, has already begun with mammoth DNA found frozen in ice. Other extinction stories had less of a happy ending. In 1987, the last dusky seaside sparrow perished in captivity through ordinary causes. Yet even here there was a bright side. Scientists had anticipated this demise and routed the demons of extinction, at least in part, by crossbreeding the last five dusky seaside males with females of the Scott's seaside sparrow—a close relative of the dusky sparrow.

The cloning craze persisted. In 1987, scientists first cloned the gene for, again, Duchenne muscular dystrophy. Several months afterwards they found just how the gene wreaked havoc in the body. The anomalous gene could not manufacture a specified protein needed by striated muscle in the body. However, this protein was not needed for cardiac or smooth muscle. Knowing this much it may soon be conceivable either to alter the gene itself, or at least to control its harmful effects.

By 1988, Philip Leder of NIH, inventor of the 'triplet binding assay,' which allowed researchers to study the relationship between codons and amino acids, and Timothy Stewart of the Harvard Medical

School, building on Philip Sharp's work on introns, had genetically engineered an entire mouse. Moreover they actually received a patent for this feat—the first time the latter had ever occurred. More possibilities loomed: why not a human? Though this has not yet occurred, the possibility is terrifyingly real. What havoc this may or may not wreak with religion and society is anyone's guess.

GENETICS IN THE SERVICE OF EVOLUTION

Still, biologists had only barely tapped the wealth of information embedded in genes. As things progressed, scientists realized that not only could an examination of DNA and RNA tell them how the body manufactures proteins, but it also could tell them how to determine when two species diverged from one another in the evolutionary pathway. For example, it explained when chimpanzees split off from monkeys to become a distinct species.

Evidence of still other information in genes came from the brilliant intellect of the Oregon chemist Linus Pauling, who, along with Emile Zuckerkandl, an associate of Pauling's at his institute, formulated the suggestion of a "biological clock" to pinpoint such divergences. A biological clock is an innate physiological rhythm that controls and monitors certain behaviors. The sleep cycle, growth, feeding, and the menstrual cycle are synchronized with solar, tidal, lunar, and seasonal cycles. Apparently the external environment "sets" biological clocks in some species, although often the clock continues to operate according to the normal rhythms even when the organism is isolated from its environment. Soon, the biologist Arthur Robinson would do innovative experimentation on this conception.

THE NEW CATASTROPHISM

Although DNA research was, arguably, the most dramatic and portentous work of recent times, there were significant things going on in other areas of biology. By the 1970s there had been a shift in evolutionary and biological thought. In the nineteenth century, there were two competing doctrines about how both the earth and animal species evolved. The dominant one, "uniformitarianism," held that alterations occurred gradually in both the earth and living species over enormous intervals. The other, "catastrophism," held that there had been rapid and overwhelming reverses periodically in the chronicles of the earth and living organisms. Although uniformitarianism held sway well into the twentieth century, the scientists Stephen Jay Gould, already known for his attack on the idea of recapitulation, and Niles Eldredge a paleobiologist at the American Museum of Natural History, broke abruptly with that teaching. In their dramatically inventive and radical hypothesis, the history of biology and geology was a narrative marked by "punctuated equilibrium." In fact, according to Gould, there are five mass extinctions in the fossil record, beyond the one at the end of the Permian period, when 90% of all living species vanished. Essentially this was at least a partial acceptance of the antiquated "catastrophism" postulate. According to "punctuated equilibrium," evolution proceeds via long periods of relative stability, intermingled with sudden change. While the speculation was stunning, it did not at that moment cause any incredible stir in the scientific community; indeed, some biologists dogmatically sympathetic to the contemporary uniformitarian view, rejected Gould's convictions.

But in 1980 something took place that regenerated interest in the revised catastrophism. The father–son team of Luis of the Lawrence Radiation Laboratory and Walter Alvarez were digging in Italy to mine the ancient Cretaceous period in the earth's history, which ended about sixty-five million years ago. But when they analyzed a sample of clay from the site, Luis Alvarez found that the amount of the metal iridium in it was immense—much larger than what one would expect according to the then-current geological thinking. Scientists could not ignore these outcomes. Luis Alvarez was one of the world's preeminent physicists and had, in 1968, captured the Nobel prize in particle physics for his studies of bargons and mesons. Other scientists straight away confirmed the iridium findings by analyzing other clay samples at several other sites on the globe.

As it turned out, the only reasonable explanation of the giant deposits of iridium was that at some stage in the earth's past, a massive object, maybe a comet or gigantic asteroid saturated with iridium, must have hit the earth. The calculations intimated that this may have ensued between the end of the Cretaceous and the opening of the Tertiary period, the so-called K–T boundary. Scientists further theorized that the gargantuan dust cloud that would surely have come after from such a collision would have obstructed all sunlight, causing such a drop in temperature that organisms which had evolved to "fit" the previous temperatures could no longer survive. Also, with the sunlight blocked, photosynthesis could no longer go on. Thus the impact would have drastically disrupted the food chain right at its core. Because such an enormous imbalance in nature would be the aftermath of something like this—such as disruptions of basic predator–prey relationships, for instance—a "snowball" effect would ensue and many other species would become extinct as well. This scenario could easily have caused the extinction of many species—the dinosaurs included. What made the belief seem all the more plausible was that paleontologists had known for a long time that the dinosaurs had, in fact, disappeared toward the end of the Cretaceous period!

Further inquiry only added to the evidence. The American paleontologist David Raup of the Geological Field Museum of National History, who does not support Darwin's idea of gradual change, and his colleague J. J. Sepkoski, Jr., began to seek out other mass extinctions that had occurred toward the end of the Cretaceous period (see Raup's essay "Conflicts Between Darwin and Paleontology," *Bulletin of the Field Museum of Natural History,* January 1979). Not only did they uncover several more but they found that these extinctions had occurred periodically, approximately every twenty-six million years. Since there was no known event on earth that could explain such extinctions, they again turned to possibilities in space. The most impressive surmise had to do with the "Oort cloud," a ring of comets known to exist in the solar system. According to Raup and Sepkoski's thesis, something may have disrupted the ring, possibly "Nemesis," a companion star of the sun, which hurtled some of the comets out of their natural orbits. If some of the comets had hit the earth, this could easily account for the global extinctions. That, in turn, could have caused something like an "acid rain" strong enough to ravage essentially any living plant or animal on earth.

Most recently, new discoveries support Alverez's general view that a gigantic asteroid hit the earth sixty-five million years ago. These findings appeared in the respected magazine *Nature* in October of 1992, co-authored by Alvarez. The article suggested that a 110-mile-wide underground crater had in fact appeared sixty-five million years ago and was caused by an asteroid. On the other hand, scientists like Dewey McLean of the Virginia Polytechnic Institute and State University argue strongly that even if such a catastrophe did happen, the paleontological record does not show that there was any sudden and dramatic loss of life. So, at present the issue appears unresolved. Even so, although some of the details and consequences of these reflections are still disputable, most scientists grant that something conceivably did collide with the earth toward the end of the Cretaceous period.

NEW FRONTIERS IN PHYSIOLOGY

Building on the verdicts of Claude Bernard in the nineteenth century, who found that the liver stores glucose which the body uses when it needs extra energy, the American biologists Carl and Gerty Cory pushed back the physiological mist surrounding scientific understanding of glycogen metabolism. Still more secrets surrendered themselves when Baron Alexander Todd of Glasgow, so helpful to Watson and Crick in showing how sugar and phosphate groups hook together, successfully created ADP and ATP (adenosine diphosphate and adenosine triphosphate) in the laboratory, the compounds that enable cells to use energy stored in glycogen.

By 1956, the biologist Earl Sutherland of Kansas had attained his ambition of isolating AMP, or adenosine monophosphate, which he consequently realized was also a significant step in the energy-manufacturing process in mammalian cells. The AMP story ultimately ended in 1971, when scientists at the National Institutes of Health in Bethesda showed that AMP allows neurons to "communicate" with one another, that is, transfer signals between them. The biological community recommended Sutherland for the Nobel prize for his trouble, which he received in 1971. Similar innovations in intercellular communication came with the triumph of Robert Michel, who, in 1975, guessed correctly that calcium "signals" between cells are a major factor in the functional integrity of cell membranes.

In this era, Bernardo Houssay of Argentina conducted his justly renowned review of the physiological functioning of the pituitary gland. Houssay graduated from the University of Buenos Aires in 1911, where he had already become interested in endocrine functioning, especially the pituitary gland. Among other things his findings proved that there were products of the pituitary gland that can cause diabetic symptoms to appear. Houssay built on previous research, this time on the results of Philip Smith of South Dakota who was the first to realize that the pituitary was the "master" gland of the entire endocrine system. Oddly, political problems interfered. When the Nobel committee gave him the prize in 1947, the Peron-controlled press made the preposterous assertion that the Nobel committee intended the award to embarrass Peron, costing Houssay his professorship at the University of Argentina. He regained it only when Peron left Argentina in the 1950s.

Researchers did not direct all their vigor toward endocrinological studies. In 1967, physiologist Haldan Hartline of Rockefeller University, already known for his work on the vision mechanisms in arthropods, George Wald of downstate New York, and the Finnish physiologist Ragnar Granit of the Nobel Institute would win the Nobel prize for adding to biological understanding of the physiology of the human eye. Wald also discovered vitamin A in the retina as an essential ingredient in pigmentation, while Granit also studied the role of the retina in color vision.

Born in 1900, Granit was still another in a long line of Swedish biologists, including Svedberg, Theorell, Euler, and many others. He received his medical degree from the University of Helsinki and served as professor of physiology at the Helsinki Institute for a number of years. The physiology of vision was his earliest pursuit and one that stayed with him throughout his days. He first disclosed that there are at least three distinct groups of "cones" on the retina, each manifesting a different sensitivity to light. In the ensuing years, he expanded his attentions to include the vertebrate nervous system generally.

ANIMAL NAVIGATION

There was some curiosity about animal "navigation" systems after the war as well. As alluded to already, ornithologists at various laboratories were observing the ways in which birds navigated, speculating that they might follow low-frequency sound, sunlight, the stars, and even the earth's magnetic field. University of Halle zoologist Karl von Frisch did some of the early research along these lines when, in 1947, he found that bees rely on polarized light to orient themselves in space. Others guessed that other species presumably did so as well.

In 1964, W. D. Hamilton's classic paper "The Genetic Evolution of Social Behavior" proved that there was a scientific basis for "altruistic" behavior among bees. Bees transmit more of their genes to the next generation by "encouraging" the queen to reproduce, while they stayed "celibate." Soon, sociobiologists would try to incorporate these decrees into human behavior. Some, like Edward O. Wilson, perhaps best known for his doctrine that any island has an optimal number of species, argued that ordinary altruism, or regard for the interests of others, in human beings had nothing to do with social conditioning, upbringing, or the like. It too was all in the genes. This was nowhere near the end of investigations into social behavior. In 1964, the psychologist Harry Harlow of the University of Wisconsin scored some points for the anti-sociobiological forces when he proved working with rhesus monkeys that rearing a monkey in complete isolation can cause it severe psychological damage. More generally, he showed that many kinds of emotional bonds—not just sex—hold human and animal societies together. Nevertheless, by 1971 Wilson was again at his desk. He published his book *The Insect Societies,* an application of some of his thoughts to the behavior of insect colonies. Further insights into animal behavior surfaced with the observations of James Bednarz in the 1980s, who found that certain species of hawks, particularly the Harris hawk native to New Mexico, will hunt in families, that is, offspring plus parents will hunt as a group.

PHYSIOLOGICAL PSYCHOLOGY

Many labored to try and fathom complex psychological phenomena during this period as well in the field now known as "physiological psychology." In 1949, for instance, Donald Hebb of McGill University, after splendid work on the consequences of sensory deprivation and brain damage on animals, published his factious book *The Organization of Behavior,* in which he suggested that science could explain memory by the fact that the brain can build neural networks "containing" memory when synapses, or junctions where one nerve cell connects with another, fire simultaneously. Later, this belief would evolve into the so-called trace theory of memory, the notion that whenever someone remembers something, the brain stores that memory as a "trace," an actual modification of the brain. In 1973, Timothy Bliss intensely reviewed Hebb's data and found that energetic bursts of electricity could strengthen the neural network and improve memory, thereby confirming much of what Hebb had said.

In 1983, G. L. Collingridge began experimenting with a group of receptors in the brain called NMDA receptors, so named because the chemical *N*-methyl D-aspartate was one of the first chemicals found in these receptors. Receptors typically are nerve endings that detect external stimuli. Collingridge, after many months of exertion, found that by chemically blocking these receptor sites, he would inhibit the normal activity of the neurons believed to store memory. By 1987, explorations into NMDA had offered some reason, therefore, for believing that such receptors were part of the machinery by which the brain stored memories. Yet these studies went beyond many raw earlier trials in showing that NMDA receptors behaved most efficiently when stimulated to match the rate of the so-called theta rhythms of the brain. The theta rhythms are a pattern of brain waves that occur at the rate of about four to seven hertz per second and are characteristic of light sleep.

Other experimentation was more controversial. In 1961, a reputable biologist actually claimed that flatworms who have cannibalized other flatworms that had previously learned a route through a maze, would themselves learn the maze even faster. It was an spellbinding fantasy. If it were sound, presumably a college undergraduate might get high grades by cannibalizing another who was getting high grades. Critics widely dispute this, to say the least. This view is possibly on the level of Arthur Jensen's supposition in the late 1960s that blacks are genetically inferior to whites, social adaptation being irrelevant.

Later, the downstate New York physiologist Eugene Aserinsky of the Jefferson Medical College discovered "rapid eye movements." Since these REM patterns correlated well with dreams, Aserinsky suggested that an exhaustive neural and behavioral analysis of dreams was probable.

Despite sustained assaults on such ideas from many quarters, the proponents of physiological philosophies of mind did not give up. Caltech psychobiologist Roger Sperry, David Hubel of Harvard Medical School, also known for his work on vision, and Torsten Wiesel of Rockefeller University began to try to "map" the brain. Sperry majored in both biology and psychology as an undergraduate and followed this with graduate education at Harvard, the Yerkes Institute, and Caltech. Through his seminal research on the amphibian nervous system, he became one of the first to note that there were huge differences between the two hemispheres of the brain. The popular conjecture today that the left half is primarily "verbal" while the right half is "emotional" and "spatial" comes directly from his efforts. He won the 1981 Nobel prize for this hypothesis though he also did significant studies of conditioned responses. Hubel enrolled at McGill University, moving from physics to biology in the tradition of F. H. C. Crick and Carl Sagan. The Harvard Medical School appointed him to the faculty in 1959, where he first began his collaboration with Wiesel. They began their inquest by watching the photosensitivity of the cerebral cortex which, in turn, led to attempts to "map" the entire brain. Their results were passable, at least on a general level. They indicated, for instance, which parts of the brain were responsible for memory, rational thought, and so forth, though they had no luck at all in making memory any more specific—that is, in locating individual memories in "traces."

Then in 1967, a ghastly experiment gave added impetus to physiological psychology. The American physician Michael Gazzaniga and his research group conducted a series of inquiries into "brain bisection." If they cut the corpus callosum, the part joining the two hemispheres of the brain, a person would split into two "personalities." So, in 'split-brain' patients an object positioned so that light hits only the left side of the retina will be invisible to them. Some argued that this "proved" that two minds could occupy the same body. For example, on occasion the left hand might "argue" with the right hand over which was going to use a pencil to draw with. (Gazzaniga's results appeared in a 1967 *Scientific American* article, "The Split Brain in Man.")

Somewhat less theatrical but much more compassionate was the work of Eric Couchesne in the 1980s on autism. For many years, scientists had conjectured that autism resulted from insufficient numbers of Purkinje cells in the brain, cells which act to help transmit "signals" through the brain and nervous system. Couchesne and his team incontrovertibly confirmed this view in 1988, by showing that autistic children all had the relevant sorts of aberrations in the brain shortly after birth.

The twentieth century was, therefore, an age of conquest—the conquest of the gene, the conquest of evolution, and the conquest, or at least the beginnings of the conquest, of a number of tragic illnesses.

CHAPTER 24

Evolution and Genetic Engineering

Inevitably, the brisk progress in both genetic engineering and evolutionary research would intersect. The American biologists Charles Sibley and Jon Ahlquist of Yale in 1984 applied DNA experimental techniques to reveal that the relationship between humans and chimpanzees was much closer than the relationships between chimpanzees or humans to other great apes. They also devised an evolutionary tree for birds in their 1986 *Scientific American* article, "Recasting bird phylogeny by comparing DNA's." In a similar way, they also applied DNA techniques to show that previous descriptions of how and from what species songbirds and other birds had evolved was not accurate. The year 1983 also saw much genetic research. The American biologist Walther Gehring and his collaborators found the so-called homeobox—a particular sequence that coordinates the general development of certain structures in mammals and segmented worms. Also in 1983, the American biologist Barbara McClintock captured the Nobel prize for her breakthroughs on the genetics of corn, which had almost at once led to spectacular increases in corn farming efficiency. At about the same time Andrew Murray and Jack Szostak oversaw the creation of the first manufactured chromosome.

In 1987, the halt on genetic research by the lawsuit mentioned previously ended: geneticists were again giving chase at full throttle.

GENETICS IN THE SERVICE OF MEDICINE

Science matched this exploit in 1990, when biologists found the gene responsible for neurofibromatosis, commonly known as "elephant man's disease" because of the gross disfigurement it causes. That disclosure led them to the protein the gene produced. On July 31, 1990, the Recombinant DNA Advisory Committee sanctioned the first successful gene therapies for human beings. These included treatment for such a malady as adenosine deaminase deficiency, a genetic ailment that quickly obliterates the immune system.

THE RISE OF BIOETHICS

In the 1990s biology witnessed the emergence of a entirely new field of biology and philosophy—bioethics. Biologists, philosophers, theologians, and scientists joined forces to probe the myriad ethical problems involved in biology and medicine. Signs of growing interest in the new field had appeared in the 1970s and 80s with the advent of recombinant DNA and other scary medical/biological technologies. Also in the 1970s, environmentalists had begun voicing mounting agitation over population growth and the destruction of the environment, whether through the dumping of chemical wastes or through nuclear pollution. In 1972, Harvard University sponsored one of the very first conferences that devoted a substantial amount of time to the ethical aspects of biological questions. In 1976, the proceedings of the conference were published by D.C. Heath/Lexington Books as *Issues in Population Education*. By the 1980s ethical concerns had expanded to encompass new horrors over the possibility that impetuous genetic technological orchestrations might inadvertently unleash a devastating new organism. Movies like *The Andromeda Strain* from 1971 only served to heighten such alarm. Soon, the assault on science broadened.

As always, political developments will affect the course of ethics. A recent example is the appointment of John Gibbons, previously the director of the Office of Technology Assessment for thirteen years, as President Clinton's chief science advisor. There is powerful evidence that Gibbons has a great deal of sympathy for animal rights groups that have often trashed scientific laboratories using animals for research. An aide to Gibbons confirmed in February of 1993 that Gibbons's wife was a member of PETA (People for the Ethical Treatment of Animals) and that Gibbons himself had supported PETA's attempts to secure their overall goal.

Science itself became the target of charges of fraud, which included allegations of scientists manipulating data. The years following showed that, in fact, corruption and fraud was and is a dark, unchecked, and pervasive malignancy in the scientific community. One of the most recent developments concerned the so-called David Baltimore case. In 1986, Professor Thereza Imanishi-Kari, then on the faculty of Tufts University, published a paper in the journal *Cell* claiming that a gene transfer into mice had had a far greater affect on their immune systems that anyone had thought it could. Baltimore was a co-author of the paper, although he was not accused of any wrongdoing. Instead, critics charged that Professor Imanishi-Kari had not in fact conducted some of the experimental trials she claimed she had done. Although her guilt or innocence has not been conclusively established, subsequent evidence appears to have cleared her, and in July of 1992, the U.S. Attorney's Office decided there was insufficient evidence to convict her. Still, the fact that such allegations keep appearing illustrates the degree of concern over the problem of scientific fraud.

ETHICAL PROBLEMS IN MEDICINE

Ethical qualms escalated with the advance of technology, as opportunities opened up for experimenting with embryos, using *in vitro* fertilization techniques to produce pregnancies for couples that might otherwise go childless, and so forth. So it was that in 1978 the first "test-tube" baby made her debut. A product of artificial insemination of the egg outside of the mother's body, her birth merely added fuel to the burgeoning controversy surrounding all such manipulations of the ordinary reproductive processes. The abortion controversy deepened and heightened trepidation over ethical aspects of many biological and medical procedures. Many became increasingly disturbed over the use of laboratory animals in medical experimentation. Surrogate parenting became an emotionally charged issue after the notorious "Baby M" case.

The very definition of "death" became a mammoth ethical problem. With the advent of imaginative technologies that allowed the brain-dead, or human beings in "pvs" (persistent vegetative state), to remain alive, even after their brains had ceased functioning, many physicians and ethicists counseled a radical modification of the definition of death. Instead of the old criterion whereby a physician could declare someone dead after the heart and lungs had stopped functioning, many now proposed that physicians should be able to declare someone legally dead when the entire brain had ceased functioning, even though machines could keep that person breathing. With such suggestions came the complex problem of euthanasia. Could it ever be ethically justifiable to end someone's life when they were still breathing, although any potential for meaningful existence had vanished? Could such actions be morally forgivable when a person was sentient, but suffering from an incurable and devastatingly painful pestilence, such as cancer or AIDS? The issue has become further tangled very recently with research appearing in various journals offering powerful reasons for believing that the concept of brain-death is radically confused.

Positively the most bizarre idea in recent history is cryonics. If dying of an incurable disease, you can, for a huge fee, arrange for scientists to freeze your head, or your entire body—even more expensive—in liquid nitrogen. The theory is that after they have "popsickled" you for a few generations, some new technological marvel of the next millennium will allow you to be thawed, cured, and to get on with your life. Predictably, this eccentric theory is extremely controversial, although it is difficult to prove that it is technologically futile. Of course, ethical and spiritual difficulties abound here as well, in such efforts to cheat the crypt.

THE HUMAN GENOME PROJECT

As always, scientists proceeded anyway. At the genetic level, again, there were a number of innovative technological advances and new projects. For example, there is the Human Genome Project, which has carried with it the fear that eugenics, or improving the race via genetic tinkering, might develop into an attractive philosophy to some. Biologist Robert Sinsheimer, reknowned for discovering a single-strand DNA and participating in the *in vitro* replication of infective DNA, then Chancellor of the University of California first instigated the recommendation in 1984. Reflecting on the large and costly requirements of biological investigations, it occurred to him that it would be enormously useful to map all of the three billion bases that constitute human chromosomes. Such a complete genetic map of human beings would make the search for defective genes that cause certain diseases much easier. A good example of this was the search by geneticist Raymond White of the Howard Hughes Medical Institute of the University of Utah. Although he had, by 1985, by studying Mormans of Utah and southern Idaho narrowed the search for the gene causing cystic fibrosis to a narrow region of the chromosome, the gene was not found until 1989. Many scientists feel that if White had had the complete "map," the gene would have been found much earlier. (Just as it is much easier to find a small town if you have a complete map of the area where the town is.) The United States Department of Energy began to fund the project, since the scientific community already knew that radiation caused genetic damage. Later, both the National Research Council and the National Institutes of Health added their forces to the project.

Perhaps most importantly, James Watson, co-winner of the Nobel prize for unraveling the structure of DNA, became the project head. Under Watson's leadership the project devised new ways of attacking the problem. One was to construct a "map" of all chromosomes by finding a "marker," or short piece of DNA that scientists can recognize indirectly, since they already know the appearance of adjacent sections. Such markers occurred about every two million bases or so.

Eventually, both the European scientific community and Japan joined in the project with geneticists everywhere, spotlighting some very specific aspects of the problem. As mentioned already, the project is not without somber critics. As always, there are concerns. Many brood over the possibility that unscrupulous people might use the facts gained through the project to weed out genetic "undesirables." Such fears notwithstanding, if all goes well, this genetic information could facilitate the determination of the evolutionary ancestry of many species of animals. Scientists may be able to tell, for instance, whether it is really true that human beings are the descendants of humanlike creatures who lived in Africa about a quarter of a million years ago, though at the moment this is still extremely debatable.

New technology will certainly figure into this. One of the most promising recent developments was the creation of the nation's first department of molecular biotechnology at the University of Washington in 1992. Its head is the brilliant and energetic Leroy Hood, who perhaps looks more like a movie star than a scientist. With better computers and even robots, he believes that the manipulation of genetic data will dramatically accelerate biological progress. If all goes well, the department expects to use its technology to finally produce a complete map of the hundred-thousand genes and the three billion nucleotides, or bases, that comprise them. That, in turn, should allow the medical profession to do such things as predict the health a given person might expect to enjoy during his or her lifetime. As always, money is an issue and many scientists have suggested that the money ought to be spent elsewhere.

An interesting recent offshoot of this is the "dog genome project." Professor Jasper Rine, a geneticist of the University of California at Berkeley, is currently trying to map the entire dog genome. If successful, this project will allow Rine and others to find the genes that cause many sorts of dog behavior, including such highly specific traits as the tendency of Newfoundlands to rescue drowning people. As Rine explained in a 1993 interview,

> "A major contribution of mouse genetics to human genetics has been . . . being able to map genes from one organism so one would know where to

look for similar genes in another organism . . . having a third mammal with a high-resolution map will aid this comparative mapping . . ."[55]

Closely related to the Human Genome Project is an old idea of the biologist Sydney Brenner of the Medical Research Council laboratories in Cambridge—the odd-sounding notion that it might be possible as well as useful to know everything there is to know about comparatively crude organisms, for reasons similar to those stated by Rine above. Brenner had recommended *Caenorhabditis elegans* for such a program, since in this and certain other nematodes (a variety of worm), there are less than 1,000 cells. An assault on this worm did in fact begin and science today often calls the enterprise "The Worm Project." In fact, there have been bonanzas. Biologists know the embryonic lineage of each of its cells, what the ancestral zygote was like, and so forth. They accomplished all of this in less than twenty years. By 1986, biologists understood thoroughly this worm's nervous system and were well on their way to constructing a "genome," or a genetic map, of this organism. Scientists envision this as a preliminary step for the Human Genome Project.

But while molecular biology has undeniably been the most bewitching sort of inquest going on in recent history, classical biology, or the biology of the entire organism, is thriving too. Among other things, paleontologists were unearthing many undiscovered species of both plants and animals. By the end of the 1980s, for instance, biologists had found two new species of lemurs, never before seen, on Madagascar, as well as an atypical species of monkey in Brazil. Most recently, a creature was found in Pakistan having both goat and cow features.

NEW TECHNOLOGY IN THE SERVICE OF FOSSIL STUDIES

New technology also revived old conundrums in paleontology, such as how the dinosaurs had vanished. Using new techniques at first intended for physicians, such as CAT scans—computerized axial tomography—paleontologists began scrutinizing fossil bones. Not surprisingly, much controversy erupted. In 1989, some scientists found hollow spaces in museum dinosaur bones. Since bird bones are hollow, they theorized that the relationship of birds to dinosaurs may be much closer than anyone had suspected. Other inquiries focused on comparatively less global issues and more on matters of detail. Examinations of *Tyrannosaurus rex*, for instance, indicated that their front limbs were far stronger than anyone had previously envisaged. Yet the most amazing find occurred in 1989, when the paleontologists Philip Gingerich of the University of Michigan Museum of Paleontology, Holly Smith, and W. L. Simons chanced upon fossils in Egypt intimating that there once lived a whale with hind legs. Supposedly, this fifty-million-year-old creature was still an ocean animal as the hind legs appeared too weak and puny to carry it about on land. One speculation was that they had been useful in reproduction.

EMBRYOLOGICAL ADVANCE

In the field of embryology, the cutting edge has, for some years, been on "feedback" mechanisms. The idea here is to explain how a fertilized egg develops the various kinds of tissues that constitute the adult organism. In the feedback thesis, a cell has something like an ordinary thermostat. When the temperature within the cell reaches a prearranged level, the "furnace" shuts off. When it reaches another level, the cell starts to form, say, nervous tissue. Another view proposes that certain hormones called "histones" may either engender or retard certain sorts of tissue synthesis. When the egg reaches a certain stage of embryonic development, a particular histone appears which "tells" the cell to start producing, say, striated muscle tissue.

Epilogue

Where biological curiosity will lead in the years to come is anyone's guess. It is, I think, inevitable that someone, someday, will clone a human being. Should that happen, our traditional concepts of "personhood," the soul, and much of our ethical tradition will behold itself in turmoil. That said, it is an unarguable truth that the field of bioethics will simply continue to expand as more and more troubled scientists, philosophers, theologians, and laypeople try to meet such crises.

In geology a spectacular claim emerged in January of 1993, adding to the continuing controversy over "catastrophism." At the annual meeting of the American Geophysical Union, the geologists Michael Rampino of New York University and Verne Oberbeck of NASA's Ames Research Center in California claimed that asteroids colliding with the earth might have initiated movements of the continents, producing debris long supposed to have been caused by glacial movement. Such collisions, some thought to have originated some 250 million years ago, could have triggered the mass extinctions the are thought to have occurred at that time. According to some estimates, close to 96 percent of all species perished in that cataclysm.

In paleontology, a remarkable find also occurred in 1993, when Paul Sereno of the University of Chicago found a complete skeleton of the most primitive dinosaur so far discovered. This adds tremendously to our knowledge of the evolution of dinosaurs. In the words of Professor Sereno, the new dinosaur "is close to what we expected the common ancestor of all the dinosaurs to look like."

On another front, looming developments in technology will inevitably cast new light on the tattered "mind–body" problem so elegantly stated by Descartes in the seventeenth century. Is the mind some sort of ethereal substance that just happens to be lodged in a physical body? This, of course, is the classic theological view of man. Or, is a doctrine such as "functionalism" true? According to the latter, a "mental state" is something that is causally connected with other mental states and with behavior of some sort. In some versions of functionalism, it is claimed that "mental states" are just states of the brain or nervous system—the so-called psycho-physical identity theory stated some thirty years ago by the English philosopher J. J. C. Smart. With the advance of computer technology and robotics, it is not unthinkable that someone, someday, will succeed in constructing a cybernetic organism that will outwardly be indistinguishable from a human being, either in appearance or behavior. Have we then created a human being? I think the answer is yes.

Genetic engineering will most likely keep scaring many in the very understandable way that it has so far. Apprehension about doomsday viruses will not easily disappear. If we can responsibly overcome such problems, both ethical and scientific, the future progress of biology and the potential for the genetic treatment of diseases is boundless.

Endnotes

1. Edwin Smith Papyrus as quoted in Eldon Gardner, *History of Biology,* (Burgess, Minneapolis, 1965), p. 14.
2. Ibid., p. 10.
3. As quoted in Charles Bakewell, *Sourcebook in Ancient Philosophy* (Scribner, New York, 1935), p. 45.
4. Aristotle, *De Partibus Animalium,* Richard McKeon, editor and translator (Random House, New York, 1941).
5. Aristotle, *De Generatione Animalium,* Richard McKeon, editor and translator (Random House, New York, 1941).
6. As quoted in Charles Bakewell, *Sourcebook in Ancient Philosophy* (Scribner, New York, 1935), p. 308.
7. From Hippocrates's *Nature of Man, Humours, Aphorisms and Regimen,* Loeb Classical Library, vol. IV, W. H. S. Jones, translator (Harvard University Press, Cambridge, 1931).
8. St. Thomas Aquinas, *In Quatuor Libros Sententiarum,* in Aquinas, *Opera Omnia,* curante Roberto Busa, S.I. (Frommann-Holzboog, Stuttgart, 1980), vol. I, p. 145.
9. From a letter written in 1717 in Adrianapole from Lady Mary Wortley

Montagu, *Letters of the Right Honorable mary Montagu*, 1793, Barrois (private).

10. Aristotle, *Historia Animalium*, Richard McKeon, editor and translator (Random House, New York, 1941).
11. William Harvey, *Exercitatio Anatomica* de Motu Cordis *et Sanguinis* (William Fitzer, Frankfort, 1628), chapter 1.
12. Alfonso Borelli, *De Motu Animalium*, second edition (A. Petrum Vander, Leyden, 1685).
13. Edwin Grant Conklin, as quoted in Robert B. Downs, *Landmarks in Science* (Libraries Unlimited, 1982), p. 137.
14. Stephen Hales, *Vegetable Statics* (London, W. and J. Innys and T. Woodward, 1727).
15. Thomas Sydenham, *The Works of Thomas Sydenham* (Greenhill, London, 1849).
16. Carolus Linnaeus, *A Dissertation on the Sexes of Plants*, James E. Smith, translator (London, 1786), p. 56.
17. Carolus Linnaeus, *Systema Naturae*, 10th edition (L. Salvii, Stockholm, 1758–59).
18. Erasmus Darwin, "Loves of the Plants, A Poem: with Philosophical Notes" (Lichfield, 1789), canto 4, lines 287–390, 399–406, pp. 164–165.
19. Erasmus Darwin, *Zoonomia, or the Laws of Organic Life*, 2 volumes (Dublin, P. Byrne and W. Jones, 1794), from the introduction.
20. Lazzaro Spallanzani, *Saggio di Osservazioni Microscopiche concernenti al Sistema della Generazione dei Signori Needham e Buffon* (Modena, 1767).
21. Conrad Sprengel, *The Secret of Nature Discovered in the Structure and Fertilization of Flowers* (Leipzig, 1894) (Washington, Saad Publications, 1975).
22. Le Bon, as quoted in Eldon Gardner, *History of Biology* (Burgess, Minneapolis, 1965).
23. Francis Galton to Karl Pearson, 26 Oct, 1901, in Pearson, *The Life, Letters and Labours of Francis Galton* (Cambridge University Press, Cambridge, 1914–1930), vol. IIIA, pp. 246–247.
24. Paper read to the Brunn Natural History Society in 1865, translation from W. Bateson, *Mendel's Principles of Heredity* (Cambridge University Press, Cambridge, 1909).
25. Walter S. Sutton, "The Chromosomes in Heredity," *Biological Bulletin*, vol. 4, 1903, pp. 231–251.
26. Jean-Louis Agassiz, *Études sur les Glaciers* (Neuchatel, Jent et Gassman, 1940) 2 vols.
27. Charles Darwin, *On the Origin of Species by Means of Natural Selection, or the Preservation of Favoured Races in the Struggle for Life* (J. Murray, London, 1859).

28. Ibid.
29. See T. H. Huxley, *Evidence as to Man's Place in Nature* (Williams and Norgate, London, 1863), as well as Huxley's *Evolution and Ethics and Other Essays* (D. Appleton, New York, 1896).
30. Ibid.
31. Ernst Haeckel, *Generelle Morphologie* (Berlin, 1866).
32. John E. Harris, "Structure and Function in the Living Cell," as quoted in *Classics in Biology*, Sir S. Zuckerman, editor (Philosophical Library, New York, 1960).
33. Louis Pasteur, "Memoir on the Organized Corpuscles Which Exist in the Atmosphere" (Paris, 1896).
34. Joseph Lister, speech delivered at the Sorbonne in 1892 on Pasteur's seventieth birthday.
35. Theodor Schwann, "Microscopical Researches on the Similarity in Structure and Growth of Animals and Plants" (London, The Sydenham Society, 1847).
36. Matthias Schleiden, "On Phytogenesis," in Johannes Müller, *Archives for Anatomy and Physiology*.
37. T. H. Huxley, "The Physical Basis of Life," in *Evolution and Ethics and Other Essays*" (D. Appleton, New York, 1909).
38. Claude Bernard, *Introduction to the Study of Experimental Medicine* (Henry Schuman, New York, 1949).
39. A. V. Hill, address to the British Medical Association at Manchester, as quoted in *Classics in Biology*, Sir S. Zuckerman, editor (Philosophical Library, 1960), p. 251.
40. Philip Eggleston, "What Can the Chemist Tell Us about the Living Cell?" as quoted in *Classics in Biology*, Sir S. Zuckerman, editor (Philosophical Library, 1960), p. 39.
41. Lecture by D. W. Ewer, "What Are Enzymes and Why Are They So Important?" as quoted in *Classics in Biology*, Sir S. Zuckerman, editor (Philosophical Library, 1960), p. 55.
42. Hugo De Vries, *Species and Varieties, Their Origin by Mutation* (Open Court, Chicago, 1976) p. 12.
43. Alexander Fleming, *Linacre Lecture of 1946 at St. John's College*, Cambridge University Press, Cambridge, 1946).
44. Ludwig Wittgenstein, *Philosophical Investigations* (MacMillan, New York, 1953).
45. William Bateson, talk before the American Association for the Advancement of Science, Toronto, 1921.
46. T. H. Morgan, "Sex-Limited Inheritance in *Drosophila*," *Science*, vol. 32, 1910, pp. 120–122.

47. H. J. Muller, "Artificial Transmutation of the Gene," *Science*, vol. 66, 1927, pp. 84–87.
48. J. D. Bernal, "The Physical Basis of Life," as quoted in *Classics in Biology*, Sir S. Zuckerman, editor (Philosophical Library, New York, 1960), p. 35.
49. John Searle, *Minds, Brains and Science* (Harvard University Press, Cambridge, 1984).
50. James Watson and Francis Crick, "A Structure for Deoxyribose Nucleic Acid," *Nature*, vol. 171, 1953, pp. 737–738.
51. Joshua Lederberg, "What Do We Seek in Space?" in *Life Beyond the Earth*, Samuel Moffat and Elie Shneour, editors, *Vistas in Science no. 2 National Science Teachers Association* (Washington, DC, 1965), pp. 153–156.
52. "Genetic Effects of the Atomic Bombs in Hiroshima and Nagasaki" in *Report of the Genetics Conference on Atomic Casualties of the National Research Council, Science*, vol. 106, 1947, pp. 331–333.
53. Rachel Carson, *Silent Spring* (Houghton-Mifflin, Boston, 1962).
54. I attempt to clarify it in my article, "Gillett on Consciousness and the Comatose," *Bioethics*, vol. 6, 1992, pp. 365–374.
55. Kim McDonald, "A Researcher Teams Up With Pet Dogs to Probe Links Between Genes, Behavior," *The Chronicle of Higher Education*, February 10, 1993.

Index

Acetylcholine, 294
ACTH, 285
Addison's disease, 277
Adenosine monophosphate (AMP), 357
Adenosine triphosphate (ATP), 274
Adrenal glands, 277
Adrenalin, 277
Adrian, Edgar, 235, 294
Agassiz, Louis, 213, 217, 225
Ahlquist, Jon, 363
AIDS, 352
Air, as fundamental material, 16, 28, 43
Albinus, Bernhard, 152–153
Alchemy, 56, 118, 133
Alcmaeon, 39
Alder, Kurt, 291
Alexander the Great, 34
Alpini, Prospero, 87
Alvarez, Luis and Walter, 355–357
Amici, Giovanni, 245
Amino acids, 287–288, 324, 334
Ampère, André, 195
Analogies, compared to homologies, 212
Anatomy
  in age of Vesalius, 76, 80–81
    and theological controversies, 81–83
  and Albinus, 152–153
  Aristotelian, 37–38
  of brain, 274
  and chemistry, 145–146
  comparative, 152
  and Da Vinci, 60
  early studies of, 27
  Galen's ideas of, 52–53
  in Hippocratic collection, 40–41
  homology and analogy in, 212–213
  and Malpighi, 121–123
  of nervous system, 265–266
  of plants, studies by Grew, 129–130
Anaximander, 19, 20
Anaximenes, 16, 27

Anemia, 323
  sickle-cell, 315, 349
Animals
  behavior studies, 306–308, 330
  navigation of, 278–279, 359
  see also Zoology
Anscombe, Elizabeth, 211
Anthrax, 250, 253, 347
Antibiotics, 284–285
  resistance to, 335–336
Antibodies, 309, 336
  monoclonal, 337
  production by cells, 339
Appert, Nicolas, 183
Arber, Werner, 283
Ardrey, Robert, 308
Arianism, 81
Arington, Charles, 321
Aristotle, 28–38
  as authority in Middle Ages, 56–57
  biology of, 32
  and Cesalpino, 98
  concepts of the soul, 49, 73
  and Democritus, 26
  and Malpighi, 121
  metaphysics of, 29–31
  as philosopher, 18, 140
  and Pliny, 47
  and Ray, 127
  on reproduction, 31, 35–36
  on spontaneous generation, 65
  taxonomy of, 32–35, 80, 127
  and Vesalius, 78
Armstrong, David, 50, 211
Artedi, Petri, 148–149
Artificial life, 341–342
Asclepiades, 48
Aserinsky, Eugene, 361
Ashkin, Arthur, 343
Astronomy, 190
Atanasoff, John, 313
Atherosclerosis, 327
Atom
  nucleus of, 286
  studies of, 145, 196
Atomic bomb, 270, 288, 317, 345
Atomists, 22, 24–25, 48–49
Autism, 362
Avery, Ostwald, 297, 331
Avogadro, Amadeo, 234
Axelrod, Julius, 295

Babylonian medicine, 9–13
Bacher, Robert, 328
Backus, John, 314
Bacteria
  anaerobic, 252
  and antibacterial substances, 283–285
  early studies of, 115
Bacteriophages, 283
Baer, Karl Ernst von, 240–242
Baird, Spencer, 209
Balfour, Francis, 228
Baltimore, David, 338, 344–345, 365
Banting, Frederick, 276
Bárány, Robert, 273
Barghoorn, Elso, 321
Barr, Murray, 319
Bartholin, Caspar, 86
Bartholin, Thomas, 99–100, 102
Bawden, Frederick, 282
Bates, Henry, 218
Bateson, William, 203, 281, 297
Barton, Derek, 336
Bauhin, Kaspar, 97, 102
Baumé, Antoine, 182
Bayliss, William, 272, 276
Beadle, George Wells, 304–305
Beaumont, William, 267
Becher, Johann Joachim, 117–118
Beckwith, Jonathan, 343
Bednarz, James, 359
Bees, behavior of, 359
Behavior
  and physiological psychology, 360–362
  social, in animals, 359
Beijerinck, Martinus, 207
Békésy, Georg von, 237
Bell, Charles, 237–238
Bell, Thomas, 277

Belon, Pierre, 67–68
Belozersky, Andrei Nikolaevitch, 297
Beneden, Edouard van, 204
Bennet, J. Claude, 336
Berg, Paul, 348
Bergstrom, Sune, 295–296
Beriberi, 323, 324
Berkeley, George, 143, 159
Bernal, John Desmond, 285, 286, 312
Bernard, Claude, 210, 264–267, 357
Beroza, Morton, 307
Berry, Clifford, 313
Best, Charles, 276
Bichat, Marie Francois, 162, 166
Big-bang theory, 328
Biochemistry, 336–337
  in Darwinian era, 239–240
Bioethics, 364–365
Biogenetic law, 241
Biological clocks, 354
Biology, meeting with chemistry, 117–124
Birds, navigational ability of, 278–279, 359
Bittner, John, 311
Blakesle, Albert, 302
Bliss, Timothy, 360
Bloch, Konrad, 317
Blood
  circulation of: *see* Circulation, of blood
  physiology of, 319
  typing of, 309
  as vital force, 142
Blood vessels, early studies of, 43, 52, 53
Blumberg, Baruch, 339
Bode, Johan, 190
Bogert, C. M., 319
Bois-Reymond, Emil, 235–236
Bolton, Edmund, 105
Bonner, James, 318–319, 330
Bonnet, Charles, 132, 142
  and embryology, 164–165
Bordet, Jules, 309
Borelli, Giovanni, 107–109, 121, 122
Bornstein, Murray, 350
Botany
  in age of Vesalius, 87–88
  in Darwinian era, 215–218
  and Grew, 129–130
  and Hales, 130–132
  and Helmholtz, 238–239
  and Malpighi, 124
  in Renaissance era, 63–64
  and rise of population theory, 191–192
  and Valerius Cordus, 73
  see also Plants
Botstein, David, 350
Bourget, Louis, 134
Boussingault, Jean, 238–239
Boveri, Theodor, 204–205
Boyer, Herbert, 344, 348
Boyle, Robert, 133–134, 135
Boylston, Zabdiel, 70
Braconnot, Henri, 182
Bragg, William, 312
Braid, James, 191
Brain, 160, 200, 236–238
  map of, 274
  early studies of, 26, 27, 37, 38, 40, 43, 53, 72, 122–123
  and physiological psychology, 360–362
Brenner, Sydney, 327–328, 330, 368–369
Bridges, Calvin, 304
Broca, Paul, 197, 200
Brown, Michael, 327
Brown, Robert, 217–218, 256–257, 258, 263
Brownian motion, 217–218
Brunfels, Otto, 63
Bubonic plague, 247
Buchner, Eduard, 266
Buckland, William, 210
Buddhism, 18, 34
Buffon, Georges-Louis, 141, 151, 154–156, 173, 177, 185
  taxonomy of, 157–158
Burnet, Frank, 309
Bush, Vannevar, 313
Butenandt, Adolf, 278, 307

Caius, John, 91
Calcar, Jan Stephan van, 76, 88
Calvin, John, 81, 82
Calvin, Melvin, 317
Camerarius, Rudolph, 128–129
Cameron, Ewan, 326
Campbell, Kevin, 350
Cancer
  gene defects in, 351
  and viruses, 311, 338
Candolle, Augustin de, 128
Cannizzaro, Stanislao, 234
Capillaries, studies of, 114–115, 122, 228
Carbon, John, 352
Carbon in plants, 239
Carbon-14, and tracer technology, 317–318
Carson, Rachel, 347
Caskey, Charles, 343
Casserius, Julius, 79
Castle, William, 281–282
Catastrophism, 19, 154, 187, 215, 355–357, 371
Cavendish, Henry, 189, 190
Cell studies, 229, 255
  antibody production in, 339
  botanical cytology, 261
  communication in, 357–358
  and Golgi, 245–246
  meiotic division, 205
  mitochondria in, 316, 318–319
  mitotic division, 204–205
    colchicine affecting, 302
  and Nägeli, 261
  nucleus in, 217, 256, 259, 263
  physiology, 318–319
  and Purkinje, 257–258
  respiration, 323–324
  and Schleiden, 259–260
  and Schwann, 257
Cell theory
  emergence of, 125–138
  and Miescher, 262
  and Virchow, 263–264
  and von Mohl, 262–263
  in zoology, 261–262
Celsius, Olaf, 146
Celsus, Aulus, 48
Cesalpino, Andrea, 97–98
Chain, Ernst, 284
Chambers, Robert, 215
Chance, Britton, 273
Change, causes of, 30–31
Charles I, 91, 105, 126
Charles II, 105
Charles VIII, 69
Chemical bonding, 234, 287
Chemistry
  and anatomy, 145–146
  and Boyle, 133–134
  continuing rise of, 189–190
  in Darwinian era, 233–234
  and Magnus, 56
  meeting with biology, 117–124
  and physics, 144–145
  in Renaissance era, 61–63
  in twentieth century, 286–288
Chill-le-Vignoble, Charles, 264
Chlorofluorocarbons, effects of, 346
Cholera, 249, 253–254
Christina, Queen, 101
Chromatography, paper, 315
Chromosomes, 204–206, 255, 282
  abnormalities in diseases, 315
  Barr bodies in, 319
  genes on, 299–300
  manufacture of, 364
Chyle vessels, 100
Circulation
  of blood, 37, 81–83, 228
    early studies of, 122
    and Harvey, 91–93
    and Leewenhoek, 114–115
    theological controversies, 81–82
  in plants, 131–132
Clarke, Louise, 352
Classification systems: *see* Taxonomy
Claude, Albert, 316
Clayton, John, 148
Cline, Martin, 353
Cloning research, 352–354

Codons, 327
Coelacanths, 321
Cohen, E. T., 272
Cohen, Stanley, 340, 344
Cohn, Ferdinand, 227
Cohn, Julius, 260
Cold elements, female, 22, 36
Collingridge, G. L., 360
Collip, James, 277
Colombo, Realdo, 79, 83
Columella, 45–46
Combustion, studies of, 117–120
Computer use, 185, 313–314, 329
Conditioned reflexes, 267
Condors, California, 347
Confucius, 21
Conklin, Edwin, 125
Controlled experiments, introduction of, 107
Copernicus, 76, 104
Cordus, Euricius, 73
Cordus, Valerius, 73
Corey, Elias, 326, 332
Coronado, Roberto, 350
Correns, Karl, 280, 281
Cortés, Hernando, 69
Cortisone, 277
Cory, Carl and Gerty, 357
Couchesne, Eric, 362
Couper, Archibald, 234
Courtois, Bernard, 182
Cowles, R. B., 319
Crick, Francis, 297, 302, 328, 332–334, 361
Cryonics, 366
Crystallography, X-ray, 285, 286, 312
Cuenot, Lucien, 281
Cumming, William, 288
Curie, Marie, 197
Cushing, Harvey, 272
Cuvier, Georges Leopold, 151–152, 154, 156, 185–187, 194, 212
Cybernetic organisms, 372
Cystic fibrosis, 367
Cytochrome, 274
Cytology: *see* Cell studies

Dale, Henry Hallet, 294, 295
Dalton, John, 133, 145
Dam, Henrik, 324, 325
Dan, Samuel, 240
Darevsky, Ilya, 308
Darwin, Charles
  and Chambers, 215
  controversy over, 153, 223–225
  The Descent of Man, 221
  and early theories of evolution, 32, 180
  and Gray, 217
  and Haeckel, 230
  on mechanistic theory, 210, 211
  and Mendel, 279–280
  on metamorphosis, 95
  on natural selection, 206, 303
  The Origin of Species, 221
  and Owen, 213
  and plant evolution, 218
  and rise of evolution, 218–221
  on sexual selection, 222–223
  and Sprengel, 194
  supported by Huxley, 224–225
  on survival of the fittest, 192
  and uniformitarianism, 187–188
  and Wallace, 226
Darwin, Erasmus, 153–154, 156–157, 158, 169
Davies, Kay, 349
Da Vinci, Leonardo, 59–61, 74
Davis, Adelle, 325
Davis, Marguerite, 322
Davis, Morris, 327
Davis, Ronald, 350
Davy, Humphrey, 182
DDT, effects of, 346–347
Death, definition of, 366
de Kruif, Paul, 285
Delbruck, Max, 335
Democritus, 24–28, 37, 49, 292
  atomic theory of, 24–25
  biological work of, 25–28

Dental practices, Sumerian, 12
Derham, William, 137
Desault, Pierre-Joseph, 162
Descartes, René, 103, 140–141
  and Borelli, 108
  mind-body concept, 72, 372
  on nervous system, 160
Desia, Manoj, 326
Determinism, 140–143
D'Herelle, Felix, 283
d'Holbach, Baron, 141
Diabetes mellitus, 276
Dickinson, Roscoe, 312
Diderot, Denis, 183–184
Diels, Paul, 291
Digestive system, 163
  Beaumont on, 267
  Bernard on, 266
  early studies of, 38, 52, 66
Dillenius, J. J., 149
Dinosaurs, 186, 211
  extinction of, 356
  primitive, 371–372
  relation to birds, 369
Dintzis, Howard, 340
Diocles, 27
Diogenes of Apollonia, 27–28
Diphtheria, 248
Disease
  bearers of, 206–207
  causes of, early concepts of, 73–74
  and cell theory, 263–264
  in Darwinian era, 246–248
  in Enlightenment era, 166–168
  and environmental medicine, 191
  germ theory of, 252–253
  hypnotism as cure for, 191
  and Spallanzani, 183
  and Sydenham, 134–135
  in twentieth century, 308–310
DNA, 206, 282, 283, 296–297, 319
  enzymes affecting, 340–341
  and genetic fingerprinting, 349
  and molecular biology, 334
  in primordial soup, 292
DNA (*cont.*)
  recombinant, 343–344
  sequences in, 327–328
  structure of, 332–334
  synthesis of facsimile, 341
Doisy, Edward, 324, 325
Dolland, John, 245
Döllinger, Ignaz, 240
Downs, Robert, 125
Dreyer, William, 336
Drug therapy
  antibacterial, 283–285
  development of, 247
  in Egyptian medicine, 5–6
  in psychiatric disorders, 296
Dualism of body and soul, 29, 72
Dubos, René, 284
Duchenne muscular dystrophy, 349–350, 354
Duggar, Benjamin, 285
Duhamel, Henri, 132
Dujardin, Felix, 178–179
Dulbecco, Renato, 337–338
Dumas, Jean, 233
Duve, René de, 316
Dyes, affecting tissues, 247
Dystrophin, 350

Ear, studies of, 23, 27, 38, 41, 79, 237, 265, 273
Earth, evolution of, 155
Eckert, John, 314
Ecology, 328–329
Edelman, Gerald, 338
Eggleston, Philip, 271
Egyptian medicine, 2–7
  papyri of, 4–7
Ehrlich, Paul, 246–247, 248
Eijkman, Christiaan, 323
Einstein, Albert, 136, 229, 269
Einthoven, Willem, 311
Eldredge, Niles, 355
Electricity
  affecting memory, 360
  in living organisms, 236

Electricity (*cont.*)
  and muscle action, 107–108, 160
  studies of, 195
Electrocardiography, 311
Electron microscope, 316
Electrophoresis, 315
Elvehjem, Conrad, 323
Embryo transplants, 326
Embryology
  advances in, 370
  and Bonnet, 164–165
  in Darwinian era, 228–230
  early theories of, 22, 26, 32, 83–84
  and evolution, 230
  and Haller, 163–164
  of invertebrates, 243
  and Pander, 178
  in plants, 255–256
  and Remak, 242–243
  in twentieth century, 326–327
  in Victorian era, 254–255
  and von Baer, 240–241
  and Wolff, 176
Empedocles, 20–24, 37, 50
Empiricism, 23, 139
Encyclopedists, 184
Enders, John, 336–337
Endocrinology, 271–278, 327
Energy
  in body, 265
    sources of, 274
  potential, 31
  production in cells, 357–358
Enlightenment age, 139–169
Entomological taxonomy, 192–193
Environmental disasters, 346–347
Enzymes, 266, 274
  affecting DNA, 340–341
  interactions with genes, 305
  restriction, 343–344, 349
  urease, 272–273
Erlanger, Joseph, 295
Estrogen, 278, 324, 327
Ethical problems in biology and medicine, 364–367
Ethology, 306–308
Etruscans, dental practices of, 12
Eugenics, 198–199
Euler, Leonhard, 134
Euler, Ulf Svante von, 295
Euler-Chelpin, Hans von, 274
Eustachio, Bartolommeo, 78–79
Evans, Herbert, 304, 318, 323
Evolution
  and Darwin, 218–225
  early theories of, 32
  of the Earth, 155
  and embryology, 230
  and genetic engineering, 363–370
  genetics in, 354–355
  and Lamarck, 171
  and mutations, 280–281
  of plants, 179–180, 218
  rudimentary ideas of, 21–22
  social reactions to, 153–154
  in twentieth century, 279–281
  type approach in, 186–187
  and Wallace, 225–227
Ewer, D. W., 273
Exobiology, 335
Exons, 352
Extinctions, 187
  and cloning of species, 353
  mass, 356, 371
Eye, studies of, 23, 27, 38, 41, 43, 358–359

Fabricius, Hieronymus, 37, 83–84, 89
Fabricius, Johann, 192
Falloppio, Gabriel, 84–86, 89
Faraday, Michael, 195
Fechner, Gustav, 236
Fermentation, studies of, 66, 251–252
Fermi, Enrico, 270
Ferrets, black-footed, 321, 347
Ferrie, David, 236
Fichte, Johann, 197
Fire, as fundamental material, 16–17, 18
Fischer, Emil, 287
Fischer, Hans, 319

Fisher, Ronald, 303, 304
Fleming, Alexander, 283
Flemming, Walther, 204
Florey, Howard, 284
Folkers, Karl, 325
Food metabolism, and energy production, 266–267, 275
Forbes, Edward, 188
Ford, C. W., 315
Fossil studies: *see* Paleontology
Foster, Samuel, 105
Fox, Walter, 291
Fracastorius, Hieronymus, 252–253
Fracastoro, Girolamo, 73
Fraenkel-Conrat, Heinz, 338, 339
Franklin, Benjamin, 158
Franklin, Rosalind, 286, 332, 333
Fraud, scientific, 365
Frederick III, 258
Free will and determinism, 140–143
French Revolution, 120, 121, 162
Freud, Sigmund, 191
Fricke, Hans, 321
Friedrich, Wilhelm, 266
Frisch, Karl von, 306–307, 308, 359
Fritsch, Gustave, 274
Fuchs, Leonard, 64
Functionalism, 372
Funk, Casimir, 322–323

Gahn, Johann, 145
Gajdusek, Daniel, 339
Galen, 51–53, 74, 77, 78, 83, 87, 98
Galileo, 103, 104, 107
Galton, Francis, 156, 198–199
Galvani, Luigi, 160–162, 236, 237
Garrod, Archibald, 297
Gasser, Herbert, 295
Gause's principle, 307
Gay-Lussac, Joseph-Louis, 239
Gazzaniga, Michael, 361
Gegenbauer, Karl, 229
Gehring, Walther, 363
Geiger, Hans, 316
Gellert, Martin, 340–341
Genes, 299–301
  and cancer, 351
  enzymes and, 305
  language of, 343
  redundant, 352
  synthesis of, 341–342
  therapy with, 350–351, 364
  transfer of, and cloning research, 352–354
Genetic code, 342
Genetic engineering, 334–335
  for commercial purposes, 348
  and evolution, 363–370
Genetics
  and Beadle, 304–305
  beneficial applications of, 349–352
  and Bridges, 304
  and cell studies, 229
  and DNA studies, 334
  and eugenics, 198–199
  and evolution, 354–355
  fears over research in, 345, 348–349
  and Fisher, 304
  and Lysenkoism, 305–306, 330
  and Mendel, 200–203
  and Morgan, 297–300
  and H. J. Muller, 301
  plant, 302–303
  and Plough, 300
  statistical, 303
  and Tatum, 305
Genome project
  dog, 368
  human, 367–368
  worm, 369
Genotype, 302
Geology, 188–189
  biology-related work in, 278–279
  and Lyell, 213–215
Germ-layer concept, 241, 242
Germ plasm, 206
Germ theory of disease, 252–253
Gesner, Konrad von, 79–80
Ghosh, Arun, 326

Gibbons, John, 365
Gibbs, J. Willard, 196, 287
Gingerich, Philip, 369
Ginkgolide B, 326
Glauber, Johann, 95
Glisson, Francis, 71–72, 101
Glucose, biochemistry of, 182
Goddard, Jonathan, 105
Goethe, Johann Wolfgang, 179–180
Goldberger, Joseph, 323
Goldstein, Joseph, 327
Golgi, Camillo, 71, 245–246, 294
Gonorrhea, 167–168
Goodpasture, Ernest, 310
Goosen, Hendrik, 321
Gorgias, 39
Goudsmit, Samuel, 145
Gould, Stephen Jay, 19, 355
Graaf, Regnier de, 84, 112
Gram, Hans, 227
Granit, Ragnar, 358-359
Gravitation, theory of, 104
Gray, Asa, 217, 224
Great chain of being, 142, 144
Greek medicine, 15–44
Grew, Nehemiah, 129–130, 132
Griffith, Fred, 302
Growth hormone, genetic engineering of, 351
Guettard, Jean, 194
Guillemin, Roger, 324
Gusella, James, 350
Gyrase, 341

Haeckel, Ernst, 174–175, 180, 211, 228, 230, 254
Haffkine, Waldemar, 254
Haldane, J. B. H., 334
Hales, Stephen, 130–132
Hall, Marshall, 160
Hall, Stephen, 348
Haller, Albrecht von, 158–160, 176
  embryology of, 163–164, 165
Hamilton, W. D., 359
Hamm, Louis, 113
Hammurabi, 10
Harden, Arthur, 271, 273, 274
Hardy, G. H., 303
Hardy-Weinberg law, 303, 330
Harlow, Harry, 359
Hartman, Frank, 277
Haruna, Iciro, 338
Harvey, William, 37, 82, 83, 89–93
Hearing, studies of, 23, 237
Heart
  early studies of, 26, 37, 52, 76, 81, 83, 92
  in Egyptian medicine, 5
Hebb, Donald, 360
Hegel, 197
Heitler, Walter, 145
Helmholtz, Hermann von, 236–237
  and botany, 238–239
  and ear studies, 237
Helmont, Jan Baptista van, 65–67, 74, 106, 133
Heme, 319
Hemoglobin studies, 286, 312
Hench, Philip, 277
Henry II, 67
Hepatitis, 337, 339, 345, 351
Heracleides, 38
Heraclitus, 16–17, 20
Herder, Johann, 180
Heredity, *see also* Genetics
  and eugenics, 198–199
  and inheritance of acquired characteristics, 155, 156–157, 174–175, 206, 215, 223, 305–306
  laws of, 202–203
  Mendelian patterns in, 281
  and mutations, 280
  in plants, 181
Hermaphrodism in plants, 129, 130
Herodotus, 11, 34
Herophilus, 39, 42–43
Herpes virus vaccine, 350, 351
Hershey, Alfred, 335
Hevesy, György von, 317, 318
Higuchi, Russell, 353

Hill, Archibald, 270–271, 274
Hillier, James, 316
Hinduism, 18, 37
Hippocrates, 38–41
Histones, 370
Hitzig, Julius, 274
Hoagland, Mahlon, 340
Hodgkin, Dorothy, 185
Hoffmann, Friedrich, 118
Hofmeister, Wilhelm, 255–256
Hokfelt, Tomas, 296
Hokin, Mabel and Lowell, 294
Holley, Robert, 334, 340, 342
Homologies, 67–68, 212–213
Homunculus, 35, 164; *see also* Preformation theory
Hood, Leroy, 368
Hooke, Robert, 125–126, 260
Hopkins, Frederick, 275, 322, 323
Hoppe-Seyler, Ernst, 266
Hormone studies: *see* Endocrinology
Hou, Ya-Ming, 342
Houpis, Ioannis, 326
Houssay, Bernardo, 358
Hsi Than, 44
Hubel, David, 361
Hume, David
  critic of theology, 137, 139–140, 143, 169
  zand Cuvier, 154
  empiricism of, 22, 197
  philosophy of, 183
"Humors" or juices of body, 41, 74
Hunter, John, 142, 163, 166–168
Hunter, William, 166, 167, 168
Huntington's disease, 350
Hutchinson, G., 329
Hutton, James, 188
Huxley, Julian, 326
Huxley, T. H., 224–225, 260
Hydrogen, 120, 121
  substitutions for, 233
Hydrometers, 182
Hylomorphism, 36, 71
Hypnotism, 191
Hypothalamus, 327
Ice age, 215, 321
Imanishi-Kari, Thereza, 365
Imhotep, 3, 4
Immunology, 309–311, 335–336, 338–339
Infusoria, studies of, 115, 179
Ingenhousz, Jan, 132
Insects, 94–96
  behavioral studies, 359
  and fertilization of flowers, 193–194
  mimicry in, 304
  reproduction of, 165–166
  sex hormones of, 278
Insemination, artificial, 311, 366
Insulin, 275–276, 285, 318
Intelligence, studies of, 198, 199–200
Interferon, 352
Introns, 352
Invertebrates, 173–174, 178–179
  embryology of, 243
  physiology of, 319
  taxonomy of, 227–228
Iodine
  in seaweed, 182
  and thyroid function, 276, 318
Iron in blood, 319
Ivanov, Ilya, 311, 330
Ivanovsky, Dmitri, 282

Jackson, Joseph, 244
Jacobs, P. A., 315
Jacobson, Ludvig, 199
Jacobson, Martin, 307
Jaenisch, Rudolf, 350
James I, 89, 91, 105
Janowsky, David, 296
Jefferson, Thomas, 158, 185
Jeffreys, Alec, 349
Jensen, Arthur, 361
Jerne, Niels, 337
Johanssen, Wilhelm, 302–303
Johne's disease, 283
Jones, W. A., 307
Jussieu, Antoine-Laurent de, 150, 173
Jussieu, Bernard de, 150

Kamen, Martin David, 318
Kang, Myung-choi, 326
Kant, Immanuel, 23, 191, 197
Karrer, Paul, 323
Katz, Bernard, 295
Keilen, David, 274
Kendall, Edward Calvin, 277
Kendall, Edwin, 276–277
Kendrew, John, 286
Kepler, Johannes, 104
Khorana, Gobin, 341–342
Kielmayer, Karl Friedrich, 151
Kim, Sung-Hou, 351
Kingston, Lord, 154
Kitasato, Shibasaburo, 247
Klinefelter's syndrome, 315
Klingenstierna, Samule, 182, 245
Koch, Robert, 249–250
Kocher, Emil, 276
Kohler, Georges, 337
Kölliker, Rudolf von, 229–230
Kolreuter, Joseph, 180–181, 193
Kornberg, Arthur, 341, 342
Kossel, Albrech, 205–206
Kovalevski, Alexander, 243
Krebs, Hans, 274–275
Krogh, Schack, 228
K–T boundary, 356
Kunkel, Louis, 350
Kuru, 339

Lamarck, Jean-Baptiste, 171–175
  and inheritance of acquired characteristics, 155, 156–157, 174–175, 206, 215, 305–306
  on invertebrates, 227–228
  and taxonomy, 173–174, 184
Lamont, Johann von, 278
Lancisi, Giovanni, 70–71, 93
Landsteiner, Karl, 308–309, 330
Langmuir, Irving, 196, 234, 269, 286
Lao-tzu, 21
Laplace, Pierre, 190
Laue, Max von, 312
Laurent, Auguste, 234
Lavoisier, Antoine, 119–120, 144, 146
Lawrence, William, 197
Le Bon, Gustave, 198
Leder, Philip, 354
Lederberg, Joshua, 334
Leeuwenhoek, Anton van, 112–116
Leibnitz, Gottfried
  on beginnings of Earth, 154
  and Buffon, 156
  concepts of the universe, 26, 137, 141
  great chain of being, 142, 144
  rationalism of, 139, 197
Leucippus, 24
Levene, Phoebus, 296, 331
Levi-Montalcini, Rita, 338–340
Levin, Michael, 200
Lewis, G. N., 234, 286–287
Li, Choh Hao, 285
Libby, Willard, 318
Liebig, Justus von, 239–240, 251, 260
Life
  artificial, 341–342
  origins of, 28, 291–292, 334
    in theory of spontaneous generation, 35, 65
Linnaeus, Carolus
  and Buffon, 154
  and evolution, 153
  and Otto Müller, 116
  and Ray, 128
  and Réaumur, 192
  taxonomy of, 33, 97, 146–150, 173, 227
Lipmann, Fritz, 274, 275
Lippershey, Hans, 111
Lister, Joseph, 253
Liver, early studies of, 26, 41, 52
Locke, John, 135, 143, 159, 197
Loeb, Jacques, 293, 308, 329
Loewi, Otto, 294, 295, 296
Lorenz, Konrad, 306, 307, 308
Lotka, A. J., 303
Love and hate, forces of, 21
Lower, Richard, 78, 81, 92
Lucretius, 21–22, 48–51

Lungs, studies of, 122
Lupines, arctic, 321
Luria, Salvador, 283, 331, 335
Lyell, Charles, 187, 213–215, 219
Lymphatic system, 99–101
Lynen, Feodor, 317
Lysenko, Trofim, 305–306, 330
Lysozyme, 283, 284

MacLeod, Colin, 297
Macleod, J. R., 276
Magendie, François, 264, 265
Magic and medicine, 4, 9–10, 57–58
Magnetic fields, and navigational ability of birds, 278–279, 359
Magnetism, 195
Magnus, Albertus, 56–57
Malaria, 71, 246, 247
Malcolm, Norman, 50, 200
Malpighi, Marcello, 114, 121–124, 132
Malthus, Thomas, 191–192, 219, 226
Mammoth, frozen, 322
Mangold, Hilda, 326
Manhattan Project, 270, 288, 317
Mantell, Mary, 211
Manure, affecting plants, 239, 240
Marcus Aurelius, 45, 51
Mariotte, Edme, 99, 133
Marsh, Othneil, 209
Marshall, Richard, 343
Marsigli, Luigi, 97
Martin, Archer Porter, 315
Mason, William, 345
Materialism, 16, 23, 139
Mather, Cotton, 166
Mauchly, John, 314
Maxwell, James Clerk, 195–196
McCarthy, Maclyn, 297
McClintock, Barbara, 363
McClung, Clarence, 301
McCollum, Elmer, 322, 330
McLean, Dewey, 357
Measles, 337
Mechanistic philosophy, 141–142, 144, 184, 210–211, 293
Medawar, Peter, 309
Meiosis, 205
Memory, trace theory of, 360
Mendel, Gregor, 181, 200–203, 279–280
Mental state, studies of, 244, 372
Mesmer, Franz, 191
Mesopotamia, 9–13
Metabolism
  early studies of, 87
  and energy production, 266–267, 275
  Krebs cycle in, 275, 323
Metamorphosis studies, 94, 95–96, 166
Meyer, Herman von, 186
Meyerhof, Otto, 270–271
Michaelis, Leonor, 272
Michel, Robert, 358
Michelson, A. A., 196
Microscopes, 111–116
  improvements in, 244–245, 315–316
  and Leeuwenhoek, 112–116
Middle Ages, 55–58
Miescher, Johann, 262
Mill, James, 157
Mill, John Stuart, 143, 157, 159
Miller, Jacques, 310
Millikan, Robert, 153, 269
Millington, Richard, 315
Milstein, Cesar, 337
Mimicry, genetic basis for, 304
Mind, nature of, 292–293
Mind–body problem, 72, 372
Minkowski, Oskar, 275
Minot, George, 323
Mitochondria, 316, 318–319
Mitosis, 204–205
Mohl, Hugo von, 258, 262–263
Molecular biology, 334
Monod, Jacque, 340
Montagu, Mary Wortley, 70
Moreau, Pierre-Louis, 153
Morgan, Thomas Hunt, 297–300, 329, 331
Mueller, Erwin, 317
Muller, Hermann J., 301

Müller, Johannes, 180, 229, 235, 236, 242, 243–244, 263
Müller, Otto, 115, 116
Mumps, 337
Murphy, William, 323
Murray, Andrew, 364
Muscle studies, 159–160, 238, 270, 274
  by Hippocrates, 40
  mechanics of action, 107–108
  microscopic, 116
  physiology, 350
Muscular dystrophy, 349–350, 354
Mutations, 197, 280–281, 301
Myoglobin studies, 286
Mysticism, 65–66, 176

Nägeli, Karl von, 203, 261
Nancy, Francois, 340
Nathans, Daniel, 343
Naudin, Charles, 203
Navigation systems, animal, 278–279, 359
Needham, John, 177
Neoplatonism, 65–66
Nerve cell studies, 229, 235, 237
Nerve growth factor, 340
Nervous system, 160, 236–238, 243–244, 246, 360
  and Bernard, 265–266
  early studies of, 38, 122–123
  in embryos, 339–340
  and physiological psychology, 360–362
  physiology of, 162, 294–296
  synapses in, 245, 294
Neurofibromatosis, 364
Neurotransmitters, 294–295, 296, 329
Neutrinos, 119
Newton, Isaac, 103–104, 131, 136, 140, 154, 156, 190
Nikolaevitch, Andrei, 282
Nirenberg, Marshall, 342, 343
Nitrogen, 190
  in cell nucleus, 261
  in plants, 238–239
Noradrenaline, 295, 329
Norris, Kenneth, 278
Northrop, John, 273, 275, 282
Nossal, Gustave, 336
Nuclear energy, fears of, 345–346
Nucleic acids, 206, 282, 331
Nucleus
  of atoms, 286
  of cells, 217, 256, 259, 263
Nufer, Jakob, 64
Nutrition
  science of, 234–235
  vitamins in, 322–326
Nuttall, George, 320

Oberbeck, Verne, 371
Ocean studies, 188
Ochoa, Severo, 341
Oersted, Hans, 195
Offroy, Julien, 141–142
Oken, Lorenz, 175, 229
Oldenburg, Henry, 112
Oncogenes, 351
Oort cloud, 356
Oparin, Alexander, 292, 329
Operons, 340
Opie, Eugene, 275–276
Oriental studies in medicine, 36, 37, 43–44
Osborne, Thomas, 234
Owen, Richard, 186, 211–213
Oxygen
  discovery of, 120, 190
  metabolism in cells, 274
Oxytocin, 324

Palade, George, 316, 342
Paleontology, 185–194
  and Agassiz, 213
  development of, 86–87
  and mass extinctions, 356
  miracles in, 320–322
  new technology in, 369–370
  and Owen, 211–212
Paley, William, 136

Pancreas, 275–276
Pander, Hans Christian, 178
Papyri, medical, 4–7
  Ebers, 6
  Edwin Smith, 5–6
  therapeutic, 6–7
Paracelsus, 61–63, 66, 133
Parasitism, 71, 96, 262
Parmenides, 20
Parthenogenesis, 165, 308
Pasteur, Louis, 134, 249, 250–253
Pauli, Wolfgang, 119
Pauling, Linus
  and biological clock, 354
  and bonding of atoms, 145
  and chemical bonding, 234, 287
  and DNA structure, 332–333
  and geometry of molecules, 336
  and Müller, 301
  and sickle-cell anemia, 315, 349
  studies in Europe, 196, 269
  and vitamins, 323, 325, 326
  and X-ray crystallography, 286, 312
Pavlov, Ivan, 267
Pearson, Karl, 198
Pellagra, 323
Pende, Nicole, 271
Penfield, Wilder, 73
Penicillin, 283–284
Penzias, Arno, 328
Perutz, Max, 286, 312
Pestilence in Renaissance era, 69–74
Peter the Great, 112–113
Pfeffer, Wilhelm, 246
Pfiffuer, J. J., 277
Phages, 338
Phenotype, 302
Pheromones, 278, 307
Philosophy
  developments in, 139–143
  in Enlightenment era, 183–184
  and science in Victorian age, 197
Phlogiston, 117–119, 144
Phosphorus
  in cell metabolism, 318
  physiological importance of, 295
Photosynthesis, 132, 238, 317
Physics, 190–191
  and chemistry, 144–145
  particle, 328
Physiology
  in age of Vesalius, 87
  Aristotelian, 37–38
  blood, 319
  cellular, 318–319
  in Darwinian era, 235–237
  of Galen, 52
  of Galvani, 160
  of Haller, 158–160
  in Hippocratic collection, 40–41
  invertebrate, 319
  lymphatic system, 99–101
  of Malpighi, 121–123
  muscle, 350
  nervous system, 162–163, 265–266, 294–296
  new frontiers in, 357–359
  plant, 182
  and tracer technology, 317–318
  in twentieth century, 270–271, 308–310
Pituitary gland, 271–272, 358
Place, U. T., 50
Plague, bubonic, 247
Plants
  cell studies, 260
  classification of, 96–99
    by Aristotle, 34
    by Camerarius, 128–129
    by Ray, 127–128
  embryology of, 255–256
  evolution of, 179–180, 218
  fertilization by insects, 193–194
  genetics of, 302–303
  physiology of, 182
  reproduction of, 180–181
Plasmids, 335
Plastics, synthesis of, 291
Plato, 16, 20, 23, 28, 29, 49, 123

Pliny, 46–48, 80
Plough, Matthew, 300
Poiseuille, Jean, 181, 182–183
Polio vaccine, 310
Polio virus, 337
Population studies, 191–192, 329
Porsild, Alf, 321
Porter, Keith, 319
Porter, Rodney, 338
Potts, Percival, 191
Pouchet, Felix, 250–251
Prebus, Albert, 316
Preformation theory, 114, 123, 164–165, 254, 262
  controversies in, 163–164
  and Spallanzani, 176–177
Priestley, Joseph, 119, 120–121, 158, 190
Prince, Derek, 313
Pringsheim, Nathanael, 215–217, 238
Progesterone, 278
Prostaglandins, 295–296
Proteins, 246, 267
  separation techniques, 315
  structural studies, 286, 287
Protoplasm, 257, 258
Proust, Joseph, 182
Psychology, physiological, 360–362
Punnet, R. C., 300, 303
Purkinje, Jan, 180, 229, 257–258
Pylarini, Giacomo, 70
Pythagoras, 17–19

Quantum mechanics
  applied to chemical bonding, 287
  and formation of molecules, 145

Rabies, 253
Race
  and social Darwinism, 223
  studies of, 197–200
Radiation, effects of, 301, 320, 345–346
Radioisotope tracers, 317–318
Ramón y Cajal, Santiago, 294
Rampino, Michael, 371
Randall, Merle, 287
Rationalism, 139, 142–143, 197
Raup, David, 356
Rawlinson, Henry, 189
Ray, John, 126–128, 131, 135–136, 150
Réaumur, René de, 95–96, 163, 164, 192
Redi, Francesco, 106–107, 109, 177
Reed, Walter, 71, 248
Reflexes, conditioned, 267
Regan, Johann, 306
Regeneration phenomenon, 179
Reichstein, Taddeusz, 277
Religion: *see* Theology and science
Remak, Robert, 229, 241, 242–243
Renaissance, 59–68
  pestilence in, 69–74
Renan, Ernest, 223
Repressor genes, 340
Reproduction
  in age of Vesalius, 83–86
  and alternation of generations, 255–256
  Aristotelian, 31, 35–36, 114
  and artificial insemination, 311, 366
  in Darwinian era, 228–230
  in insects, 165–166
  and Leeuwenhoek, 113–114
  plant, 180–181
  and preformation: *see* Preformation theory
  and sexual selection, 222
  and studies of egg or ovum development, 123
Respiration
  in cells, 323–324
  early theories of, 22, 26–27
  in plants, 124, 132
Respirator, artificial, 134
Retzius, Anders Adolf, 199–200
Reverse transcriptase, 344–345
Rh factor, 309
Ribosomes in cells, 316, 342
Rickets, 71, 74
Rine, Jasper, 368
RNA, 206, 282, 319, 327
  genetic code in, 342
  messenger, 340, 343

RNA (*cont.*)
synthesis of facsimile, 341
transfer, 334, 340
Robbins, Fred, 336–337
Roberts, Richard, 352
Robinet, Jean Baptiste, 175
Robinson, Arthur, 325–326, 355
Roman medicine, 45–53
Rosetta stone, 188–189
Ross, Ronald, 71
Roth, Thomas, 319
Roux, Wilhelm, 254–255, 262
Rowland, Henry, 190, 196
Ruber, Max, 274
Rudbeck, Olof, 100–101
Rutherford, Daniel, 189, 190
Rutherford, Ernest, 189–190, 196
Ruzicka, Leopold, 278

Sabin, Albert, 310
Sacks, J., 318
Sachs, Julius von, 238, 280
Sagan, Carl, 292, 361
Salk, Jonas, 310
Samuelsson, Bengt, 295
Sanctorius, 87
Sanger, Frederick, 276
Saussure, Nicholas de, 133
Schally, Andrew, 324
Scheele, Carl Wilhelm, 145–146, 190
Schick, Bela, 248
Schimmel, Paul, 342
Schlegel, Friedrich von, 197
Schleiden, Matthias, 239, 242, 259–260, 261, 263
Scholasticism, 60–61, 71, 74
Schomaker, Verner, 332
Schreiber, Johann, 201
Schwann, Theodor, 242, 244, 251, 257, 261
Science
escalating fears of, 345–348
organizations of, 105–106, 158
professionalism in, 209–210
Scientific method, evolution of, 106–107
Scopes trial, 306, 330
Scott, K. J., 323, 327
Searle, John, 313–314
Secretin, 276
Sedgwick, Adam, 219
Selye, Hugo, 277
Semipermeable membranes, 246
Sensation
early theories of, 26
studies of, 235
Sense organs, 78–79
Sepkoski, J. J., Jr., 356
Sereno, Paul, 371–372
Servetus, Michael, 81–83
Sex hormones, 277–278
Sexual selection processes, 222
Sharp, Philip, 352
Shemin, David, 317
Sherman, Henry, 235
Sherrington, C. S., 272, 293, 294
Shockley, William, 200
Shoenheimer, Rudolf, 317
Shultes, Johann, 7, 64–65
Sibley, Charles, 363
Sickle–cell anemia, 315, 349
Siebold, Karl von, 261–262
Simons, W. L., 369
Sinsheimer, Robert, 367
Skepticism, arguments for, 139
Skolnick, Mark, 350
Slater, John, 196
Sleep studies, 361
Smallpox, 69–70, 166
Smart, J. J. C., 50, 372
Smith, Hamilton, 343
Smith, Holly, 369
Smith, James, 321
Smith, Philip, 271, 358
Smith, William, 189
Snow, John, 253
Social behavior in animals, 359
Social Darwinism, 223
Sociobiology, 329

Socrates, 39
Soul
  concepts of, 72–73, 80, 141, 142, 143–144, 244, 293
  early ideas of, 17, 20, 35, 49–50, 66
Spallanzani, Lazzaro
  and disease, 183
  and invertebrate zoology, 179
  and ovism, 176–177, 184
  and preformation theory, 142
  and spontaneous generation, 177–178
  studies of digestion, 163
Spemann, Hans, 326
Spencer, Herbert, 223, 224
Sperry, Roger, 361
Sphygmomanometers, 182–183, 228
Spiegelman, Sol, 338
Spinoza, 108, 139, 141
Spontaneous generation theory, 35, 65, 262, 292
  and Spallanzani, 177–178
  testing of, 106–107, 250–251
Sprengel, Christian Conrad, 193–194
Stahl, George, 118–119
Stanborough, Walter, 282
Stanley, Wendell, 207, 282
Stannius, Friedrich, 261
Starling, Ernest, 272, 303
Staudinger, Hermann, 291
Steenbock, Harry, 322
Steenstrup, John, 256
Steinmetz, Charles, 345
Steno, Nicholas, 86–87, 155
Sterling, Ernest, 276
Steroids, 336
Stewart, Timothy, 354
Strasburger, Eduard, 204, 206, 256
Strauss, David, 223
Streptomycin, 285
Strong, J. A., 315
Su, Wei-guo, 326
Sumerian medicine, 9–13
Summers, Jesse, 345
Sumner, James, 272–273, 282
Sutherland, Earl, 357
Sutton, Walter, 205
Suze, Konrad, 313
SV40 virus, 339
Svedberg, Theodor, 311–312, 315
Swammerdam, Jan, 93–95, 102, 111–112, 129
Swingte, W. W., 277
Sydenham, Thomas, 58, 134–135
Sylvius, Jacob, 75
Sylvius, Jacques Dubois, 74
Synapses, 245, 294, 360
Syphilis, 69, 73, 167–168, 247
Syrian medicine, 7
Szent-Györgyi, Albert, 323

Takamine, Jokichi, 277
Taoism, 36, 37, 44
Tatum, Edward, 305, 334
Taxonomy
  Aristotelian, 32–35, 80, 127
  of Buffon, 157–158
  of coral reefs, 221
  of Cuvier, 151–152
  entomological, 192–193
  and *International Rules of Zoological Nomenclature*, 320, 330
  invertebrate, 227–228
  of Jussieu, 150–151
  of Lamarck, 173–174
  of Leeuwenhoek, 115–116
  of Linnaeus, 146–150
  and plant classification, 34, 96–99, 127–129
  Ray's contributions to, 127–128
  of Renaissance zoologists, 79–80
  vertebrate, 228
Technology
  in Darwinian era, 244–245
  in Enlightenment era, 182–183
  and fears over science, 345–348
  in paleontology, 369–370
  in twentieth century, 270, 285–286, 311–318
Teleological change, 30
Temin, Howard, 338, 344–345
Terman, Lewis, 198

Testosterone, 278, 324
Tetracyclines, 285
Thaddeus, Alderotti, 48
Thales, 16, 66
Theology and science, 135–137, 210
  in anatomical studies, 81–83
  and controversy over Darwinism, 223–225
  and development in philosophy, 139–143
  and theories of evolution, 153–154
Theophrastus, 42
Theorell, Axel, 323–324
Theorell, Hugo, 358
Thevenot, Melchisedec, 94
Thomas Aquinas, St., 51, 55, 56, 136
Thompson, John, 95–96
Thomson, Charles, 188
Thucydides, 42
Thymus gland, 309–310
Thyroid gland, 276–277, 318
Tinbergen, Nikolaas, 308
Tiselius, Arne, 314–315
Titian, 76
Tjio, J., 348
Todd, Alexander, 357
Tolman, Richard, 286
Tonegawa, Susumu, 338–339
Tracer technology, 317–318
Transcriptase, reverse, 344–345
Trypanosomiasis, 247
Tschermak von Seysenegg, Erich, 280
Tso, Paul, 318
Tuberculosis, 250, 285
Turing, Alan, 313–314
Turner's syndrome, 315
Twort, Frederick, 283
Tyler, Stanley, 321
Type theory, in history of evolution, 186–187
Typhus, 337

Ultracentrifuge, 311–312
Uniformitarianism, 187–188, 355
Urease, 272–273

Vaccines
  anthrax, 253
  cholera, 254
  development of, 247–248
  hepatitis, 351
  herpes virus, 350, 351
  measles, 337
  polio, 310
  rabies, 253
  smallpox, 70, 166
Valenstein, Eliot, 73
Valentin, Gabriel, 129, 229
Van Vleck, J., 196
Vane, John, 295
Vavilov, Nikolai, 306
Vertebrate taxonomy, 228
Vesalius, 75–78, 91
Victorian era, 195–207
Vigneaud, Vincent du, 324
Virchow, Rudolf, 244, 254, 263–264
Viruses, 206–207, 282–283, 337–340
  and cancer, 311, 338
  as living organisms, 338
  reproduction of, 335
  in vaccines, 310
Vision, physiology of, 359
Vital force, concepts of, 142, 176, 236
Vitamins, 234–235, 322–326, 330
Vivisection, 78, 265, 270–271
Voit, Karl von, 266–267, 275
Volta, Allesandro, 195
Vries, Hugo de, 197, 279, 280–281

Waksman, Selman, 285
Wald, George, 358
Wallace, Alfred Russell, 225–227
Warburg, Otto Heinrich, 274
Ward, Peter, 187
Warm elements, male, 22, 36
Water, as fundamental material, 16
Watson, James, 297, 302, 332–334, 367
Watt, James, 158
Weber, Ernst, 235
Wedgwood, Hannah, 219

Weiditz, Hans, 63
Weinberg, W., 303
Weismann, August, 206
Weissmann, Charles, 352
Weller, Thomas, 336–337
Wharton, Thomas, 101
Whipple, George, 319
Whistler, Daniel, 74
White, Gilbert, 191
White, Raymond, 367
Whytt, Robert, 101, 143–144
Wiener, Alexander, 309
Wiesel, Torsten, 361
Wilberforce, Bishop, 225
Wilkins, Maurice, 332–334
Wilkis, John, 105
William, Robert, 349
William of Occam, 3, 55
Williams, Robert, 323, 324
Williams, Robley, 316
Willis, Thomas, 72–73
Willughby, Francis, 97, 127
Wilson, Allan, 353
Wilson, Edmund, 255
Wilson, Edward O., 359
Windaus, Adolf, 322
Wittgenstein, Ludwig, 98, 211, 244, 293, 313
Witzenhausen, Karl, 210, 228
Wolf, Etienne, 326–327
Wolff, Kaspar, 123, 176, 184
Wollaston, William, 163
Woodbury, Frank, 276
Woodger, J. H., 304
Woodward, Robert, 325
Wren, Christopher, 72
Wurttemberg, Eugen, 276
Wyville, Charles, 188

X-ray crystallography, 285, 286, 312
Xenophanes, 19

Yalow, Rosalyn, 324
Yamazake, Junio, 307
Yanofsky, Charles, 327
Yellow fever, 71, 248
Yersin, Alexandre, 247

Zeno, 20
Zipser, David, 342
Zoology
  cell theory in, 261–262
  invertebrate, 173–174, 178–179
  in Renaissance era, 79–80
  taxonomy of
    by Aristotle, 33–34
    Ray's contributions to, 127
Zsigmondy, Richard, 315
Zuckerkandl, Emile, 354
Zworykin, Vladimir, 316